Reproduction of Amphibians

Biological Systems in Vertebrates

Series Editors
Hiran Dutta and Douglas Kline
Department of Biological Science
Kent State University
Kent, Ohio, USA

Books in this Series

J.N. Maina—Functional Morphology of the Vertebrate Respiratory Systems
Hans Ditrich—Renal Structure and Function in Vertebrates
Seth Kisia—Muscular System of Vertebrates
Maria Ogielska—Reproduction of Amphibians

Reproduction of Amphibians

Editor

Maria Ogielska
Zoological Institute
University of Wrocław
Poland

CRC Press
Taylor & Francis Group
Boca Raton London New York

CRC Press is an imprint of the
Taylor & Francis Group, an **informa** business
A SCIENCE PUBLISHERS BOOK

CRC Press
Taylor & Francis Group
6000 Broken Sound Parkway NW, Suite 300
Boca Raton, FL 33487-2742

First issued in paperback 2018

© 2009 reserved
CRC Press is an imprint of Taylor & Francis Group, an Informa business

No claim to original U.S. Government works

ISBN 13: 978-1-138-11771-6 (pbk)
ISBN 13: 978-1-57808-307-7 (hbk)

This book contains information obtained from authentic and highly regarded sources. While all reasonable efforts have been made to publish reliable data and information, neither the author[s] nor the publisher can accept any legal responsibility or liability for any errors or omissions that may be made. The publishers wish to make clear that any views or opinions expressed in this book by individual editors, authors or contributors are personal to them and do not necessarily reflect the views/opinions of the publishers. The information or guidance contained in this book is intended for use by medical, scientific or health-care professionals and is provided strictly as a supplement to the medical or other professional's own judgement, their knowledge of the patient's medical history, relevant manufacturer's instructions and the appropriate best practice guidelines. Because of the rapid advances in medical science, any information or advice on dosages, procedures or diagnoses should be independently verified. The reader is strongly urged to consult the relevant national drug formulary and the drug companies' and device or material manufacturers' printed instructions, and their websites, before administering or utilizing any of the drugs, devices or materials mentioned in this book. This book does not indicate whether a particular treatment is appropriate or suitable for a particular individual. Ultimately it is the sole responsibility of the medical professional to make his or her own professional judgements, so as to advise and treat patients appropriately. The authors and publishers have also attempted to trace the copyright holders of all material reproduced in this publication and apologize to copyright holders if permission to publish in this form has not been obtained. If any copyright material has not been acknowledged please write and let us know so we may rectify in any future reprint.

Visit the Taylor & Francis Web site at
http://www.taylorandfrancis.com

and the CRC Press Web site at
http://www.crcpress.com

Library of Congress Cataloging-in-Publication Data
Reproduction of amphibians/editor, Maria Ogielska.
     p. cm. -- (Biological systems in vertebrates)
Includes bibliographical references and index.
ISBN 978-1-57808-307-7 (hardcover)
1. Amphibians--Reproduction. I. Ogielska, Maria.

QL669.3.R47 2009
573.6'178--dc22
                                        2009011134

Cover illustrations
Developing ovary of the grass frog *Rana temporaria*.

Photograph provided by Renata Augustyńska and Agnieszka Kotusz
Illustration of female grass frog *Rana temporaria*.
Photograph courtesy of Maria Ogielska.

This book is dedicated to my
Parents

Preface

This book is devoted to reproduction of amphibians (subclass: Lissamphibia) belonging to three extant orders: caecilians (infraclass: Gymnophiona, order: Apoda), salamanders (infraclass: Batrachia, superorder: Urodela, order: Caudata), and frogs and toads (superorder: Salientia, order: Anura). The current number of amphibian species is more than 6,400. Caecilians are represented by 176 species classified into 6 families; salamanders are represented by more than 570 species classified into 10 families, and anurans are the most numerous group with almost 5,700 species classified into 34 families (www.amphibiaweb, Duellman, 2003; Larson, 2003).

Caecilians, salamanders, and frogs differ not only in external morphology, but also in reproductive modes. Caecilians are oviparous or viviparous. Viviparity is always connected with adaptations of oviducts and other parts of the female reproductive system. Fertilization in caecilians is internal and males are equipped with a special copulatory organ called phallodeum.

Salamanders are mainly oviparous with external or internal fertilization. Male salamanders have no copulatory organs and internal fertilization takes place when females actively collect sperm packages (spermatophores) laid by males on the background on land or in water. Females are able to store the sperm in special cloacal pockets (spermathecae). Larviparity and viviparity in salamanders are very rare and known only in a few species. Urodela are exceptional among vertebrates in respect of neoteny. This phenomenon is characteristic of some species or entire families and consists of sexual maturity of permanent larvae that do not complete metamorphosis.

Anurans are the best known group of amphibians. Fertilization is external, with exception of vivparous species and two species of the tailed-frogs. Reproduction of frogs and toads is very diverse and classified into several modes including terrestrial, arboreal, and aqueous eggs, parental care performed by males or females, foam nests, and viviparity. This topic, however, was described elsewhere (Duellman and Trueb, 1986; Duellman, 2003) and is not covered by our book.

We focused on anatomy and histology of reproductive systems, as well as cytology and molecular mechanisms that regulate gametogenesis. The content of the volume is divided into eight chapters describing various aspects of amphibian reproduction. Chapter 1 is devoted to the origin of somatic and germ cells during formation of gonads at larval stages and juvenile period. Chapters 2, 3, and 4 cover male reproduction system, testis development and structure, urogenital connections and reproductive tract, spermatogenesis, and regulation of reproductive cycle in caecilians, salamanders, and anurans. The same schedule was applied to chapters 5, 6, and 7 that are devoted to females. Chapter 8 deals with early embryonic and postembryonic development, direct development, and neoteny; special attention, however, has been paid to modifications of gametogenesis and meiosis in hybrids and polyploids (hybridogenesis and gynogenesis).

The final shape of this volume is a result of a good cooperation between the authors, although it took a long time to finish the whole volume, because all of us were really busy with teaching, administration, and research at our home universities.

Jean-Marie Exbrayat from the Universite Catolique de Lyon, France, presented valuable and original data on oogenesis and spermatogenesis of caecilians that is the least known group of amphibians.

Mari Carmen Uribe Aranzábal from the Universidad Nacional Autónoma de Mèxico coverd the topic in salamanders. Jolanta Bartmańska and I represent the University of Wrocław, Poland; we contributed to the largest portion of the volume devoted to frogs and toads that are the best studied amphibians.

I am deeply grateful to the series editors, Doug Kline and Hiran Dutta, for their encouragement and invaluable help in language correction, because none of the authors is a native English speaker.

My former PhD students Agnieszka Kotusz, Renata Augustyńska, and Beata Rozenblut provided several beautiful photos of various aspects of anuran ovaries and oogenesis. Special thanks are reserved for my former PhD student and now my co-worker Piotr Kierzkowski, who arranged the photos and created all line drawings and schemes that illustrate the texts of mine and Jolanta Bartmańska; he and Beata Rozenblut also corrected figures provided by other authors.

Financial support of the University of Wrocław, Zoological Institute (DS 1018) has facilitated the preparation of the volume.

References

Larson A, Weisrock DW, Kozak KH. 2003. Phylogenetic systematics of salamanders (Amphibia: Urodela), a review. [In]: Reproductive Biology and Phylogeny of Urodela. Ed DM Sever. Science Publishers, Inc. 2003. pp. 31-108.

Duellman WE. 2003. An overview of anuran phylogeny, classification and reproductive modes. [In]: Reproductive Biology and Phylogeny of Anura. Ed BGM Jamieson, Science Publishers, Inc. 2003. pp. 1-18.

Duellman WE, Trueb L. 1986. Biology of Amphibians. McGraw-Hill Book Company, USA. pp. 670.

Maria Ogielska
Wrocław, January 2009

Contents

1

The Undifferentiated Amphibian Gonad

Maria Ogielska

Maria Ogielska

FORMATION OF GONADS IN AMPHIBIANS

In amphibians, formation of gonad takes place during larval stages. The differentiation of gonads into ovaries or testes usually occurs around the time of metamorphosis, although there are some exceptions to this rule. The genetic sex is established at the time of a zygote formation, when the male and female pronuclei of an ovum and a spermatozoon fuse. However, gonads are in undifferentiated stages of development before the first morphological features characteristic of a male or a female gonad are detectable. During that time the somatic cells of a gonad start to differentiate according to the sex-determining genes and under their influence the germ cells will transform into oogonia or spermatogonia.

Origin and Differentiation of Somatic Cells

The process of gonad formation in Anura, Urodela, and Gymnophiona are similar and for this reason they will be described together. However, most of our knowledge comes from the studies on Anura. (For more detailed information on formation of larval gonad in Gymnophiona see Chapter 'Spermatogenesis and Male Reproductive System in Gymnophiona' in this volume).

During larval development of amphibians, two longitudinal folds of coelomic epithelium are formed in the dorsal part of body cavity along both sides of gut mesentery (Figs. 1 and 2). The formation of genital ridges were described in *Rana silvatica* (Witschi, 1929); *Rana pipiens* and *Xenopus laevis* (Merchant-Larios and Villalpando, 1981; Iwasawa and Yamaguchi, 1984);

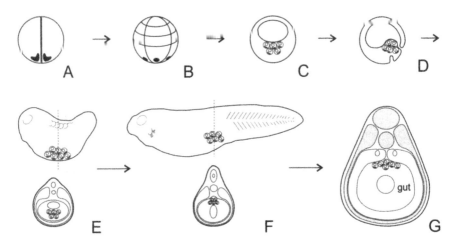

Fig. 1. Schematic illustration of specification and migration of presumptive primordial germ cells (pPGCs) in Anuran amphibians. A—Early cleavage; small islets of germ plasm (black dots) fuse and form few bigger aggregates distributed to four vegetal blastomeres. B—Early blastula. Germ plasm is still distributed to four vegetal blastomeres after the asymmetric mitotic divisions. C—Late blastula; PGCs are located in the endoderm. D—Gastrulation; PGCs are a part of the archenteron. E—Tail-bud stage; PGCs are localized in the forming gut. F, G—Sagittal and cross section of an early larva after hatching; PGCs begin to migrate actively and gather close to the region of the forming mesentery and genital ridges (G). For further details, see Figure 2.

Rana nigromaculata (Tanimura and Iwasawa, 1988); *Rhacophorus arboreus* (Tanimura and Iwasawa, 1989). In both Anura and Urodela (Nieuwkoop, 1950), the folds—called genital ridges—develop independently of the germ cells, as indicated by the differentiation of sterile gonads in absence of the germ cells (Wylie *et al.*, 1975, Nieuwkoop and Sutasurya, 1979; Ogielska and Wagner, 1993). These genital ridges soon become invaded by primordial germ cells (PGCs) (Figs. 2 and 3A-C). In Anura, PGCs are originally distributed along the entire length of genital ridges, but soon gather in the central parts, leaving both ends devoid of germ cells. This re-arrangement gives rise to 3 parts of a gonad (Fig. 5A): the central part forms the proper gonad; the anterior (progonad or progonium) forms the fat body; and the posterior (epigonad or epigonium) forms a sterile appendage, which sometimes can produce a small additional fat body (Witschi, 1929). The three parts are also called *pars progonalis, pars gonalis* and *pars epigonalis*, respectively. In Urodela and Gymnophiona, the progonad and epigonad transform into a fat body (*corpus adiposum*), whereas in Anura, the epigonad degenerate and only progonad transforms into a fat body. In *Pleurodeles waltl* (Urodela), shortly before sex differentiation, the fat body

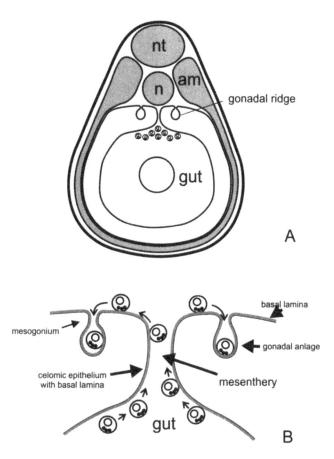

Fig. 2. Active migration of primordial germ cells (PGCs) in Anuran amphibians. A—Cross section of an early larva. PGCs are situated in the dorsal part of the gut. nt, neural tube; am, axial mesoderm and somatopleura; gonadal ridges are formed as folds of the peritoneal epithelium close to mesentery; n, notochord. B—Enlarged area of the gonadal ridges, the PGCs (each PGC contains germinal granules marked as black dots) migrate across the mesentery toward the gonadal ridges; the migration is guided by fibronectin in the basal lamina of the coelomic epithelium. See also Figures 4A and B.

is located on the median side of gonadal anlagen, and the gonad proper is located distally (Dournon *et al.*, 1990). The morphology and formation of the fat body in Anura is described in sections devoted to the structure of ovaries and testes in this volume.

The formation of the primordial gonad starts when migrating PGCs invade a somatic gonadal ridge; either one by one in *Xenopus laevis*, or in groups in *Rana pipiens* (Merchant-Larios and Villalpando, 1981). It usually

4

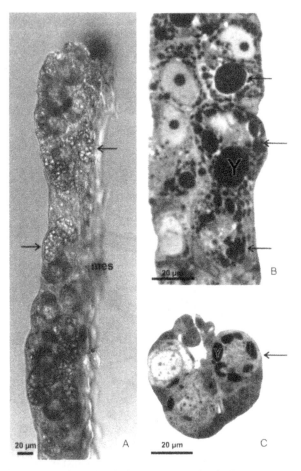

Fig. 3. Gonadal anlage of a tadpole beginning to feed (Gosner stage 25 and 26). Primordial germ cells (PGCs) are indicated by arrows. A—*Rana temporaria*; total view. B—*Rana ridibunda*; longitudinal section. Iron hematoxylin staining C—*Rana temporaria*; cross section. Toluidine blue staining Y, yolk platelets.

takes place in a tadpole just before beginning to feed (Gosner stage 24; Nieuwkoop and Faber stage 44). The undifferentiated gonad is attached to body cavity wall by specialized mesentery, called the mesogonium, which is formed by two parallel sheets of peritoneal epithelium of a gonad (Figs. 2B; 3A; 4C,D; 5A,B). The basal membrane of the gonadal epithelium is continuous with the basal membrane of the coelomic epithelium, and is situated inside a gonadal ridge (Fig. 4A). At the Gosner stage 26 in Anura (early limb bud), the epithelial cells give rise to somatic parts of a gonad, both future cortex, and medulla; the proliferating epithelial cells interrupt the basal membrane and migrate to the interior of the gonadal ridge,

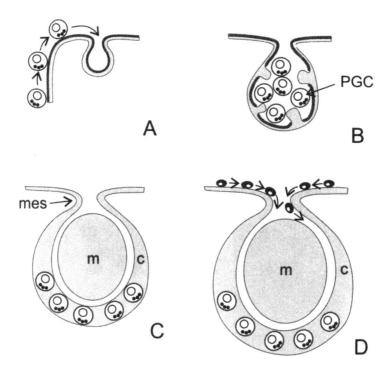

Fig. 4. Origin of somatic and germ cells of differentiating amphibian gonad. The figures represent right gonadal anlage shown in Figure 2B. Parts of gonads formed as a result of proliferation of coelomic epithelium are marked grey. A— Genital ridge is formed by invagination of coelomic peritoneal epithelium and its basal lamina (dark line). Here, pPGCs migrate across the basal lamina of the inner side of mesentery toward the gonadal ridge. B—PGCs are localized inside genital ridges, thereby forming the gonadal anlagen. Cells of the epithelial wall (shaded) of the gonad start to proliferate, disrupt the basal lamina, and invade the interior of gonadal anlage. C—Progenitor cells of peritoneal epithelium form the cortex (C) containing PGCs, and the medulla (M). D—Undifferentiated gonad immediately before sexual differentiation. Mesenchymal cells (shown in black, with arrows) start invading the space between the cortex and the medulla through the mesogonium.

thereby forming a medullar part (Fig. 4B) (Merchant-Larios and Villalpando, 1981; Iwasawa and Yamaguchi, 1984; Tanimura and Iwasawa, 1988, 1989). In a sexually undifferentiated gonad, the gonial cells derived from PGCs are present in its cortical part (Figs. 4C; 5B). Somatic cells of a gonadal anlage will give rise to follicular cells of the future ovary, or to seminiferous tubules, *rete testis,* and efferent ducts of the future testis. Also, at least some of the parenchymal cells of the testicular stroma originate from the coelomic epithelium.

6

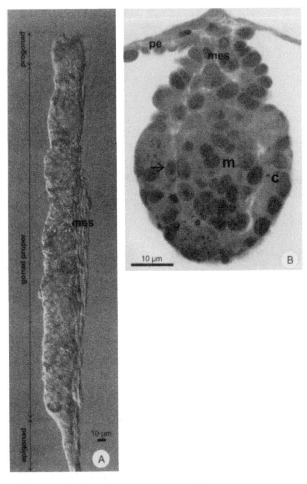

Fig. 5. Undifferentiated gonad. A—*Rana temporaria*, total view of 3 parts of a gonad: progonad, gonad proper, and epigonad. B—*R. ridibunda*, cross section of the gonad proper. Iron hematoxylin staining. c—cortex; m—medulla; mes—mesogonium; pe—peritoneal epithelium. The arrow indicates immigrating mesenchymal cells.

Additional somatic cells from other sources appear soon after sexual differentiation of gonads, when fibroblasts and blood vessels reach a gonad by a narrow space between the epithelial sheets of mesogonium (Figs. 4D; 5B). These cells originate from mesonephros, and invade the space between the cortex and the medulla through the mesentery (mesogonium). The descendants of these cells give rise to connective tissue cells (theca) in an ovary, and to interstitial Leydig cells in a testis. Thecal cells in an ovary are the main source of testosterone (T), which is converted into estrogen

(E2) by α-aromatase produced by follicle cells; interstitial cells in a testis produce androgens.

The problem of the origin of the somatic cells of the cortex and the medulla has been studied since Witschi (1931, 1957, 1959) postulated that the sexual differentiation of gonads was induced by the antagonistic action of two somatic cell-lines of different origin: the cortex, which originates from the coelomic epithelium, and the medulla, which originate from the mesonephros. As was described above, both the cortex, and the medulla have the same origin, i.e., the coelomic epithelium, and cells from the mesonephros invade the gonad *after* sexual differentiation. The only study, which supports the extragonadal origin of medullar cells, is devoted to *Bombina orientalis* (Lopez, 1989). However, this study is not detailed, and the origin of medullary cells in these species needs further investigation.

Origin and Differentiation of Germ Line Cells

Generally all cells, which will finally give rise to sperm or ova, are called germ cells. However, various authors do not use the same terminology for the same phases of germ cell formation. Germ cells from undifferentiated gonads are collectively known as primordial germ cells (PGCs) or gonocytes. Nieuwkoop and Sutasurya (1979) proposed the following nomenclature for germ cells: (1) presumptive primordial germ cells (pPGCs), for cells during extragonadal period, usually before active migration. In Anura they have asymmetrically localized germ plasm resulting from unequal mitotic division. This class of cells is characteristic of Anura, but absent from Urodela; (2) primordial germ cells (PGCs) for cells during migration, colonization and clonal multiplication within gonadal anlage and indifferent gonad; (3) oogonia or spermatogonia after sexual differentiation of gonads.

Primordial germ cells form at least two morphologically distinct classes of cells and it seems to be reasonable to name them differently. I strongly recommend using the term "gonia" or "gonial cells", as was originally suggested by Witschi (1929) for germ cells that are inside indifferent gonads and are derived from primordial germ cells. These cells are easily distinguished from PGCs because they are smaller and are devoid of yolk due to vitellolysis. In other words, gonial cells are the descendants of PGCs that proliferate inside undifferentiated gonad, and have no yolk. Until a gonad is sexually differentiated, it is not possible to distinguish oogonia from spermatogonia according to their morphology. The nomenclature and differentiation of germ cells is summarized in Figure 6. For sake of clarity, in the following sections I will use only the name PGCs, without classifying germ cells into pPGCs or PGCs.

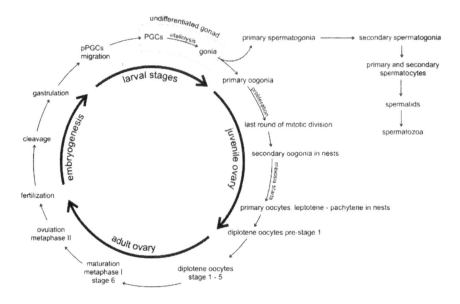

Fig. 6. Nomenclature of the germ-line cells in amphibians. pPGCs, presumptive primordial germ cells; PGCs, primordial germ cells.

Morphology and Differentiation of Germ-line Cells

PGCs are big cells, about 25-30 μm in diameter, with densely packed yolk platelets, a spherical or ovoid nucleus with a smooth nuclear envelope, mitochondria arranged in a group ("cloud"), centrioles, and germ plasm containing germinal granules (Figs. 3A-C; 7A,C,D). During their migration, PGCs multiply mitotically. There are four blastomeres in the blastula stage of *Xenopus laevis* that contain germ plasm (Whitington and Dixon, 1975); at the end of the tail-bud stage the number of cells carrying germ plasm (PGCs) increases to about 60 (Kamimura *et al.*, 1976). Other amphibian species have a similar number of PGCs (Hardisty, 1967). Vitellolysis, i.e., the process of yolk utilization, starts when PGCs reach genital ridges.

Primordial germ cells are the precursors of all germ cells. Among amphibians their origin is best known in Anura. Anuran PGCs contain unique granules (sex determinants, germinal granules, germ plasm) containing localized RNAs of maternal origin, *i.e.*, inherited from the vegetal hemisphere of egg cytoplasm (Blackler, 1970; Smith and Williams, 1979; Dixon, 1981; King, 1995; Matova and Cooley, 2001). The germ-cells specific RNAs (*Xcat2, Xwnt, Xpat, Xdazl, Xlsirts, fatVg, DEAD South, XFACS* and others) are localized to the germ plasm in the vegetal hemisphere and are transferred to germinal granules. The best-studied RNAs are *Xcat2* (Mosquera *et al.*, 1993), *Xpat* (Hudson and Woodland, 1998), *Xdazl*

(Houston *et al.*, 1998; Houston and King, 2000), and *Xfatvg* (Chan *et al.*, 1999, 2001). The more detailed information of the molecular characteristics of these RNAs is available in several reviews (King, 1995; Kloc and Etkin, 1995a,b; Kloc *et al.*, 1998, 2001).

The information on germ cell specific mRNAs in anurans other than *Xenopus laevis* is scanty and preliminary. Ogielska *et al.* (1998) detected the presence of *Xcat2* mRNA in small diplotene oocytes of *Rana pipiens* and *R. catesbeiana*. Marracci *et al.* (2004, 2007) isolated and characterized other genes that play a crucial role in germ line differentiation: *Rpum1, RV1, Vasa/ PL10*, and *RYB* (homologous to *Pumilio, Vasa, PL110*, and *Y-box* genes in *Drosophila* and mouse), respectively, in oocytes of *R. ridibunda, R. lessonae*, and *R. esculenta*.

Germ plasm forms in early larval gonia from the dense material produced inside the nucleus and emanating from the nuclear pores. The dense material forms aggregations, and becomes surrounded by mitochondria, thereby forming the "intermitochondrial cement" (Fig. 8B). In growing diplotene oocytes, the contact between the cement and mitochondria is gradually lost, and the cement transforms into granulo-fibrillar "nuage" material, which soon gives rise to the germinal granules (Williams and Smith, 1971; Kloc *et al.*, 2000, 2002).

The lack of germ plasm results in sterile gonads. This was demons-trated in experiments with embryos irradiated with UV in a dose, which does not damage embryogenesis, but destroys germ granules (Smith, 1966; Tanabe and Kotani, 1974; Ikenishi *et al.*, 1974; Williams and Smith, 1984). When PGCs are transplanted from normal to irradiated embryo, they will differentiate into the donor gametes. Such an experiment was carried out by Blackler (1962, 1970), who transplanted PGCs between *Xenopus laevis* and *Xenopus laevis victorianus*. These species produce eggs, which differ in color and size, thus it was possible to recognize whether the host embryo produces ova of its own, or donor type. Additionally, this experiment revealed that germ cells are not species specific, at least in closely related species.

The area of the germ plasm is surrounded by pigment granules (Ikenishi and Nakazato, 1986), and the granules of germ plasm are accompanied by various organelles, mitochondria being the most prominent (Ikenishi and Kotani, 1975). In the genus *Rana* and *Rhacophorus* there is a specific class of mitochondria accompanying germ plasm, which have paracrystalline inclusions inside cristae (Fig. 8A) (Kress and Spornitz, 1974; Ogielska, 1990).

Before fertilization, the germ plasm is distributed in the vegetal pole in form of small and numerous islets, which change their position during the cortical rotation following fertilization, then come back again and fuse,

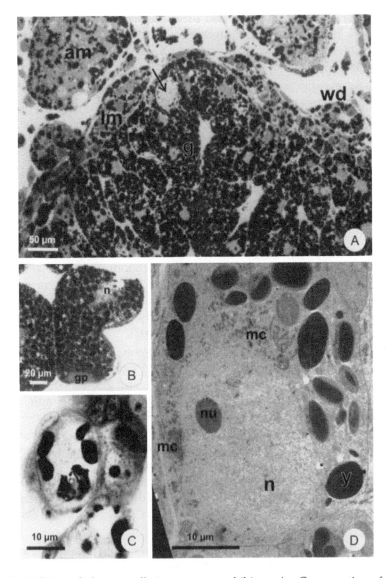

Fig. 7. Primordial germ cells in anuran amphibians. A—Cross section of a larva of *Rana lessonae* at Gosner stage 23 (compare with Fig. 2A,B). Hematoxylin and eosin (HE) staining. Arrow indicates a migrating PGC. B—Two blastomeres of *R. ridibunda* blastula during cleavage showing germ plasm (gp). Semi-thin section stained with toluidine blue. C—PGC from *R. lessonae* stained with iron hematoxylin. D—Transmission electron micrograph (TEM) showing the morphology of a PGC. am—axial mesoderm; g—gut; gp—germinal plasm; mc—mitochondrial cloud; lm—lateral mesoderm; n—nucleus; nu—nucleolus; wd—Wolffian duct; y—yolk.

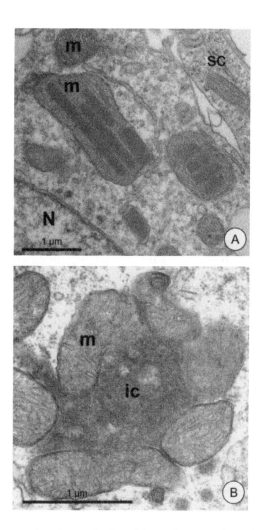

Fig. 8. TEM images showing mitochondria and germ plasm in gonial cells of *Rana temporaria*. A—Mitochondria containing paracrystalline inclusion bodies inside cristae. B—Another view of a gonial cell showing the intermitochondrial cement (germ plasm) surrounded by mitochondria. N—nucleus; m—mito-chondria; sc—somatic cells; ic—intermitochondrial cement.

thereby forming few bigger islets (Fig. 9A,B) (Savage and Danilchik, 1993); the movement is dependent on microtubules and their motor kinesin-like protein Xklp1 (Robb *et al.*, 1996). Beginning with the first cleavage furrow in Anura, the germ plasm is distributed to blastomeres and resulting endodermal cells according to a specific pattern (Figs. 1A; 9C) (Whittington and Dixon, 1975; Kamimura *et al.*, 1976; Riccio *et al.*, 1998). Up to 32-cell stage the germ plasm is restricted to 4 blastomeres as a result of asymmetric cell divisions. Then the germ plasm changes its position from peripheral in blastomeres (Fig. 7B) to juxtanuclear in PGCs (MacArthur et al., 1999). During the time period between blastula and early tadpole stages, the migrating PGCs undergo usually four symmetrical divisions, giving rise to about 60 cells containing germ plasm; these cells will invade gonadal anlagen (Dziadek and Dixon, 1975, 1977). This process was recently studied in *Xenopus laevis* by MacArthur et al. (1999). They revealed that one of the germ plasm mRNAs, *Xcat2*, is translated to protein, and accumulates in the germ plasm within a narrow window of time between the late blastula, and the beginning of neurula stages. The *Xcat2* protein most probably functions only during germ cell determination. In adults, the *Xcat2* expression is restricted to the maternal germline. It is detectable also in juvenile testes and ovaries in the first generation of pachytene spermatocytes and oocytes, respectively. Then the germ plasm disappears, and in females is restored again during diplotene stage as granules extruded from the nucleus, which soon aggregate to form the intermito-chondrial cement. In this way the germ cell-determinant is restored as the maternal information for the next generation. The cycle of formation, storage, and loss of the germ plasm was also described for water frogs (Ogielska, 1990), and is summarized in Figure 9.

The presence of sex determinants allows detection and localization of PGCs during their migration to the gonadal ridges. As was shown for *Rana dalmatina*, nuclei of PGCs selectively bind agglutinin, apparently having intranuclear lectin receptors. Lectin binding is restricted to PGCs during migration and is lost after settlement inside gonads. Generally, it is believed that the presence of germ plasm, and/or lectin binding receptors inside nuclei, protects PGCs from undergoing differentiation until they reach their final localization inside gonads (Riccio et al., 1998).

In Anurans, presumptive PGCs are localized in the endoderm, and— as members of the endoderm—are passively transferred to the forming gut. As was shown for *Rana pipiens*, the separation of the PGCs from endoderm occurs in two phases (Subtelny and Penkala, 1984). The first involves the penetration of lateral plate mesoderm cells into the upper portion of the vegetal blastomeres, and then the endoderm (Fig. 1C,D). The second phase involves the translocation of the PGCs from the endoderm to the dorsal root of the mesentery ("passive migration"). This is accomplished by passive displacement when the rudiment of dorsal mesentery is formed. These two phases of PGC movement are dependent upon the morpho-genetic events associated with the formation of dorsal mesentery, and do not involve the active migration. The active migration of PGCs was studied thoroughly in *Xenopus laevis*. At the beginning of the tail bud stage PGCs start to migrate actively and are localized in the dorsal part of the gut tube (Fig. 1E,F). At the time when the coelomic cavity is formed, the PGCs leave

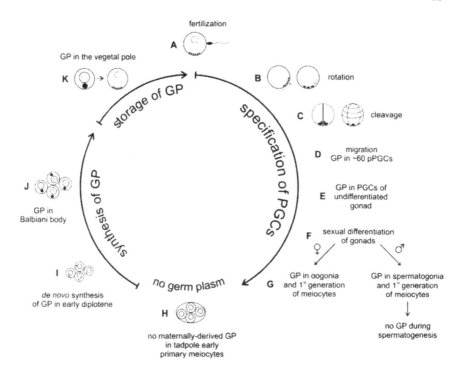

Fig. 9. Synthesis, storage, and utilization of germinal plasm in anuran amphibians. A—Fertilization; germ plasm (GP) represented by black dots is localized in the vegetal pole of an oocyte. B—Translocation of small inlets of GP during cortical rotation, and their fusion during come back to the original position. C—During cleavage up to 32-blastomere stage GP is localized to 4 vegetal blastomeres. D—During tail bud and early larval stages GP id distributed to about 60 PGCs in result of 4 cycles of mitotic divisions of GP-containing blastomeres. E—Translocation of PGCs ("passive migration") and active migration of PGCs from the gut, through mesentery to the genital ridges. F—After sexual differentiation of gonads, GP is present in oogonia and spermatogonia, and in the first generation of early primary oocytes and spermatocytes. G and H—Germ plasm diminishes from early oocytes arranged in nests. I—*De novo* synthesis of germ plasm at early diplotene. J—Localization of germ plasm in the Balbiani body. K—Migration of germ plasm from the Balbiani body to the vegetal pole of the diplotene oocyte.

the gut, migrate upward along the mesentery, and eventually to gonadal ridges (Fig. 2A,B) (Whitington and Dixon, 1975; Wylie and Heasman, 1975; Wylie *et al.*, 1975). The migration of PGCs from the gut to the gonadal ridges takes about a week, during Nieuwkoop and Faber stages 44–47 (Heasman and Wylie, 1978). PGCs form lobopodia and can move actively, as was evidenced both *in vitro* and *in vivo* (Wylie and Ross, 1976; Wylie, 1980).

PGCs move by producing filipodia on the cellular substrate prepared from a piece of mesentery. A PGC undergoes a series of contractions, which pushes the cytoplasm toward the leading filipodium. Then the next cycle starts and the cell moves forward. The movement is guided by fibronectin and is directed along the long axes of a substrate cells (Heasman et al., 1981). In the mesentery, PGCs move along the defined path over the inner surface (basal membrane) of the epithelial cells, and then are oriented toward the site of gonadal ridges formation (Fig. 2B).

In Urodela (Ambystoma, Triturus, Hynobius, Pleurodeles), precursors of PGCs are not distinguishable at early stages of development. Maufroid and Capuron (1977, 1985) studied germ line formation in Pleurodeles waltl and they stated that PGCs are the product of inductive interaction, which takes place after the 8-cell stage embryo. The PGCs originate from the ventral marginal zone (mesoderm) of gastrula. These cells migrate over the ventral blastopore lip together with presumptive posterior lateral plate mesoderm (Nieuwkoop, 1950; Nieuwkoop and Sutasurya, 1979; Perma, 1984). In Urodela (Pleurodeles waltl) primordial PGCs undergo usually 3 mitotic cycles during 2 weeks until they reach genital ridges, resulting in an average number of 97 per individual. During next 2 weeks they stop proliferating and migrate into genital ridges (Dournon et al., 1989).

Nieuwkoop and Sutasurya (1979), and Dixon (1981) summarized information available for PGCs and gonad formation in Urodela. Humphrey first described PGCs morphology in Hemidactylium and other species during 1925-1928 (reviewed by Nieuwkoop and Sutasurya, 1979). Kotani (1958) and Amanuma (1958) described primordial germ cell morphology and formation in Triturus pyrrhogaster. Ikenishi and Nieuwkoop (1978) described ultrastructure of PGCs at larval stages of the axolotl Ambystoma mexicanum, and Hamashima and Kotani (1977), and Nieuwkoop and Sutasurya (1979) of the newt, Triturus. Their results reveal that the general morphology, including the presence of intermitochondrial cement, in urodelan PGCs strongly resembles the morphology of PGCs in anurans. As Nieuwkoop and Sutasurya (1979) concluded "in the urodeles, PGCs do not develop from predetermined elements, but arise strictly epigenetically from common, totipotent cells of the animal moiety of the blastula, as a part of regional induction of the mesoderm by vegetal yolk endoderm". The origin of urodelan germ line cells from the animal hemisphere, or from ectoderm, is supported by the morphological observation of pigment granules, which are present within the cytoplasm of PGCs of Triturus pyrrhogaster (Hamashima and Kotani, 1977).

Apparently germ cell specification in Urodela is regulative, i.e., epigenetically controlled by signaling of neighboring cells on some cells not predisposed (preformed) to germ cell fate (Wakahara, 1996). This resembles the germ cell specification in mammals. As it was described for mice, the molecular signaling from the neighboring cells of the extra-embryonic ectoderm is essential for germ cell determination (reviewed by McLaren, 1999).

Germ plasm in Urodela is first detectable at larval stages, when PGCs are in close vicinity of gonadal ridges. One of maternal mRNAs charac-teristic of germ line of various animal species belongs to the *DAZ* family of genes. In *Xenopus laevis* the germ plasm protein transcribed from maternal *Xdazl* mRNA is necessary for the exit of PGCs from endoderm and migration toward mesentery and gonadal ridges (Houston and King, 2000). Johnson *et al.* (2001) described a *DAZ*-like sequence in the axolotl, *Axdazl*, homologous to *Xdazl*, and used it as a molecular marker of PGCs in this urodelan species. They detected product of this gene in PGCs as they approach and enter gonads. *Axdazl* is not localized in oocytes and early embryos up to tail-bud stage. Maternal *Axdazl* RNA is found in the animal cap, and is present in equatorial region of early embryos. Up to tail-bud stage *Axdazl* RNA is widely distributed. Concomitantly with *Axdazl* RNA being localized to PGCs, germ plasm is morphologically recognized. In adults, *Axdazl* is expressed only in testes and ovaries, and not in somatic tissues (Johnson *et al.*, 2001).

Summarizing, amphibians represent two different modes of PGCs formation known in animals: predetermined (preformed) with continuous germ line inheriting germ plasm from generation to generation in Anura, and not determined (epigenetic), characterized by a germ line that arises *de novo* in each generation in Urodela. As was pointed out by Nieuwkoop and Sutasurya (1976), the different origin of PGCs strongly support the hypothesis of polyphyletic origin of the two orders of amphibians. No data concerning the origin of PGCs are available for Gymnophiona.

Sexual Differentiation of Gonads

Morphological changes accompanying sexual differentiation of gonads are in many cases difficult to detect and may vary slightly among species. This causes difficulties in generalization and classification of possible differences. The main debate concerns the question, whether testes differentiate directly from undifferentiated gonads, or they pass through a transition phase, displaying ovary features during some time, and how long is this period.

The most common hypothesis is that differentiation of a gonad into an ovary is a primary process, and a female sex XX should be considered as plesiomorphic. This hypothesis was confirmed by experiments on a number of species of amphibians, in which ovaries of genetic female larvae underwent transformation into testes following parabiosis with genetic male larvae, or after transplantation of testis grafts. Opposite experiments did not cause transformation of testes into ovaries indicating that male gonad must produce substance(s) having the ability to sex reversal in females (reviewed by Hayes, 1998).

Gramapurohit et al. (2000) and Saidapur et al. (2001) suggested three major patterns of gonad differentiation in Anura: (1) differentiated type in which indifferent gonad directly differentiate either into ovary or testis, (2) undifferentiated type in which indifferent gonad either differentiate into ovary or remain indifferent for some time and then differentiates into testis, and (3) semidifferentiated type in which gonads differentiate initially into ovaries irrespective of the genetic sex, and later the ovary in a genetic male transform into testis following degeneration of oocytes and reorganization of gonads. Essentially, types (1) and (2) may be considered as differentiated type, because the difference concerns the time of sex differentiation of males, but not the sequence of morphological changes. The semidifferentiated type is not common, and was described for *Rhacophorus arboreus* (Tanimura and Iwasawa, 1989) and *Rana curtipes*, the endemic species to southern India (Gramapurohit et al., 2000). These authors revealed that gonads initially differentiate into ovaries in all individuals but later oocytes degenerate and testicular differentiation occurs in genetic males. The differentiation of seminiferous tubules from seminiferous cords was not shown, and the tubules were seen after the degeneration of diplotene oocytes. The presence of diplotene oocytes in testes (testis-ova) was described in the genus *Rana* and *Rhacophorus* (Iwasawa and Kobayashi, 1976; Kobayashi and Iwasawa, 1988; Tanimura and Iwasawa, 1989; Hsu et al., 1977, 1979) (see 'Spermatogenesis and Male Reproductive System in Anura' in this volume). Testis-ova were also occasionally observed in *Rana lessonae*, but it cannot be connected with semidifferentiated type of gonadal sex differentiation, because this species displays a typical differentiated type of gonad differentiation, described herein as type (1) (Ogielska and Bartmańska, 1999).

In earlier studies (Iwasawa, 1959, 1960; Hsu and Wang, 1981), gonads of some species belonging to the families Ranidae and Rhacophoridae (*Rana temporaria ornativentris, Rana catesbeiana, Rana japonica, Rana nigromaculata, Rhacophorus schlegelii*) were also described as developing first as ovaries, thus representing type 3. Under the influence of androgens produced by somatic cells of a gonad, they transform into testes, according to the genetic sex. However, more recent comparative studies on ovary development in 12 species of anuran amphibians belonging to 6 families, including 6 species of the family Ranidae, (*Rana lessonae, Rana ridibunda, Rana temporaria, Rana arvalis, Rana pipiens, Rana catesbeiana, Bombina bombina, Hyla arborea, Bufo bufo, Bufo viridis, Xenopus laevis* and *Pelobates fuscus*) (Ogielska and Kotusz, 2004) revealed that the first stages of gonad differ-

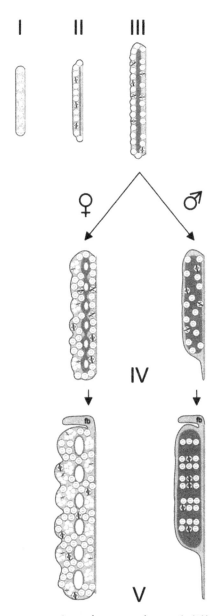

Fig. 10. Schematic representation of stages of sexual differentiation of anuran amphibian gonads. Stage I-III, the undifferentiated gonad with a central medulla and cortex containing PGCs. Stage IV and V, the early ovary (left) is formed from the entire anlage, a lumen is formed in the core, and germ-line cells are located in the cortex; the early testis (right) is shorter than its anlage, and germ-line cells migrate toward the center of the gonad; no lumen is formed, and the cortex becomes thin and devoid of germ-line cells. fb—fat body.

entiation are the same for both sexes. After the same time, usually 5 weeks, gonads start to differentiate either as testes, or ovaries, with no transition phase (Fig. 10). In undifferentiated gonads, the cortex, containing PGCs and gonia, is clearly visible and remains distinct in females, whereas disappears gradually in males. Probably it was the reason why undifferentiated gonads might have been mistaken with ovaries by some authors. In the light of above study, the type (3) pattern of gonadal differentiation, *i.e.*, the transformation of a differentiated ovary into a testis, should be abandoned. In most anuran species, sexual differentiation of gonads takes place when tadpoles are an age of about 6 weeks (timing starts with the onset of cleavage) (Ogielska and Kotusz, 2004). In *Pleurodeles waltl* (Urodela), sexual differentiation starts after 11 weeks of development (Dournon *et al.*, 1990).

The role of PGCs in sexual differentiation of gonads has long been a matter of discussion. Experiments with germ cell transplantation in *Xenopus laevis* (Blackler, 1962, 1965, 1970) indicate that PGCs have no effect on sexual differentiation of gonads, which differentiate into ovaries or testes according to the host, not the donor sex. Similar results were obtained by Dournon *et al.* (1990) for an urodele *Pleurodeles waltl*. On the other hand, Shirane (1987) studied the role of PGCs in sexual differentiation of gonads in *Rana japonica* and *R. nigromaculata*. He showed that PGCs have molecules displaying peanut-lectin affinity, which act as trigger for the expression of genes controlling sexual differentiation of somatic cells of a gonad. He suggested that somatic cells, which respond to the trigger, might differ in terms of the threshold for such a response between males and females. His findings suggest that somatic cells of a gonad are not sexually predetermined, but sexual determination follows the response to signal molecules produced by PGCs.

Genetic Control

There is no doubt that one or several genes control sexual differentiation of gonads in amphibians. However, the sex-determining genes still remain unknown, although several sex-linked genes were described (Masayuki and Nishioka, 2000; Sumida and Nishioka, 2000). The differences between species indicate that there is no common ancestral or conserved sex-linkage group in amphibians. The most convincing results, supporting the existence of genetic control of gonad differentiation is the presence of sex chromosomes in some amphibian species, although in most species sex chromosomes are not distinguishable from autosomal chromosomes (for review see Green and Sessions, 1991). Data about sex chromosomes in Gymnophiona are still lacking. However, even in the limited sample, a great variety of sex chromosomes have been described. The available data reveal that genetic sex determining systems have evolved independently in various taxa. For this reason, the study of amphibian sex chromosome evolution is challenging.

Two systems of chromosome sex determination exist in amphibians: XX/XY (male heterogametic XY), or ZZ/ZW (male homogametic ZZ) (for reviews see Gallien, 1965; Beçak, 1983; Schmidt et al., 1991). Sex chromosomes evolved by gradual loss of functional genes on a chromosome bearing sex determining genes, and subsequent heterochromatinization of remaining non-coding DNA. In amphibians, Y or W sex chromosomes are at various stages of evolution, from no detectable sex chromosomes in *Triturus helveticus* and hetrochromatin bands in *T. alpestris* and *T. vulgaris* (Schmid et al., 1979), late replicating bands in *Rana esculenta* (Schempp and Schmid, 1981), two different forms of Y chromosome in *Gastrotheca pseustes*, up to well defined sex chromosomes of various length in *Gastrotheca riobambae, Necturus maculosus, Centrolenella antisthenesi* and others (Schmid et al., 1991). Generally, Y and W chromosomes have more constitutive heterochromatin than autosomes, although *Gastrotheca walkeri* and *G. ovifera* contain less heterochromatin than autosomes (Schmid et al., 2002).

The basic system in amphibians seems to be ZZ/ZW. The coexistence of the XX/XY and ZZ/ZW systems is known only in one species, *Rana rugosa*, from various populations in Japan (Miura et al., 1996, 1998). The specimens of *R. rugosa* from western Japan have the XX/XY system, whereas the specimens from northeastern Japan have the ZZ/ZW system. A unique system 0W/00 was described in *Leiopelma hochstetteri* (Green, 1988a), which most probably evolved from ZZ/ZW found in closely related species *L. hamiltonii* (Green, 1988b). Sex determination in *L. hochstetteri* depends on the supernumerary univalent W chromosome characteristic of females, and absent from males.

The system of sex determination can also be studied experimentally by production of sex-reversed progeny, which, after reaching sexual maturity, are backcrossed to normal counter sex. This way of study sex determination is still in use, especially in species with no differentiated sex chromosomes. First proofs of sex-determining mechanisms resulted from experimental studies on gonad primordia excision and grafting during early larval stages in *Ambystoma mexicanum* and *A. tigrinum* were performed by Humphrey in years 1942-1957 (see Gallien, 1965; Schmid et al., 1991). When the primordium of ovary in a genetic female was unilaterally excised and replaced by a testis primordium from genetic male, the graft modifies the original ovary left in the host body, and reversed it into functional testis. After removal of the grafted testis, the remaining one (originally ovary) retained a male character and transformed into a functional testis. In that way the individual became a sex-reversed neo-male. After analysis of sex ratio in the progeny (male to female 1:3) of the neo-male mated with normal female, one could state that the genetic female was heterogametic ZW. Similar experiments were carried out also for other members of the ZZ/ZW system (*Pleurodeles waltl, Xenopus laevis*), and XX/XY system (*Rana japonica, R. temporaria*) (reviewed by Gallien, 1965).

The experimental sex-reversal can also be induced by hormonal treatment of larvae at critical stages of development. Such experiments were carried out in *Xenopus laevis*, in which all progeny obtained from estradiol-treated tadpoles were functional females (reviewed by Hayes, 1998). Half of the neo-females (*i.e.*, inverted genetic males), when mated to normal males gave male-only progeny. This result supported the hypothesis, that female was a heterogametic sex. Similar experiments with frogs (*Rana nigroma-culata, Rana japonica, Rana brevipoda, Hyla arborea* and *Bombina orientalis*) revealed the existence of the XX/XY system in these species (Kawamura and Nishioka, 1977).

Hormonal Control

Although sex determination in amphibians is controlled genetically, hormonal regulation of sexual differentiation of gonads is possible, and the role of hormones in sex differentiation is still discussed. The results concerning the role of steroid hormones, thyroid and pituitary glands, as well as time of their action differ among species (for review, see Hayes, 1998).

Mechanisms underlying sexual gonad differentiation are still not clear, either because of a variety of processes existing among species, or because of diverse and not unified methodology used in experiments. The problem has been studied since Witschi (1931, 1957, 1959) postulated that the sex of gonads is regulated by antagonistic system of substances produced by somatic cells of different embryonic origin. According to his theory, cells of the medulla are formed by the mesenchymal cells of the extragonadal origin i.e., the renal blastema. These cells secret the "medullarin", whereas cells of the cortex which are formed by the coelomic epithelial cells secret "cortexin". Medullarin was believed to promote differentiation of germ cells into spermatogonia after they move into the medulla, whereas cortexin was believed to promote differentiation of cells into oogonia if they stay in the cortex. According to this hypothesis, sex-determining genes would be active only during short and definite period, thereby starting a chain reaction resulting in sexual differentiation of gonads. The nature and way of action of cortexin and medullarin was tested mainly in the first half of 20th century in a number of experiments with parabiotic larvae of different sex, trans-plantation of gonads at various stages of development, and exogenous estrogen treatment. These experiments are summarized and reviewed by

Gallien (1965), and Hayes (1998). Although most of them are now of historical meaning, there is no doubt that both the medulla and the cortex are composed of cells originated from the coelomic epithelium. Witschi's hypothesis of the substances produced and acting locally within gonadal tissue, and not transported by blood circulation, was the first that dealt with the role of somatic cells in sexual differentiation of germ line cells.

Sex steroids, estrogens and androgens, have important functions due to interference with other endocrine factors, which facilitates normal sexual differentiation and development of gonads. Estrogens always lead to feminization, whereas masculinization caused by androgens depends on species. Exogenous estrogen treatment of larvae results in significant bias of sex ratio caused by sex reversal in a number of amphibian species, both Anura and Urodela. In addition, also other steroids, such as corticoids, have similar effect (summarized by Hayes, 1998). Data on Gymnophiona are lacking.

The main hypothesis concerning sex differentiation in amphibians deals with the primary genetic sex determination leading to sexual differentiation of larval gonads, which in turn produce sex steroids responsible for secondary sexual features (summarized in Fig. 11) (Hayes, 1998). To validate this hypothesis, Bögi et al. (2002) provided interesting data from experiments with administration of various steroid hormone doses during embryonic development of Xenopus laevis. They studied both the role of estrogen (17β estradiol, E2) and androgens (testosterone, T and dihydrotestosterone, DHT), and the level of mRNAs of their receptors. One of the surprising results was that both estrogen (E2), and androgens (T and DHT), were present in high concentrations in eggs, embryos, and hatched larvae. The hormones were probably transferred from female to eggs during oogenesis, and stayed accumulated in embryos. Level of the hormones decreases dramatically about 2 weeks after hatching and their endogenous production starts immediately after sex differentiation of gonads. The level of androgens is higher in males and the level of estradiol is much higher in females. Neither steroid hormone receptors, nor their mRNAs were detected in eggs and embryos, but were detected in larvae after hatching. Maternal sex steroids accumulated in eggs and embryos may probably induce expression of zygotic genes responsible for receptor mRNAs transcription, and finally to receptor synthesis. Experiments, in which Xenopus laevis were exposed to estrogens, antiestrogens, testosterone and antitestosterone gave the same results that are well known for mammals: estrogens biased sex ratio to feminization, anti-estrogens lead to neutralization caused by underdeveloped gonads, without changing sex ratio; testosterone did not change sex ratio, DHT induced significant masculinization, whereas anti-androgen caused feminization (Bögi et al., 2002).

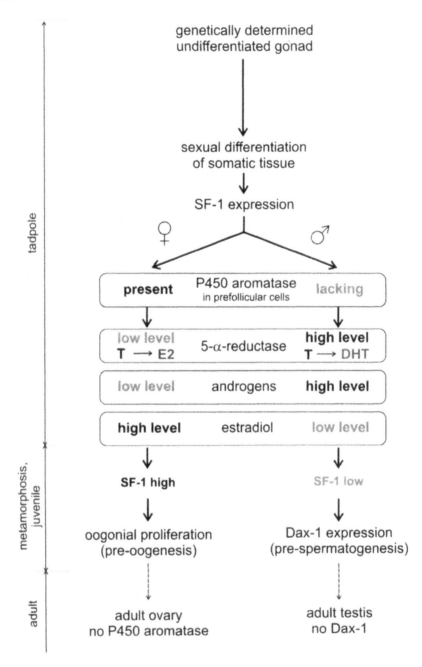

Fig. 11. A generalized model of genetic and hormonal control of gonad differentiation in anuran amphibians. E2—17β estradiol; T—testosterone; DHT—dihydrotestosterone; SF-1—steroidogneic factor 1.

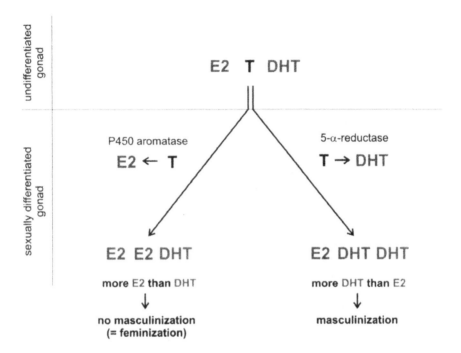

Fig. 12. Details of the hormonal control of sexual differentiation of gonads in amphibians with special reference of the fate of maternally inherited testosterone. E2—17β estradiol; T—testosterone; DHT—dihydrotestosterone.

The observations of Bögi *et al.* (2002) described above for the anuran species *Xenopus laevis* may be also valid for other amphibians. Maternally derived testosterone in genetic males has to be converted endogenously into DHT by the enzyme 5 α-reductase, which is considered to be a key in producing more active forms of androgens during testis differentiation (Fig. 12). Zaccanti *et al.* (1994) showed that larval *Bufo bufo* and *Rana dalmatina* treated with inhibitor of 5α-reductase result in more pronounced ovary development. Indirect evidence was obtained that a pronounced activity of 5 α-reductase preferentially expressed in genetic males might be a natural trigger for induction of sexual differentiation toward masculinization.

Still open are questions about the role of maternal E2 and conversion of maternal T into DHT by the aromatase. The results obtained for *Xenopus laevis* by Myiata *et al.* (1999) and Myiashita *et al.* (2000) revealed that the level of maternally inherited E2 is sufficient for ovary differentiation. However, the results obtained for other amphibian species indicate that the additional amounts of E2 are necessary for normal ovary differentiation. Chardard *et al.* (1995) revealed high level of aromatase (an enzyme converting androgens to estrogens) in larvae of an urodelan *Pleurodeles waltl* differentiating into females. Recently, Kato *et al.* (2001) cloned P450 aromatase cDNA (*P450arom*) and studied its expression in the gonads of *Rana rugosa* female, and female-to-male reversed tadpoles. They observed that the *P450arom* expression starts at the same time when sexual differentiation of gonads started; its level was high during ovarian differentiation and remained high in young ovaries 2 month after metamorphosis when small diplotene oocyte were present. After the injection of testosterone to the tadpole's body cavity, the level of *P450arom* decreased rapidly and the ovaries were transformed into testes. The transformation was demonstrated by the dramatic morphological change of the ovaries, degeneration of early diplotene oocytes, and differentiation of testicular tissues. *P450arom* was also absent in the testes and adult ovaries. Immunocytological studies of histological sections revealed that expression of *P450arom* takes place in the follicle cells of the growing ovary, thus indicating *P450arom* as an enzyme involved in ovary differentiation. These experiments also showed that the period of P450 aromatase activity is relatively short. The P450 aromatase expression is preceded by the expression and synthesis of steroidogenic factor 1 (SF-1), an evolutionary conserved nuclear receptor belonging to the steroid receptor superfamily. It was cloned and studied in *Rana catesbeiana* (Mayer *et al.*, 2002). Larval females had significantly higher levels of SF-1 than males. It is postulated that higher level of SF-1 promotes aromatase expression, and as a consequence, estradiol production. The other member of the nuclear receptors, Dax-1, is thought to act as a transcriptional factor to repress testicular differentiation in vertebrates. It is not an ovarian determining gene, but has a critical role in spermatogenesis. It was also cloned and detected in an amphibian, *Rana rugosa* (Sugita *et al.*, 2001). The *Dax-1* expression starts at the mid-blastula transition and becomes stronger before sexual differentiation of gonads. At the beginning it is expressed in both sexes, but after metamorphosis it is still present in testes, but not in ovaries. In adult gonads it is not expressed. This can be interpreted as a factor important during prespermatogenesis.

The possible existence of a "steroid sensitive window" during gonadal sex differentiation was discussed by Saidapur *et al.* (2001). As is suggested by Bögi *et al.* (2002) and Kloas (2002), maternally inherited steroids and their derivatives (see Fig. 11) may induce the expression of their corresponding receptors and thereby lead to the existence of short-term sensitivity to steroids. Villalpando and Merchant-Larios (1990) demonstrated experimental sex reversal in developing gonads in *Xenopus laevis* tadpoles treated with exogenous estradiol benzoate. The success in sex reversal depended on the stage of gonad differentiation. One hundred percent of sex reversal into females was obtained in genetic males when estradiol was administered before migration of gonial cells from the cortex of undifferentiated gonad into medulla. Ambiguous gonad was obtained when estradiol was added during migration of gonial cells into medulla, and no effect was observed when the hormone was administered when migration of gonial cells into medulla was already finished. The hypothesis proposed by Villalpando and Merchant-Larios (1990) is that genotypic females start to produce estradiol earlier than males, and this mechanism inhibits the differentiation of medullary cells. As a result, germ cells remain in the cortex and an ovary is formed. The results, however, may vary for different species; in *Rana curtipes* such a "window" was not detected. Treatment of *R. curtipes* larvae with testosterone and estradiol-17β during larval period shows no apparent effect on the gonadal sex differentiation (Gramapurohit *et al.*, 2000). The question of endogenous production of steroid hormones during larval development is still open.

The problem of global amphibian decline is now in focus (for review see Alford and Richards, 1999), because the cause of this phenomenon is still obscure. One of the causes may be chemical pollution of the environment. Some of contaminations can alter or mimic the action of steroid hormone metabolism, thereby influencing the sex ratio in amphibian larvae living in contaminated water (reviewed by Kloas, 2002). Such contaminations, known as environmental endocrine disruptors, have been shown to change the sex ratio in progeny of *Acris crepitans* (Reeder *et al.*, 1998) and *Rana rugosa* (Othani *et al.*, 2000).

Data about Urodela and Gymnophiona are scarce or lacking. Experiments elucidating hormonal regulation of sexual differentiation must be continued with different groups of amphibians. It is also worth noting that comparative studies on gonad development and differentiation are usually carried out according to somatic stages of larval and tadpole development. As was shown by Ogielska and Kotusz (2004), gonads have their own timing of differentiation, almost independent of somatic growth. Thus, some contradictory data may be a result of administration of hormones to tadpoles staged accordingly to the somatic, not gonadal development. Moreover, the time of sexual differentiation of gonads differ among species.

There are also a number of experiments with the role of other hormones, such as thyroid or pituitary, in sexual differentiation of gonads. Although these hormones can alter the sex of progeny in Anurans (reviewed by Hayes, 1998), the mechanism is unclear. On the other hand, thyroid and pituitary hormones are not needed for sexual differentiation and further development of gonads, as was shown in hypo- and thyroidectomized tadpoles of *Rana catesbeiana* (Chang and Hsu, 1987; Hsu *et al.*, 1971, 1982). The best example, however, is the existence of neoteny, i.e., sexual maturity of urodelan larvae. Generally, neoteny is a result of insufficiency of hormones or their receptors during metamorphosis. It can be caused by the lack of thyrotropin-releasing hormone (TSH-RH), as in *Ambystoma gracile* or *A. tirgrinum*, lack of thyroid-stimulating hormone (TSH), as in *A. mexicanum*, or defective receptors of TSH as in *Siren, Necturus* and *Eurycea neotens* (Gilbert, 2000), which is a good evidence that neoteny has evolved independently in various families of Urodela.

Role of Temperature

Temperature is an environmental factor, which can control sex determination in cold-blooded vertebrates. The best known are examples of most species of turtles and all crocodilians, among which sex of progeny is established during development according to temperature, but always within natural ranges. In the European turtle *Emys orbicularis* incubation of eggs in temperature below 25°C leads to all-male progeny, whereas temperature above 30°C produces all-female progeny. It was revealed that in turtles estrogen is essential for ovary differentiation. The enzyme aromatase, which converts androgens into estrogens, displays temperature-dependent activity (low in 25°C, and dramatically increased in 30°C). It is not known, however, whether this mechanism is controlled at the level of transcription, or protein activity (Gilbert, 2000).

The influence of temperature on sex of progeny has been studied since the 1930s, both in anurans (*Bufo, Rana*) and urodeles (*Pleurodeles, Hynonbius*) (reviewed by Hayes, 1998). These studies showed that the temperature of water can affect the process of sexual differentiation of gonads in reared larvae. However, in all cases experiments were carried out in high temperatures, close to lethal extremes (up to 36°C). Recently such experiments were carried out for *Hyperolis viridiflavus, Bufo boreas* and *Pyxicephalus adspersus*, but no differences in the sex ratio of progeny were observed (Hayes, 1998). Temperature-dependent sex differentiation of gonads was studied recently also in an urodele *Pleurodeles waltl* (Dournon et al., 1990; Chardard and Dournon, 1999; Chardard et al., 1995). In this species of newt, males are homogametic ZZ and females heterogametic ZW. Under natural temperature conditions (about 20°C) sexual differentiation conforms strictly the sexual genotype. However, when temperature is raised to 32°C during a thermosensitive period at the beginning of sexual differentiation of gonads, larvae of female ZW genotype develop and mature as fertile neo-males.

Summarizing, it should be stated that temperature seems to convert genetic sex only in abnormally high ranges, which probably alter natural action of hormones or hormone regulating enzymes. Within natural temperature ranges, sexual differentiation of gonads is controlled genetically.

The role of temperature on sex determination in amphibians is still not well understood, and any generalization is possible at the moment. However, temperature is still taken into consideration as a possible epigenetic factor in sex determination in these poikilothermic animals, but evidence from natural populations is lacking (reviewed by Eggert, 2004).

Acknowledgements

I would like to thank my former Ph.D. students: Agnieszka Kotusz for providing photos 3A, 5A, 8A,B of developing gonads of *Rana temporaria*, and Piotr Kierzkowski for preparing schematic drawings and helpful discussions.

References

Alford RA, SJ Richards. 1999. Global amphibian declines: A problem in applied ecology. Annu. Rev. Syst. 30: 133-165.

Amanuma A. 1958. On the role of the dorso-caudal endoderm in the formation of the primordial germ cells. J. Inst. Polytechnics Osaka City Univ. Series D 9: 211-216.

Beçak W. 1983. Evolution and differentiation of sex chromosomes in lower vertebrates. Differentiation 23 (Suppl.): S 3-S 12.

Blackler AW. 1962. Transfer of primordial germ cells between two subspecies of *Xenopus laevis*. J. Embryol. Exp. Morph. 10: 641-651.

Blackler AW. 1965. Germ-cell transfer and sex ratio in *Xenopus laevis*. J. Embryol. Exp. Morph. 13: 51-61.

Blackler AW. 1970. The integrity of the reproductive cell line in the Amphibia. Current Topics in Dev. Biol. 5: 71-87.

Bögi C, G Levy, I Lutz, W Kloas. 2002. Functional genomics and sexual differentiation in amphibians. Comp. Biochem. Physiol. B 133: 559-570.

Chan AP, M Kloc, L Etkin. 1999. fatvg encodes a new localized RNA that uses a 25-nucleotide element (FVLE1) to localize to the vegetal cortex of *Xenopus* oocytes. Development 126: 4943-4953.

Chan AP, M Kloc, S Bilinski, L Etkin. 2001. The vegetally localized mRNA *fatvg* is associated with germ plasm in early embryo and is later expressed in the fat body. Mech. Dev. 100: 137-140.

Chang L-T, CY Hsu. 1987. The relationship between age and metamorphic progress and the development of tadpole ovaries. Proc. Natl Sci. Counc. B ROC 11: 211-217.

Chardard D, G Desvages, C Pican, C Dournon. 1995. Aromatase activity in larval gonads of *Pleurodeles waltl* (Urodele Amphibian) during normal sex differentiation and during sex reversal by thermal treatment effect. Gen. Comp. Endocrinol. 99: 100-107.

Chardard D, C Dournon. 1999. Sex reversal by aromatase inhibitor treatment in the newt *Pleurodeles waltl*. J. Exp. Zool. 283: 43-50.

Dixon KE. 1981. The origin of the primordial germ cells in the Amphibia. Netherlands J. Zool. 31: 5-37.

Dournon CH, CH Demassieux, D Durand, M Lesimple. 1989. Primordial germ cell proliferation in the salamander *Pleurodeles waltl*: genetic control before gonadal differentiation. Int. J. Dev. Biol. 33: 477-485.

Dournon CH, D Durand, CH Demassieux, M Lesimple. 1990. Differential germ cell proliferation in the salamander *Pleurodeles waltl*: controls by sexual genotype and by thermal epigenetic factor before differentiation of sexual phenotype of gonads. Int. J. Dev. Biol. 34: 365-375.

Dziadek M, KE Dixon. 1975. Mitoses in presumptive primordial germ cells in post-blastula embryos of *Xenopus laevis*. J. Exp. Zool. 192: 285-291.

Dziadek M, KE Dixon. 1977. An authoradiographic analysis of nucleic acid synthesis in the presumptive primordial germ cells of *Xenopus laevis*. JEEM 37: 13-31.

Eggert C. 2004. Sex determination: the amphibian models. Reprod. Nutr. Dev. 44: 539-549.

Gallien LG. 1965. Genetic control of sexual differentiation in Vertebrates. In: Organogenesis, (Eds) De Haan RL, Ursprung H. Holt, Reinhart, and Winston, New York, Chicago, San Franciso, Toronto, London. Pp. 583-610.

Gilbert S. 2000. Developmental Biology, Sinauer Associates, Inc. Sixth edition.

Gosner LK. 1960. A simplified table for staging anuran embryos and larvae with notes on identification. Herpetologica 16: 513-543.

Gramapurohit NP, BA Shanbhag, SK Saidapur. 2000. Pattern of gonadal sex differentiation, development and onset of steroidogenesis in the frog, *Rana curtipes*. Gen. Comp. Endocrin. 119: 256-264.

Green DM. 1988a. Cytogenetics of the endemic New Zealand frog, *Leiopelma hochstetteri*: extraordinary supernumerary chromosome variation and a unique sex-chromosome system. Chromosoma 97: 55-70.

Green DM. 1988b. Heteromorphic sex chromosomes in the rare and primitive frog Leiopelma *hamiltoni* from New Zealand. J. Heredity 79: 165-169.

Green DM, SK Sessions. 1991. *Amphibian Cytogenetics and Evolution*, Academic Press. pp: 393-430.

Hamashima N, M Kotani. 1977. Ultrastructure of the primordial germ cells in the newt, *Triturus pyrrhogaster*. Zool. Magazine 86: 239-245.

Hardisty MW. 1967. The number of vertebrate primordial germ cells. Biol. Rev. 42: 265-287.

Hayes TB. 1998. Sex determination and primary sex differentiation in amphibians. Genetic and developmental mechanisms. J. Exp. Zool. 281: 373-399.

Heasman J, CC Wylie. 1978. Electron microscopic studies on the structure of motile primordial germ cells of *Xenopus laevis in vitro*. J. Embryol. Exp. Morphol. 46: 119-133.

Heasman J, RO Hynes, AP Swan, V Thomas, CC Wylie. 1981. Primordial germ cells of *Xenopus laevis* embryos: the role of fibronectin in their adhesion during migration. Cell 23: 427-447.

Houston DW, J Zhang, JZ Maines, SA Wasserman, ML King. 1998. A *Xenopus DAZ*-like gene encodes an RNA component of germ plasm and is a functional homologue of *Drosophila boule*. Development 125: 171-180.

Houston DW, ML King. 2000. A critical role of *Xdazl*, a germ plasm-localized

RNA, in the differentiation of primordial germ cells in *Xenopus*. Development 127: 447-456.

Hsu CY, NW Yu, HM Liang. 1971. Age factor in the induced metamorphosis of thyreoidectomized tadpoles. J. Embryol. Exp. Morphol. 25: 331-338.

Hsu CY, CH Chiang, HM Liang. 1977. A histochemical study on the development of hydroxysteroid dehydrogenases in tadpole ovaries. Gen. Comp. Endocrinol. 32: 272-278.

Hsu CY, LH Hsu, HM Liang. 1979. The effect of cyproterone acetate of the activity of Δ^5-3β-hydroxysteroid dehydrogenase in tadpole sex transformation. Gen. Comp. Endocrinol. 39: 404-410.

Hsu CY, HJ Wang. 1981: Production of sterile gonads in tadpoles by 17β-ureide steroid. Proc. Nat. Sci. Council China, part B: Biological Science 5: 322-327.

Hsu CY, KL Li, MH Lu, HM Liang. 1982. The presence of intramitochondrial yolk crystals in oocytes of hypophysectomized bullfrog tadpoles. Develop. Growth Differ. 24: 319-325.

Hudson C, HR Woodland. 1998. *Xpat*, a gene expressed specifically in germ plasm and primordial germ cells of *Xenopus laevis*. Mech. Dev. 73: 159-168

Ikenishi K, M Kotani. 1975. Ultrastructure of the 'germ plasm' in *Xenopus* embryos after cleavage. Dev. Growth Diff. 17: 101-110.

Ikenishi K, M Kotani, K Tanabe. 1974. Ultrastructural changes with UV irradiation in the "germinal plasm" of *Xenopus laevis*. Develop. Biol. 36: 155-168.

Ikenishi K, PD Nieuwkoop. 1978. Location and ultrastructure of primordial germ cells (PGCs) in *Ambystoma mexicanum*. Develop. Growth Differ. 20: 1-9.

Ikenishi K, S Nakazato. 1986. A natural marker for blastomeres of the germ line in cleavage stage *Xenopus* embryos. Dev. Biol. 113: 259-262.

Iwasawa H. 1959. Effects of thiourea on the gonadal development of *Rana temporaria ornativentris* larvae. Zool. Magazine 68: 6-11.

Iwasawa H. 1960. Inhibitory effect of *para*-hydroxypropiophenone on the development of gonad in thyroidectomized frog larvae. Endocrinol. Jap. 7: 181-186.

Iwasawa, H, M Kobayashi. 1976. Development of the testis in the frog *Rana nigomaculata*, with special reference to germ cell maturation. Copeia 1976: 461-476.

Iwasawa H, K Yamaguchi. 1984. Ultrastructural study of gonadal development in *Xenopus laevis*. Zool. Sci. (Japan), 1: 591-600.

Johnson AD, RF Bachvarova, M Drum, T Masi. 2001. Expression of axolotl *DAZL* RNA, a marker of germ plasm: widespread maternal RNA and onset of expression in germ cells approaching the gonad. Dev. Biol. 234: 402-415.

Kamimura M, M Ikenishi, M Kotani, T Matsuno. 1976. Observations on the migration and proliferation of gonocytes in *Xenopus laevis*. JEEM 36: 197-207.

Kato T, K Matsui, M Takase, M Kobayashi, M Nakamura. 2001. Expression of P450 aromatase protein in developing and in sex-reversed gonads of the XX/XY type of the frog *Rana rugosa*. Gen. Comp. Endocrinol. 137: 227-236.

Kawamura T, M Nishioka. 1977. Aspects of reproductive biology of Japanese anurans. In: The Reproductive Biology of Amphibians (Ed.) DH Taylor and SI Guttman. Plenum Press, New York. pp. 103-139.

King M. 1995. mRNA localization during frog oogenesis. In: Localized RNAs,

(Ed.) HD Lipshitz, Springer, pp. 137-148.

Kloas W. 2002. Amphibians as a model for the study of endocrine disruptors. Int. Rev. Cytol. 216: 1-57.

Kloc M, LD Etkin. 1995a. Two distinct pathways for the localization of RNAs at the vegetal cortex in *Xenopus* oocytes. Development 121: 287-297.

Kloc M, LD Etkin. 1995b. Genetic pathways involved in the localization of RNA in *Xenopus* oocytes. In: Localized RNAs, (Ed.) H.D. Lipshitz, Springer, pp: 149-156.

Kloc M, C Larabell, APY Chan, LD Etkin. 1998. Contribution of METRO pathway localized molecules to the organization of the germ cell lineage. Mech. Dev. 75: 81-93.

Kloc M, S Bilinski, LD Etkin. 2000. The targeting of Xcat2 mRNA to the germinal granules depends on a cis-acting germinal granule localization element within the 3'UTR. Dev. Biol. 217: 221-229.

Kloc M, S Bilinski, AP Chan, LH Allen, NR Zearfoss, LD Etkin. 2001. RNA localization and germ cell determination in *Xenopus*. Int. Rev. Cytol. 203: 63-91.

Kloc M, MT Dougherty, S Biliñski, APY Chan, E Brey, ML King. 2002: Three dimensional ultrastructural analysis of RNA distribution within germinal granules of *Xenopus*. Dev. Biol. 241: 79-93.

Kotani M. 1958. The formation of germ cells after extripation of the presumptive lateral mesoderm of *Triturus* gasrtrulae. J. Inst. Polytechnics Osaka City Univ Series D9: 195-209.

Kress A, UM Spornitz. 1974. Paracrystalline inclusions in mitochondria of frog oocytes. Experientia 30: 438-456.

Lopez K. 1989. Sex differentiation and early gonadal development in *Bombina orientalis* (Anura: Discoglossidae). J. Morphol. 199: 299–311.

MacArthur H, M Babunenko, DW Houston, ML King. 1999. Xcat2 RNA is a translationally sequestered germ plasm component in *Xenopus*. Mech. Dev. 84: 75-88.

Marracci S, C Casola, S Bucci, M Raghianti, M Ogielska, G Mancino. 2004. Characterization of genes coding for regulators of germline development in the hybridogenetic *Rana esculenta* complex. Abstracts of the 50th Meeting of the Italian Embryological Group. Pavia, Italy, June 2-5, 2004.

Marracci S, C Casola, S Bucci, M Raghianti, M Ogielska, G Mancino. 2007. Differential expression of two *vasa/PL10*-related genes during gametogenesis in the special model system Rana. Dev. Genes Evol. DOI.1007/s00427-007-0143-6.

Masayuki S, M Nishioka. 2000. Sex-linked genes and linkage maps in amphibians. Comp. Biochem. Physiol. B126: 257-270.

Matova N, L Cooley. 2001. Comparative aspects of animal oogenesis. Dev. Biol. 321: 291-320.

Maufroid J-P, A Capuron. 1977. Recherches recentes sur les cellules germinales primordiales de *Pleurodeles Waltlii* (Amphibien Urodele). Mem. Soc. Zool. Franc. 41 Symp. L. Gallien: 43-60.

Maufroid J-P, AP Capuron. 1985. A demonstration of cellular interactions during the formation of mesoderm and primordial germ cells in *Pleurodeles waltlii*. Differentiation 29: 20-24.

Mayer L, SL Overstreet, CA Dyer, CR Propper. 2002. Sexually dimorphic expression of steroidogenic factor 1 (AS-1) in developing gonads of the American bullfrog, *Rana catesbeiana*. Gen. Comp. Endocrinol. 127: 40-47.

McLaren A. 1999. Signaling for germ cells. Genes Dev. 13: 373-376.

Merchant-Larios H, I Villalpando. 1981. Ultrastructural events during early gonadal development in *Rana pipiens* and *Xenopus laevis*. Anat. Rec. 199: 349-360.

Miura I, H Ohtani, A Kashiwagi, H Hanada, M Nakamura. 1996. Structural differences between XX and ZW sex lampbrush chromosomes in *Rana rugosa* females (Anura: Ranidae). Chromosoma 105: 237-241.

Miura I, H Ohtani, M Nakamura, Y Ichikawa, K Saitoh. 1998. The origin and differentiation of the heteromorphic sex chromosomes Z, W and Y in the frog *Rana rugosa*, inferred from sequences of a sex-linked gene, ADP/ATP translocase. Mol. Biol. Evol. 15: 1612-1619.

Mosquera L, C Forristal, Y Zhou, ML King. 1993. A mRNA localized to the vegetal cortex of *Xenopus* oocytes encodes a protein with a nanos-like zinc finger domain. Development 117: 377-386.

Myiashita K, N Shimizu, S Osanai, S Myiata. 2000. Sequence analysis and expression of the P450 aromatase and estrogen receptor genes in the *Xenopus* ovary. J. Steroid Biochem. Mol. Biol. 75: 101-107.

Myiata S, S Koike, T Kubo. 1999. Hormonal reversal and genetic control of sex differentiation in *Xenopus*. Zool. Sci. 15: 335-340.

Nieuwkoop PD. 1950. Casual analysis of the early development of the primordial germ cells and the germ ridges in urodeles. Arch. Anat. Microscpique Morphol. Exp. 39: 257-268.

Nieuwkoop PD, J Faber. 1967. Normal table of *Xenopus laevis* (Daudin): A Systematical and Chronological Survey of the Development from the Fertilized Egg Till the End of Metamorphosis. 2nd ed. Amsterdam: North Holland.

Nieuwkoop PD, LA Sutasurya. 1976. Embryological evidence for a possible polyphiletic origin of the recent amphibians. J. Embryol. Exp. Morphol. 35: 159-167.

Nieuwkoop PD, L Sutasurya. 1979. Primordial Germ Cells in the Chordates. Cambridge University Press, pp. 187.

Ogielska M. 1990. The fate of intramitochondrial paracrystalline inclusion bodies in germ line cells of water frogs (Amphibia, Anura). Experientia (Basel) 46: 98-101.

Ogielska M, E Wagner. 1993. Oogenesis and ovary development in natural hybridogenetic water frog, *Rana esculenta* L. 1. Tadpole stages until metamorphosis. Zool. J. Physiol. 97: 349-368.

Ogielska, M, J Bartmańska. 1999: Development of testes and differentiation of germ cells in water frogs of the *Rana esculenta*—complex (Amphibia, Anura). Amphibia Reptilia 20: 251-263.

Ogielska M, M Kloc, LD Etkin. 1998. Molecular analysis of localized RNAs in oogenesis of *Rana* sp. (Amphibia, Anura). Folia Morphol. (Warszawa) 57: 46.

Ogielska M, A Kotusz. 2004. Pattern and rate of ovary differentiation wit h reference to somatic development in anuran amphibians. J. Morphol. 259: 41-54.

Othani H, I Miura, Y Ichikawa. 2000. Effects of dibutyl phthalate as an environmental endocrine disruptor on gonadal sex differentiation of genetic males of the frog *Rana rugosa*. Environ. Health Perspect. 108: 1189-1193.

Perma M. 1984. Are the primordial germ cells (PGCs) in Urodela formed by the inductive action of the vegetative yolk mass? Dev. Biol. 103: 109-116.

Reeder AL, GL Foley, DK Nichols, LG Hansen, B Wikoff, S Faeh, J Eisold, MB Wheeler, R Warner, JE Murphy, VR Beasley. 1998. Form and prevalence of intersexuality and effects of environmental contaminants on sexuality in cricket frogs (*Acris crepitans*). Environ. Health Perspect. 106: 261-266.

Riccio M, T Telò, PP Giorgi, M Pirazzini, S Santi, F Di Grande. 1998. Nuclear labeling in primordial germ cells of *Rana dalmatina* embryos by Dolichos biflorus agglutinin. Ital. J. Zool. 65: 241-248.

Robb DL, J Heasman, J Raats, C Wylie. 1996. A kinesin-like protein is required for germ plasm aggregation in *Xenopus*. Cell 87: 823-831.

Saidapur SK, NP Gramapurohit, BA Shanbhag. 2001. Effect of sex steroids on gonad differentiation and sex reversal in the frog, *Rana curtipes*. Gen. Comp. Endocrin. 124: 115-123.

Savage RM, MV Danilchik V. 1993. Dynamics of germ plasm localization and its inhibition by ultraviolet irradiation in early cleavage *Xenopus* embryos. Dev. Biol. 157: 371-382.

Schempp W, M Schmid. 1981. Chromosome banding in Amphibia. VI. BrdU replication patterns in Anura and demonstration of XX/XY sex chromosomes in *Rana esculenta*. Chromosoma 83: 697-710.

Schmid M, J Olert, CH Klett. 1979. Chromosome banding in Amphibia. III. Sex chromosome in *Triturus*. Chromosoma 71: 29-55.

Schmid M, I Nanda, C Steinlein, K Kausch, T Haaf, JT Epplen. 1991. Sex-determining mechanisms and sex chromosomes in amphibia. In: Amphibian Cytogenetics and Evolution, (Ed.) DM Green and SK Sessions. Academic Press. pp. 393-430.

Schmid M, W Feichtinger, C Steimein, I Nanda, C Mais, T Haaf, RV Garcia, AF Badillo. 2002. Chromosome banding in Amphibia. XXII. Atypical Y chromosomes in *Gastrotheca walkeri* and *Gastrotheca ovifera* (Anura, Hylidae). Cytogenet. Genome Res. 96: 1-4.

Shirane T. 1987. Role of peanut-lectin-affinity molecules (PALM) on primordial germ cells in the initial determination of sex in Anura. J. Exp. Zool. 243: 495-502.

Smith LD. 1966. The role of "germinal plasm" in the formation of primordial germ cells in *Rana pipiens*. Dev. Bio. 14: 330-347.

Smith LD, M Williams. 1979. Germinal plasm and germ cell determinants in anuran amphibians. In: Maternal Effects in Development, (Eds) Newth DR and Balls M. Cambridge University Press, pp. 167-197.

Subtelny S, JE Penkala. 1984. Experimental evidence for a morphogenetic role in the emergence of primordial germ cells from the endoderm in *Rana pipiens*. Differentiation 26: 211-219.

Sumida M, M Nishioka. 2000. Genetic linkage groups in the Japanese brown frog (*Rana japonica*). J. Hered. 91: 1-7.

Sugita J, M Takase, M Nakamura. 2001. Expression of DAX-1 during gonadal development of the frog. Gene 280: 67-74.

Tanabe K, M Kotani. 1974. Relationship between the amount of "germinal plasm" and the number of primordial germ cells in *Xenopus laevis*. J. Embryol. Exp. Morphol. 31: 89-98.

Tanimura A, H Iwasawa. 1988. Ultrastructural observations on the origin and differentiation of somatic cells during gonadal development in the frog *Rana nigromaculata*. Develop. Growth Diff. 30: 681-691.

Tanimura A, H Iwasawa. 1989. Origin of somatic cells and histogenesis in the primordial gonad of the Japanese tree frog, *Rhacophorus arboreus*. Anat. Embryol. 180: 165-173.

Villalpando I, H Merchant-Larios. 1990. Determination of the sensitive stages for gonadal sex-reversal in *Xenopus laevis* tadpoles. Int. J. Dev. Biol. 34: 281-285.

Wakahara M. 1996. Primordial germ cell development: is urodele pattern closer to mammals than to anurans? Int. J. Dev. Biol. 40: 653-659.

Whitington P, KE Dixon. 1975. Quantitative studies of germ plasm and germ cells during early embryogenesis of *Xenopus laevis*. JEEM 33: 57-74.

Williams MA, LD Smith. 1971. Ultrastructure of the "germinal plasm" during maturation and early cleavage in *Rana pipiens*. Dev. Biol. 25: 568-580.

Williams MA, LD Smith. 1984. Ultraviolet irradiation of *Rana pipiens* embryos delays the migration of primordial germ cells into genital ridges. Differentiation 26: 220-226.

Witschi E. 1929. Studies on sex differentiation and sex determination in amphibians. I. Development and sexual differentiation of the gonads of *Rana silvatica*. J. Exp. Zool. 52: 135-265.

Witschi E. 1931. Studies on sex differentiation and sex detrmination in amphibians. V. Range of cortex-medulla antagonism in parabiotic twins of *Ranidae* and *Hylidae*. J. Exp. Zool. 58: 149-165.

Witschi E. 1957. The inductor theory of the sex differentiation. J. Fac. Sci. Hokkaido Univ. Series VI, 13: 428-439.

Witschi E. 1959. Age of sex determining mechanisms in vertebrates. Science 130: 372-375.

Wylie CC. 1980. Primordial germ cells in Anuran embryos: Their movement and guidance. BioScience 30: 27-31.

Wylie CC, J Heasman. 1975. The formation of the gonadal ridge in *Xenopus laevis*. I. A light and transmission electron microscope study. JEEM 35: 125-138.

Wylie CC, M Bancroft, J Heasman. 1975. The formation of the gonadal ridge in *Xenopus laevis*. II. Scanning electron microscope study. JEEM 35: 139-148.

Wylie CC, TB Ross. 1976. The formation of the gonadal ridge in *Xenopus laevis*. III. The behaviour of isolated primordial germ cells in vitro. JEEM 35: 149-157.

Zaccanti F, S Petrini, ML Rubatta, AM Stagni, PP Giorgi. 1994. Accelerated female differentiation of the gonad by inhibition of steroidigenesis in amphibia. Comp. Biochem. Physiol. A107: 171-197.

2

Spermatogenesis and Male Reproductive System in Amphibia—Anura

Maria Ogielska and Jolanta Bartmańska

Structure of Testes in Anura

Structure of Testes in Adults

Testes in Anura are paired spherical or ovoid organs lying on the ventral side of kidneys. They are usually situated near the medial anterior part of the kidneys. Each testis is enveloped by peritoneal epithelium, beneath which the connective tissue capsule (*tunica albuginea*) is formed. In the dorsal part of a testis, the two sheets of peritoneum form a double-layered *mesorchium* that contains a system of efferent ducts (*vasa efferentia; ductuli efferentes*), which transport sperm from testes to kidney (Rugh, 1951; Witschi, 1956; Hiragond and Saidapur, 2000). The *tunica albuginea* contains blood vessels, smooth muscles and nerves. In some species, such as *Bombina bombina, B. variegata, B. orientalis* (Madej, 1964; Golmann *et al.*, 1993), or *Nectophrynoides occidentalis* (Zuber-Vogel and Xavier, 1965), the *tunica albuginea* contains various amounts of melanophores. In *Physalemus cuvieri* (de Oliveira *et al.*, 2002) cells containing dark pigment are also present in interstitial tissue.

Testes are filled with a mass of convoluted seminiferous tubules, which are separated from each other by interstitial tissue and loose connective tissue. Spermatogenesis occurs inside additional compartments of seminiferous tubules formed by a number of spherical vesicles known as cysts. Walls of cysts are composed of one or few Sertoli cells. Cysts are closed until germ cells achieve a stage of elongated spermatids. Then cysts open and differentiating spermatozoa gain contact with the lumen of the seminiferous tubule. During breeding season spermatozoa are released into the lumen of seminiferous tubule and are transported by a system of efferent ducts to mesonephros, and finally to Wolffian ducts and cloaca. With very few exceptions, fertilization in Anura is external, *i.e.*, it takes place outside a female body. During copulation, known in amphibians as *amplexus*, spermatozoa are released directly on eggs released by a female.

Development of Testes and Differentiation of Seminiferous Tubules

The schematic illustration of stages of testes development and differentiation in anurans is shown in Fig. 1. The stages of undifferentiated gonads (Fig. 1, stages I-III) are described elsewhere (see 'The Undifferentiated Amphibian Gonad' in this volume). The first morphological sign of sexual differentiation of a gonadal anlage is the displacement of germ cells from the cortical region into medulla (Fig. 1, stage IV; Fig. 3A). This rearrangement takes place at Gosner (1960) stage 28 in *Rana lessonae*, *R. ridibunda* (Ogielska and Bartmańska, 1999), and at stage 43 in *Rhacophorus arboreus* (Tanimura and Iwasawa, 1989).

The structure of an early testis was described in *Rana silvatica* and *R. temporaria* (Witschi, 1929, 1956); *R. nigromaculata* (Iwasawa and Kobayashi, 1976; Iwasawa *et al.*, 1987; Tanimura and Iwasawa, 1988); *R. lessonae* and *R. ridibunda* (Ogielska and Bartmańska, 1999); *Xenopus laevis* (Iwasawa and Yamaguhi, 1984); *Rhacophorus arboreus* (Tanimura and Iwasawa, 1989). Sexual differentiation of testes is characterized by formation of seminiferous tubules and ductules of the *rete tetsis*, which will form a system of sperm transportation. The former cortex of undifferentiated gonad is devoid of germ cells, becomes poorly detectable, and soon disappears. Primary spermatogonia become encapsulated by proliferating somatic cells deriving from the peritoneal epithelium, thereby forming the primordial seminiferous cords, *i.e.*, anlagen of further seminiferous tubules (Fig. 1, stage VI; Fig. 3B,C; 4A). The developing seminiferous cords join the *rete testis*, which at that time is the most conspicuous part of a gonad (Fig. 1, stage VIII). Forming tubules of the *rete testis* become connected to primordial seminiferous cords by short canals (*tubuli recti*). Along the dorsal side of a testis, the proliferating cells of the medulla soon form rudiments of a system of efferent ducts (Fig. 1, stage IX; Fig. 4A,B).

At the time of sexual differentiation, other somatic cells, *i.e.*, mesenchymal cells and blood vessels invade the gonad (marked black in Fig. 1, stages IV-X). They penetrate a thin space between the cortex and the medulla and then proliferate inside a mass of already existing medullar somatic cells originating from coelomic (peritoneal) epithelium. In juvenile males after metamorphosis, some of mesenchymal cells will differentiate

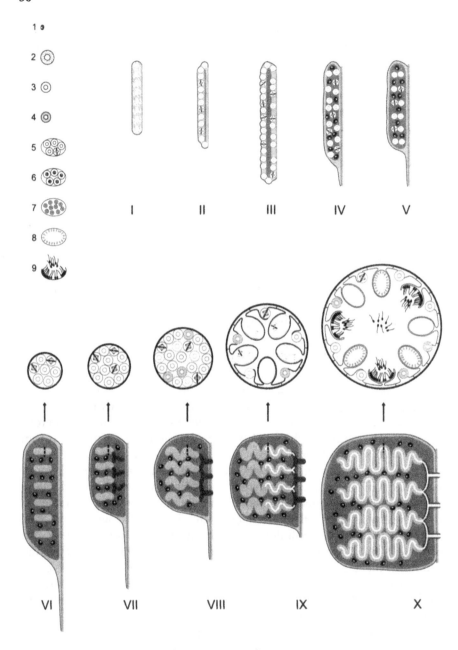

Fig. 1. See caption on the next page.

into interstitial (Leydig) cells producing steroid hormones (Hsu and Wang, 1981). Primordial seminiferous cords transform into seminiferous tubules with a lumen inside. Seminiferous tubules are elongated and convoluted blind sacs, which form a continuous connection with tubules of the *rete testis* (Fig. 1, stage X) and *ductuli efferentes*, which join the lateral canal, the Bowman's capsule of the kidney canal, and finally open to the Wolffian duct (Rugh, 1951).

Seminiferous tubules are originally short, but become longer during the course of testis development. During the juvenile period, seminiferous tubules grow in size mainly as a result of increasing number of spermatogenic cells (Fig. 5A), but also due to increasing diameter of the tubule. In the frog *Rana nigromaculata* after metamorphosis the diameter of tubules is 80 μm; whereas one month later it increases to 104 μm. After hibernation,

---

Fig. 1. Schematic illustration of stages of development of anuran testes and differentiation of seminiferous tubules. Stages I-X are shown as longitudinal sections; beginning with stage VI, cross sections of differentiating seminiferous tubules are shown in a row above corresponding stages. Stages I-III show undifferentiated gonad; stages IV-X show the testis. Stage I—gonadal anlage filled with primordial germ cells (PGCs). Stages II and III—PGCs undergo vitellolysis, multiply and transform into gonial cells distributed in the cortical part, whereas the central part become filled with somatic medullar cells (deep gray). Stage IV—Early testis; the cortical part disappears and gonial cells (now called primary spermatogonia) migrate into the medulla. At the same time mesenchymal cells (marked in black) invade the space between the cortex and the medulla, and will give rise to connective tissue and Leydig cells. Testis differentiates from the anterior part of the gonad, whereas the distal part becomes devoid of germ cells and forms a thread-like process, which eventually disappears (stages IV-VIII). Stage V—Primary spermatogonia form aggregations. Stages VI-X show further development of the testes cross-sections show above the arrows. Stage VI—Primary spermatogonia are arranged into sex cords separated from the interstitial tissue by a thin sheet of epithelium shown in cross sections. Stage VII—Along with formation of sex cords, which now grow in diameter (cross section), ductuli of *rete testis* (marked in darker gray) differentiate from the somatic cells of the medulla. Stage VIII—Ductules of *rete testis* penetrate the mesorchium and leave the testis toward mesonephros. Stage IX—About the time of metamorphic climax *rete testis* is already formed, and sexual cords transform into seminiferous tubules (shown in cross section) filled with stem-cell primary spermatogonia, cysts with secondary spermatogonia, and cysts with early spermatocytes. Stage X—Adult testis with well-developed seminiferous tubules connected with efferent ductules.
Legend: 1—mesenchymal cell; 2—primordial germ cell; 3—primary spermatogonium; 4—dark primary spermatogonium; 5—cyst with secondary spermatogonia; 6—cyst with primary spermatocytes; 7—cyst with secondary spermatocytes; 8—cyst with spermatids; 9—open cyst with spermatozoa.

i.e. during the second year of life, the diameter achieves about 170 μm. Seminiferous tubules become a dominant part of the testis and the *rete testis* is barely detectable (Iwasawa and Kobayashi, 1976, Iwasawa *et al.*, 1987). The basal lamina of the wall epithelium lines the lumen of a seminiferous tubule (Cavicchia and Movglia, 1983).

During differentiation the testis develops from the anterior part of the gonadal anlage. The posterior part becomes narrower and devoid of germ cells, until it transforms into a thin, thread-like process, which soon disappears (Fig. 1, stages IV-VIII; Fig. 2A) (Witschi, 1929; Ogielska and Bartmańska, 1999). The testis, which developed from the anterior part, starts to enlarge; the size of the testis reflects the size, length, and number of differentiating and elongating seminiferous tubules. Jørgensen and

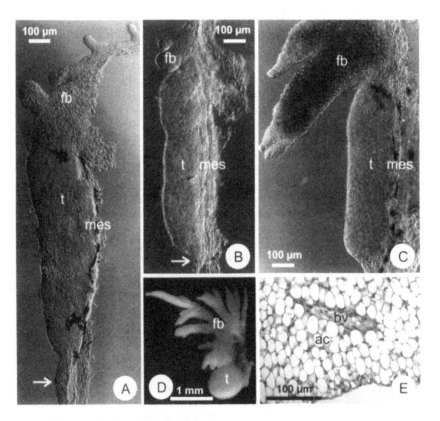

Fig. 2. Total view of developing testes and fat bodies. A-C—*Rana temporaria* during metamorphosis (Gosner stages 42-46). Arrows in A and B indicate the posterior part of the testis that becomes devoid of germ cells. Nomarski optics. D—Juvenile *R. lessonae*. E—Section of fat body. ac—adipose cells; bv—blood vessel; fb—fat body; mes—mesorchium; t—testis.

Billeter (1982) did not exclude formation of new seminiferous tubules in sexually active males of the toad *Bufo bufo*; however formation of new seminiferous tubules in sexually active males should be examined.

Sertoli Cells and Cyst Formation

In a full-grown testis (Fig. 1, stage X) seminiferous tubules are filled with cysts, in which spermatogenesis occurs (Fig. 5B,C). A cyst is formed by one or few somatic cells, which differentiate from the epithelium of the early gonad together with walls of seminiferous tubules, and engulf each of the primary spermatogonia. These cells are referred as the cyst or Sertoli cells. Sertoli cells were described in a number of anuran species: *Rana temporaria, R. esculenta, R. pipiens* (reviewed by Pudney, 1993); *Rana catesbeiana* (Sprando and Russell, 1988); *Bufo arenarum* (Cavicchia and Movgilia, 1983); *Pachymedusa dacnicolor* (Rastogi *et al.*, 1988); *Odontophrynus cultripes* (Báo *et al.*, 1991); *Xenopus laevis* (Reed and Stanley, 1972). The primary spermatogonium inside a cyst forms filipodia, which attach to the membrane of the Sertoli cell (Pudney, 1995). During spermatogenesis, the spermatogonium inside a cyst starts to multiply, thereby giving rise to a few generations of the secondary spermatogonia, which enter meiosis. After the second meiotic division, the resulting spermatids transform into spermatozoa during the process called spermiogenesis (spermio-histogenesis). Acrosomal regions of spermatids are inserted in numerous foldings of the apical part of the Sertoli cell processes, whereas axonemes are directed to a cyst lumen. A cyst is formed by the thin processes of the apical part of Sertoli cells, whereas their basal part is apposed to the basal lamina of a tubule wall. Processes of Sertoli cells are stabilized by microtubules, arranged in bundles longitudinal to the long axis of spermatids (Báo *et al.*, 1991). In Anura, cysts remain closed until spermatids start to elongate. Then cysts open and release spermatozoa to the lumen of a tubule. Open cyst resembles a cup attached to the tubule wall, with bundles of 60-150 early spermatozoa with flagellae stretching toward the lumen (Witschi, 1956; Roosen-Runge, 1977; Báo *et al.*, 1991; Pudney, 1993, 1995). During spermiation, *i.e.*, release of spermatozoa to the tubule lumen, a vacuolized apical part of Sertoli cell is also detached. The basal part remains intact and can form a new cyst (Burgos and Vitale-Carpe, 1976b; Lofts, 1974; Grier, 1993; Pudney, 1995).

Sertoli cells also mediate the hormonal action of gonadotropins, *i.e.*, hormones regulating spermatogenesis. Sertoli cells are big flat cells, with multiple processes connected by desmosomes and tight junctions (Cavicchia and Movgilia, 1983; Bergman *et al.*, 1983; Tanimura and Iwasawa, 1988; Rastogi *et al.*, 1988; Guraya, 2001). The junctions ensure isolation of the spermatogenic compartment from the environment of

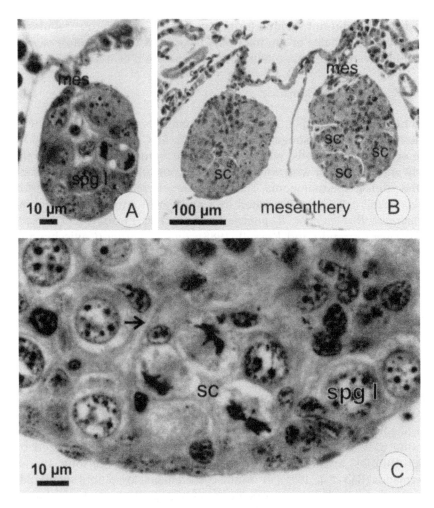

Fig. 3. Cross-sections of developing testes of *Rana lessonae*. A—Gosner stage 30 (compare with Fig. 1, stage V). B—Gosner stage 35 (compare with Fig. 1, stages VI and VII). C—Gosner stage 33 (compare with Fig. 1, stage VII). The arrow in C indicates the forming seminiferous tubule wall. mes—mesorchium; spg I—primary pale spermatogonia; sc—sex cords.

seminiferous tubule, thus forming a blood-testis barrier. The permeability of seminiferous tubule epithelium was described by Cavicchia and Movgilia (1983), who demonstrated the formation of blood-testis barrier by injection of lanthanum hydroxide to the circulation system of *Bufo arenarum*. They revealed that Sertoli cells forming a cyst around primary spermatogonia up to leptotene spermatocyte stage are permeable for lanthanum hydroxide, *i.e.*, are not protected by the barrier. Cysts containing germ cells at more advanced stages than leptotene are thereby isolated from the seminiferous lumen microenvironment. Formation of the blood-testis barrier is connected with the differentiation of specific tight junctions between cell membranes of interdigitated processes formed by neighboring Sertoli cells. Tight junctions of Sertoli cells surrounding early stages of spermatogenesis (primary spermatogonia up to leptotene stage) are much less extensive than those present in more advanced cysts. Sertoli cells are also referred as follicle, supporting, sustentacular or nurse cells (Witschi, 1956; Pudney, 1993). Their cytoplasm is filled with tubular mitochondria, glycogen (Rastogi *et al.*, 1988; Sprando and Russel, 1988; Guraya, 2001), and their nuclei have evenly distributed pores (Cavicchia and Movgilia, 1983). Lysosomes of Sertoli cells of *Odontophrynus cultripes* contain acid phosphatase, and glucose-6-phosphatase was detected in endoplasmic reticulum and Golgi complex (Fernandes and Báo, 1998).

Studies of the anuran, *Rana cyanophylcis*, indicate that the number of Sertoli cells does not change with age, nor does the size of their nuclei (Pancharatna *et al.*, 2000). A peculiar situation was described in *Bombina variegata*, where spermatogenesis proceeds without Sertoli cells and spermatozoa do not form bundles, but a mass of cells embedded in medullar tissue (Obert, 1976). However, our recent results do not support this fact (Pecio *et al.*, 2007). We reexamined the testis structure in *B. variegata* in breeding males, tadpoles and metamorphosed juveniles where Sertoli cells were readily identified. They separated germ cells from the basement membrane and formed the walls of cysts.

Interstitial Tissue

The somatic tissue spread between seminiferous tubules is called interstitial tissue (Fig. 5B,C). It is composed of steroidogenic Leydig cells, myoid cells, nerve fibers, blood vessels, fibroblasts and collagen fibrils. Leydig cells display changes in morphology, strictly related to a stage of spermatogenesis.

Interstitial cells appear between seminiferous tubules soon after sexual differentiation of testes, as was described for *Rana sylvatica* (Witschi, 1929). In *Rana cyanophlyctis*, the number of Leydig cells increase with male age (Pancharatna *et al.*, 2000), being the lowest in one-year-old specimens, the highest (twice as numerous) in 2-4 year old sexually active males, and decreasing again in the oldest, five-years-old animals. The interstitial cells are involved in hormonal control of spermatogenesis. The greatest number of Leydig cells match well with the period of active spermatogenesis, indicting steroidogenic potency and the role of these cells in sexual activity of breeding males. Leydig cells in juvenile testes of the common toad *Bufo bufo* are spindle-shaped with small nuclei, whereas in adult males with active spermatogenesis they are voluminous cells rich in cytoplasm and with large nuclei; islets of well developed Leydig cells can be found in varying numbers among poorly differentiated interstitial tissue (Jørgensen and Billeter, 1982).

Bidder's Organ

In species belonging to the family Bufonidae, the most anterior part of the gonads, both ovaries and testes, differentiate into the structure known as the Bidder's organ. It is much easier to distinguish in males than in females, because its structure resembles rather an ovary than a testis. For this reason the Bidder's organ has been primarily described as characteristic of males. Although the presence of the Bidder's organ in male bufonids has been known since the 19th century, its role, differentiation and structure are still not fully understood. Studies on the Bidder's organ still remain scarce.

The Bidder's organ develops from the anterior part of the gonad proper, next to the fat body (Viertel and Richter, 1992). Its differentiation, with comparison to the gonad proper, was studied in *Bufo japonicus formosus* by Tanimura and Iwasawa (1986, 1987). In this species the Bidder's organ starts to differentiate in larvae 6 days after hatching, at stage corresponding to Gosner's 24/25. At this time the germ cells were scattered among the somatic cells of a gonad. During metamorphosis the Bidder's organ gradually increases and rapidly enlarges at the time of metamorphic climax. Neither definite cortico-medullary structure, nor the primary gonadal cavity is observed throughout the further development, although a small space appears in the center of the Bidder's organ at the same time, when the secondary ovarian cavity is formed in ovaries. At the premetamorphic stages, the germ cells grow rapidly in volume, resembling young diplotene oocytes. Their number grows until metamorphosis and does not change thereafter. Normal meiocytes appear at the periphery of the Bidder's organ later, *i.e.*, about the same time as in developing ovary (in females, and in corresponding time in males). At the time when ovarian diplotene oocytes appear, the Bidderian oocytes start to degenerate. The role and nature of these cells are unknown.

Recently Farias *et al.* (2002) reported the histological and ultrastructural study of adult males of the Brazilian species *Bufo ictericus*. The Bidder's organ in this species is composed of the cortex containing diplotene oocytes at various stages of development and the medulla with large blood vessels. The connective tissue of the medulla is rich in acid carboxylates and collagen fibers. The other class of the collagen fibers are localized around the follicles, as was also reported for *Nectophrynoides malcolmi* (Wake, 1980), *Bufo japonicus formosus* (Moriguchi *et al.*, 1991; Tanimura and Iwasawa,

1992), and *B. woodhousii* (Pancak-Roessler and Norris, 1991). The latter authors studied the relationship between the testes and the Bidder's organs in adult males of *Bufo woodhousii*. An experimental group of animals had bilaterally removed testes (orchidectomy) and was treated with gonadotropins; in result the oocytes in the Bidder's organ of the castrated males begun to develop. Normally the oocytes are in pre-vitellogenic stages and never grow further; in orchidectomized males, even without administration of gonadotropins, the oocytes started the uptake of yolk and the weight of the Bidder's organ increased. As was suggested by Pancak-Roessler and Norris (1991), the Bidder's organ probably posses receptors for gonadotropin and is able to respond in a manner similar to ovary, but this function is normally inhibited by the testes. The experiments with orchidectomized *Bufo marinus* were also carried out by Brown *et al.* (2002). They revealed that the maximum increase of the bidderian oocyte growth occurs 1 month after testes removal.

Fat Body

During larval development, the anterior part of gonad (the progonad) looses its germinal function and differentiates into fat body (Witschi, 1929) (Fig. 2A-D). It is a finger-like structure, and the axis of each "finger" contains an artery and a vein (Zancanaro *et al.*, 1996). Studies on fat body development, as well as its physiological role in males are scarce. The first molecular studies on fat body differentiation were performed by Chan *et al.* (1999, 2001), who isolated the *fatvg* gene from *Xenopus laevis* germ plasm. This gene is a homologue of a mammalian gene expressed in adipose tissue and in a tadpole gonad is expressed in the fat body. *fatvg* is originally associated with the germ plasm located in germ cell precursors, but before the colonization of germinal ridges the two cell lineages segregates. One of them forms PGCs, whereas the other forms precursors of the fat body. The fat body precursor cells, which form two masses of cells at the dorso-anterior part of the endodermal margin, colonize the anterior part of the genital ridges and later transform into fat bodies.

An unusual feature of amphibian gonad is its role in overall energy metabolism as a main organ of fat deposition, storage and release. Schlaghecke and Blum (1978) studied changes in fat body mass in the frog *Rana esculenta* and stated that fat body mass changes during seasonal cycle, whereas the body mass remains relatively constant. The frog's fat body is composed of cells, which resemble adipocytes of mammalian fat tissue (Zancanaro *et al.*, 1996). The entire organ is composed of large adipocytes of various sizes (15-140 μm in diameter), each filled with a large drop of fat (Fig. 2E).

Sperm Collecting Ducts and Urogenital Connections

The sperm collecting system in Anurans, as in other amphibians, is composed of a network of intro-testicular tiny canals (*rete testis*) collecting spermatozoa from seminiferous tubules (Fig. 1, stages VII and VIII; Fig. 4A, B), a system of extra-testicular efferent ducts (*vasa efferentia, ductuli efferentes*) passing through the mesorchium, and a pair of Wolffian ducts (*ductus Wolffi*), which open to the cloaca (Fig. 6). During embryonic development of a male vertebrate, genital ridges develop in close contact with kidneys, and soon anatomical connections between gonads and excretory system are formed. The proximal part of a kidney is either entirely or partly functionally connected with the genital system. In amphibians, the proximal part of mesonephros transports sperm, whereas the distal part transports urine. Thereby, the proximal part plays a role of epididymis (Blüm, 1985).

In amphibians, as in most cold-blood vertebrates, kidney ducts serve also as ways for sperm transportation. Amphibian kidneys, both prone-phros in larvae and mesonephros (opistonephros) in adults, have a primary urinary duct—the Wolffian duct. Primordia of the primary urinary ducts are situated at the level of the fifth somite of an early embryo. The primary urinary duct originates from two parts: the anterior mesodermal primordium and a small distal part of ectodermal cloacal diverticulum. The primary urinary ducts leave each kidney in the anterior region and converge immediately behind. In the majority of anuran species the Wolffian duct is connected to the outer edge of kidney along its whole length. A functional pronephros is essential for the development and maintenance of the Wolffian duct; if the pronephros is experimentally removed, the anterior part of the duct degenerate (Cambar, 1949; Viertel and Richter, 1992; Hiragond and Saidapur, 2000). On the other hand, the anterior part of the Wolffian duct in *Rana dalmatina* is essential for mesonephros growth and differentiation (Cambar, 1947a,b, 1948). The other duct connected with gonad development and differentiation is the Müllerian duct. In females it differentiates into an oviduct, whereas in males it usually degenerates or remains vestigial. In freshly metamorphosed male *Rana nigromaculata* the Müllerian ducts are composed of solid cords of cells. During second year of life the Müllerian ducts are reduced and form only longitudinal thickenings of the peritoneum.

Intratesticular Sperm-collecting Ducts (rete testis)

In Anura, seminiferous tubules of testes open to a network of sperm-collecting ducts known as the *rete testis*. At larval stages of male amphibians the *rete testis* originate entirely from medullar cells, already differentiated

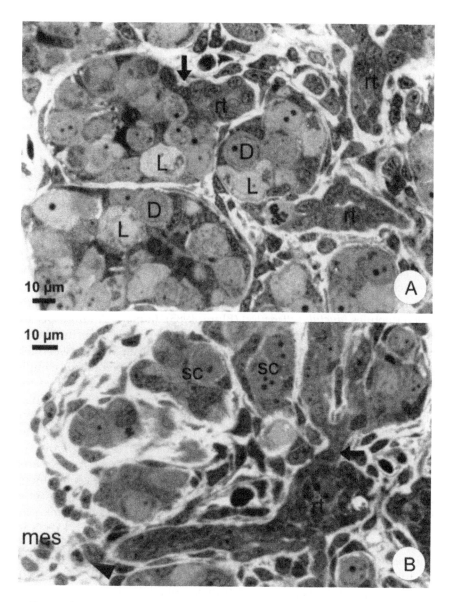

Fig. 4. Cross sections of developing testes (stage VIII) of *Rana lessonae* at Gosner stage 40. Well-differentiated sex cords (sc) filled with primary pale (L) and dark (D) spermatogonia are shown in A—Connections between *rete testis* (rt) and differentiating seminiferous tubule (arrows) are shown in A and B. The forming efferent ductule (arrowhead in B) grows toward mesorchium (mes). sc—sex cords; D—dark primary spermatogonia; L— pale primary spermatogonia; rt—rete testis.

from coelomic epithelium of a gonadal anlage (see 'The Undifferentiated Amphibian Gonad' in this volume). Collecting ducts are situated in the center or at one side of a testis, depending on a species. Their walls are composed of epithelial cells, which are cuboidal in Rana pipiens (Rugh, 1951). Collecting ducts of the rete testis leave the testis and are called efferent ducts (ductules) when outside a gonad.

Efferent Duct (ductus efferens)

The formation of efferent ducts during organogenesis was described in *Rana nigromaculata* (Tanimura and Iwasawa, 1988) and *Rhacophorus arboreus* (Tanimura and Iwasawa, 1989). The cells of the dorsal portion of medullar cell mass of *rete testis* proliferate toward mesorchium and form medullar cell cords, which differentiates into tiny efferent ductules (*ductuli efferentes*) (Fig. 4A,B). The cell cords invade mesonephros and soon transform into rudimentary efferent ducts. Efferent ducts in adults traverse the mesorchium and join the kidney. The number and length of the ducts differ among species and in some cases, when they are tiny structures, they are called ductules (*ductules efferentes*) (Blüm, 1985). In many Anuran species, efferent ducts join the lateral kidney canal (*canalis lateralis*). The number of efferent ducts per gonad ranges from 3 to 30 and can vary between left and right gonad (Fig. 6A-D). The length and shape of efferent ducts (simple or branched) is species specific. Bhaduri and Basu (1957) described the anatomy of efferent ducts in 8 species of African anuran families (Ranidae, Hyperolidae, Hemisotidae and Rhacophoridae). Efferent ducts in some species are hardly visible, because of their short length and narrow mesorchium. In these species (*Rana mascareniensis mascareniensis, Phrynobatrachus natalensis, Afrixalus fulvovittatus, Leptopelis karissimbiensis, Arthroleptis sylvaticus*) testes are tightly juxtaposed to kidneys. In others, testes are loosely attached to kidneys by broad mesorchia. In *Rana subsigillata* there are 4 efferent ductules, in *Hemisus marmoratum guineense* 3 or 4; in *Phrynobatrachus versicolor* and *Cacosternum boettgeri* 4 or 5; in *Chrysobatrachus cupreonitens* and *Kassina argyreivittis ruandae* 6. Testes in *Chiromantis rufescens* are extremely large and 14-16 branched efferent ductules join kidney along its whole length (Bhaduri and Basu, 1957). Recently Hiragond and Saidapur (2000) described 7-9 ductules per gonad in *Rana cyanophlyctis*; 6-8 in *Rana limnocharis*; 8-12 in *Polypedates maculates*; 3-4 in *Microhyla rubra*; 25-30 in *Bufo melanosticus*, and 14-18 in *Bufo fergusonii*.

After passing through efferent ducts (and lateral canal), sperm enter the Bowman's capsule of the Malpighian corpuscles in the kidney via short (epididymal) canals (*ductus epididymis*). Then the sperm passes by kidney tubules into the Wolffian ducts. In majority of anuran species the

Wolffian duct is united to the outer edge of kidney along its whole length. However, in some species the canals run outside and the Wolffian duct is situated in a distance from the kidney. The number of the kidney tubules outside the kidney varies among species: 13-20 in *Rhacophorus arboreus*; 12-14 in *R. schlegelli*; 10-16 in *R. buergeri* (Iwasawa and Michibata, 1972); about a dozen in *Chiromatis rufescens*, and 3 in *Kassina argyreivittus ruandea* (Bhaduri and Basu, 1957).

Wolffian Ducts

In male anurans, the Wolffian ducts serve both as ureter and sperm duct, and for this reason is also called urogenital duct. The Wolffian ducts are attached to the lateral margin kidney and leave it at various levels, according to the species. As reported by Bhaduri and Basu (1957), Wolffian ducts leave kidneys at the anterior (*Chiromantis rufescens, Kassina argyreivittis ruandae*), middle (*Afrixalus fulvovittatus, Chrysobatrachus cupreonitens, Hemisus marmoratum guineense*), one-third posterior (*Rana mascareniensis mascareniensis, Phrynobatrachus natalensis, Phrynobatrachus versicolor, Leptopelis karissimbiensis*), or posterior (*Cacosternum boettgeri, Rana subsigillatai, Artheloptis sylvaticus, Hyperolius viridiflavus coerulescens*) part.

After leaving kidneys, Wolffian ducts run posterior toward the cloaca, where they enter, usually as two separate openings, in some cases separated by a papilla. In *Hemisus marmoratum* they fuse before reaching cloaca and enter the cloaca by one opening. The urinary bladder is situated at opposite side of Wolffian ducts and is not involved in sperm evacuation (Bhaduri and Basu, 1957). According to Hiragond and Saidapur (2000), the lumen of the Wolffian duct is lined by epithelium, which can be pseudostratified (*Rana cyanophlyctis*), stratified (*R. limnocharis*), columnar (*Microhyla rubra, Polypedates maculates*), ciliated columnar (*Bufo melanosticus*) or cuboidal (*Bufo fergusonii*).

The posterior parts of Wolffian duct close to their openings to cloaca are enlarged and are lined by glandular epithelium that is more columnar than that lining the anterior duct. These enlargements serve as sperm storage sites (Rugh, 1951; Bhaduri and Basu, 1957; Blüm, 1986; Viertel and Richter, 1992; Hiragond and Saidapur, 2000). From sperm storage sites, sperm is released outside via the cloaca during amplexus.

Generally four types of sperm storage sites are distinguished: ampulla, seminal vesicle, coiled Wolffian duct, and urogenital sinus. Iwasawa and Michibata (1972) provided a detailed anatomical and histological study of the storage sites in 17 anuran species belonging to 4 genera (*Bufo, Hyla, Rana*, and *Rhacophorus*).

Ampullae are characteristic of *Bufo bufo japonicus, Hyla arborea japonica*,

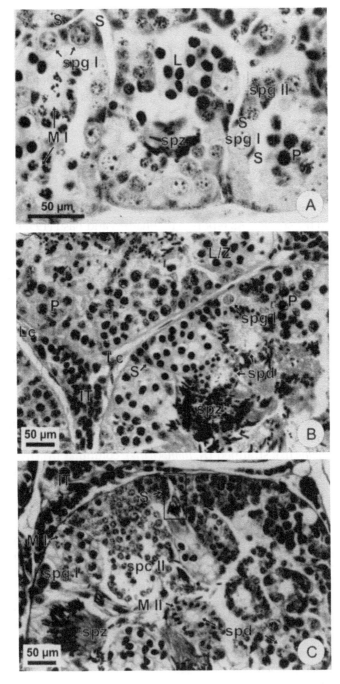

Fig. 5. See caption on the next page.

Rana nigromaculata, Rana brevipoda, Rana limnocharis, Rana rugosa, Rana catesbeiana, Rhacophorus buergeri and *Rhacophorus japonicus*. The ampulla is an enlargement of the distal part of Wolffian duct situated at the level of the posterior half, or in the rear of kidney. The ampulla is the largest during breeding period and decreases rapidly soon after. The wall of ampulla during breeding period is thin; its connective tissue is stretched and lined by a single layer of flat or cuboidal epithelium, which becomes columnar when the lumen becomes smaller. In most species the epithelium is folded and forms papillae. In *Hyla arborea japonica* the epithelium lining the ampulla lumen during breeding season is ciliated. After the breeding season the size of ampulla decreases and its wall thickens due to folding of connective tissue. The ampulla in *Rana catesbeiana* and *Rhacophorus japonicus* has numerous small or about 10 large outpouchings, respectively. Bhaduri and Basu (1957) described ampullae in *Phrynobatrachus versicolor, Cacosternum boettgeri, Arthroleptis sylvaticus, Hyperolius viridiflavus, Leptopelis carissimbiensis*, although they did not use the name ampulla, but dilatation. They did not find any enlargements of Wolffian ducts in one species of *Rana* (*R. subsigillata*). Iwasawa *et al.* (1987) studied the development and differentiation of the ampulla in *Rana nigromaculata* during the first and second year of life. During the first year there is no difference between the structure of the ampulla in males and females. The epithelium of this portion of the Wolffian duct is smooth and unilayer. During the second year of life the difference becomes remarkable. In males the epithelium becomes stratified, the diameter of the ampulla increase, and is about 1.6 times larger than during first year of development; the inner surface of the ampullar wall become folded.

Seminal vesicles are present in *Rana japonica, R. chensinensis, R. ornativentris, R. tagoi, R. tsushimensis,* and *R. okinavana* (Iwasawa and Michibata, 1972). The seminal vesicle is a large, knob-like outgrowth of a distal part of Wolffian duct. The lumen of the Wolffian duct is connected with each seminal vesicle via several canals. The lumen of seminal vesicle

---

Fig. 5. Cross-sections of testes of *Rana lessonae*. A—Juvenile male before I hibernation (stage IX). Seminiferous tubules are filled with primary spermatogonia (spg I), cysts with secondary spermatogonia (spg II), cysts with primary spermatocytes during first meiotic division (MI), and few spermatozoa (spz). B and C—Sexually mature male (stage X). Seminiferous tubules are well developed and contain all stages of spermatogenesis, and interstitial tissue with Leydig cells. spg I—primary spermatogonia; spg II—secondary spermatogonia; L—primary spermatocytes in leptotene; L/Z primary spermatogonia in leptotene/zygotene; P—primary spermatocytes in pachytene; spd—spermatids; spz—spermatozoa; Lc—Leydig cell; S—Sertoli cell outlined by a frame; IT—interstitial tissue.

has numerous outpouchings, which can be also recognizable externally. At the breeding season the lumen of seminal vesicle and lumen of Wolffian duct at the level of seminal duct increases and is covered by flat or cuboidal epithelium and papillae. Bhaduri and Basu (1957) also reported seminal vesicles in *Phrynobatrachus natalensis*, *Phrynobatrachus versicolor* and *Rana mascareniensis*.

Coiled Wolffian ducts were described in *Rhacophorus arboreus*, *R. schlegelii* (Iwasawa and Michibata, 1972), *Chiromantis rufescens*, and *Kassina argyreivittus ruandea* (Bhaduri and Basu, 1957). In these species Wolffian ducts arise from the anterior part of kidney and run posteriorly increasing its diameter and become coiled. Both the Wolffian duct and associated ductules change during breeding season, when the lumen is lined by columnar epithelium with numerous active secretory cells, walls are thin and stretched with no papilla, and the lumen is filled with spermatozoa. After breeding season the lumen shrinks, walls become thicker, and papilla appear again. In two other *Rhacophorus* species (*R. buergeri* and *R. japonicus*) there is no coiled Wolffian duct and sperm-collecting apparatus is of the ampulla type. Complicated coiled Wolffian ducts in *Rhacophorus arboreus* and *R. schlegelii* are probably connected with the mode of oviposition. These two species lay eggs in foam nests, whereas the other lay eggs in water. Foam nest are formed of excrections of reproductive tracts of both female and male. It is suggested that structure of male reproduction system (coiled Wolffian ducts or ampulla) in Rhacophoridae is related more to breeding habits than to systematic status.

The forth type of sperm storage site, the urogenital sinus, is a cloacal derivative, and Wolffian ducts do not open directly, but in close vicinity to it. Urogenital sinuses were described in *Ascaphus truei*, *Microhyla ornata*, *Rhinophrynus dorsalis*, *Arthelopus sylvaticus*, *Phrynobatrachus natalensis*, *Rana sylvatica* (Iwasawa and Michibata, 1972), as well as in *Phrynobatrachus natalensis*, *Arthroleptis sylvaticus* and *Kassina argyreivittus* (Bhaduri and Basu, 1957).

Sperm-transporting Ducts

In some Anuran species, genital tracts are partly or entirely separated from urinary tracts. Although studies of urogenital connections in Anura are scarce, they show that organization of sperm collecting ducts in Anura is diverse and phylogenetic relationships are unclear at this time. Blüm (1986) distinguished 3 types of organization of sperm transporting ducts. The first type was described in *Bombina* (formerly *Bombinator*), where the lateral kidney canal is separated from the mesonephros and transverse canals budding from the lateral canal make contact with cranially situated nephrons, and lead to the Wolffian duct (see Fig. 6). The second type is known from *Discoglossus*, where the mesonephros is not involved in sperm transport. Efferent ducts and lateral kidney canal are absent and replaced by a direct connection between testis and Wolffian duct. The Wolffian duct also collects urine from the proximal part of mesonephros. Nephrons of the distal part of mesonephros are connected with a secondary urinary duct, *i.e.*, the ureter. The third type of sperm transport system was described in *Alytes*, where genital duct is entirely separated from urinary tract. In this Anuran species the Wolffian duct leaving the testis acts only as sperm transporting canal, and urine from the entire kidney is collected by secondary urinary duct (ureter) (see Fig. 6D).

In more advanced genera (*Rana, Polypedates, Microhyla* and *Bufo*) studied recently by Hiragond and Saidapur (2000), the system of sperm collecting ducts does not resemble either of those described by Blüm (1986). In these amphibians sperm is transported from testes to the Bowman's capsule of the kidney and further to the Wolffian duct, which serves as urogenital duct. No secondary urinary duct was found. A lateral kidney duct is present only in *Rana cyanophlyctis*. In other species studied by Hiragond and Saidapur (2000) (*Rana limnocharis, Polypedats maculatus, Microhyla rubra, Bufo melanosticus* and *B. fergusonii*) the lateral kidney duct is absent and urogenital connections are direct. In these species the proximal part of kidney is involved in transport of sperm from the testis to Wolffian duct, and the distal part drains urine. However, in some species sperm was also observed in distal nephrons, but most probably only for temporary storage.

Cloaca

The cloaca is the common receptacle for the alimentary tract, urinary bladder and Wolffian ducts that enter the cloaca dorsally. The lumen of the cloaca is lined, in many cases, by columnar ciliated epithelium. Connective tissue underlies the epithelium and contacts with circular and longitudinal layers of muscles, which are continuous with muscle of intestine. A sphincter surrounds the external opening of the cloaca. In anurans, the cloaca has a simple structure and serves as the last portion of reproductive system, by which the sperm is transported to the exterior (Duelmann and Trueb, 1986).

A peculiar adaptation of cloaca to internal fertilization is known from a tailed frog *Ascaphus truei* and its sibling species *Ascaphus montanus* (Nielson *et al.*, 2001), although it is known in more detail for the former species. The cloaca is elongated and forms an external part, known as a tail, which forms a copulatory male organ. It is bulbous in shape and physiologically acts as a penis of higher vertebrates. The detailed description of mating as well as anatomy and histology of the *Ascaphus*

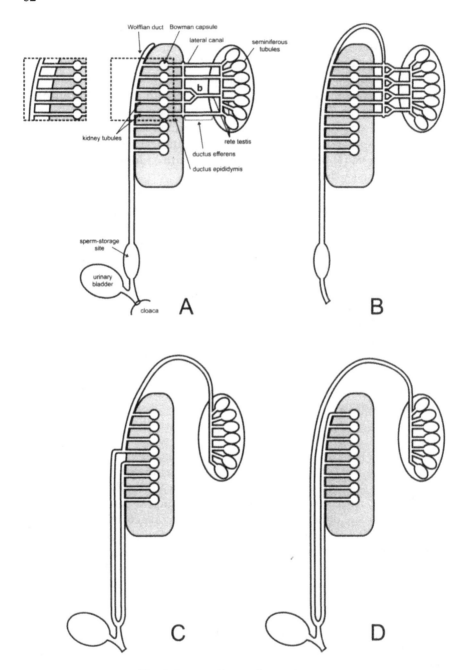

Fig. 6. See caption on the next page.

truei penis was recently provided by Sever *et al.* (2003). The penis has its own musculature, which enables erection. The cloacal opening at the top of penis forms a groove. The lumen of the penis-part of the cloaca is lined by ciliated columnar epithelium with goblet cells. Together with underlying connective tissue containing blood vessels, nerves and collagen fibers, the internal epithelium forms numerous folds. Cloacal glands release their proteinous secretion near the cloaca opening. The penis has specialized tissue consisting of large blood sinuses, which are accompanied by lymph sacs. This structure, which resembles *corpora cavernosa* in higher vertebrates, enables erection and copulation. Additionally two cartilaginous rods support an erected penis. Connective tissue capsules enclose the rods. During the sexual act the penis is inserted into female cloaca and for this reason Sever *et al.* (2001) suggested that this kind of copulation should be named "copulexus", instead of amplexus.

Although *Ascaphus truei* is not the only amphibian species, in which internal fertilization occurs, it is unique because of at least two reasons. The first is the presence of a penis, whereas other in species with internal fertilization sperm is transferred by cloacal apposition during amplexus. The second is the way, by which spermatozoa are stored, *i.e.*, in oviducts.

Spermatogenesis in Anura

The process of spermatogenesis in Amphibians, similarly to Mammals, can be divided into two phases: prespermatogenesis and active spermatogenesis (Fig. 7). Prespermatogenesis is restricted to tadpole and early juvenile stages, and consists of multiplication of primary spermatogonia and formation of a pool of "juvenile" primary spermatogonia. During that time no other stages of spermatogenetic cells are present in a testis. Juvenile primary spermatogonia transform into spermatogenic stem cells ("adult"

Fig. 6. Schematic illustration of urogenital connections in Anura. A—In most species seminiferous tubules are connected with a system of tiny canals known as *rete testis* when inside a testis, or efferent ductules, when outside the gonad. Efferent ductules traverse the mesorchium, and can be simple or branched (b). They fuse with the lateral canal of the kidney, which has several epididymal canals, which join the Bowman's capsule. The Bowman's capsules connect to Wolffian ducts via kidney tubules. The Wolffian ducts can be juxtaposed on kidney or join with kidney tubules by additional extragonadal canals (shown in the inset). B—In *Bombina* the Wolffian duct has direct connection with the testis, whereas in *Discoglossus* (C) and *Alytes* (D) urinary and genital systems are separated. According to Bhaduri and Basu (1957) and Hiragond and Saidapur (2000) (A), and Blüm (1986) (B-D).

primary spermatogonia), which give rise to many generations of spermatozoa during the entire adult life of a male.

Active spermatogenesis is usually staged according to phases of meiotic division, into primary and secondary spermatogonia, primary and secondary spermatocytes, spermatids, and spermatozoa. Each of these stages is characterized by various morphological features, which are described below. Additional sub-stages are usually applied for primary spermatocytes undergoing a sequence of phases of the first meiotic division (leptotene, zygotene, pachytene, diplotene, diakinesis). More recently Manochantr *et al.* (2003) proposed even more detailed sub-staging based on chromatin organization in spermatogonia and spermatocytes, and acrosome and tail formation in spermatids of *Rana tigerina*.

Morphology of Germ Cells

Primary spermatogonia. Primary spermatogonia are big (about 15 µm in diameter) single cells enclosed by 1–3 Sertoli cells. In the genus *Bufo* (Poska-Teiss, 1933; Cavicchia and Moviglia, 1983), *Pachymedusa* (Rastogi *et al.*, 1988), *Xenopus* (Al-Mukhtar and Webb, 1971), and *Scinax* (de Oliveira *et al.*, 2003) primary spermatogonia have big, highly lobulated nuclei. In the genus *Rana* nuclei are generally smooth in outline (Rastogi *et al.*, 1985; Ogielska and Bartmańska, 1999; Manochanter *et al.*, 2003), but in some cells nuclei may contain deep finger-like invaginations of cytoplasm containing clusters of mitochondria (Ogielska and Bartmańska, unpublished). The nuclear envelope of primary spermatogonia has unevenly distributed nuclear pores, forming aggregations and pore-free areas, as was described in *Bufo arenarum* (Cavicchia and Moviglia, 1983). The nuclei contain mainly euchromatin with small, rather evenly scattered heterochromatin blocks formed by aggregations of tiny 30 nm chromatin fibers (Manochantr *et al.*, 2003).

Usually two, but in some spermatogonia more than two nucleoli are observed. Kalt and Gall (1974) revealed in *Xenopus laevis* that during this

---

Fig. 7. Stages of spermatogenesis in amphibians. Prespermatogenesis is restricted to the juvenile period (gonadal stages I-VII), and "active" spermatogenesis occurs during the whole reproductive life span of a male. Light primary spermatogonia multiply and renew the pool of stem-cell spermatogonia; dark spermatogonia are committed to enter the unique cell cycle, which gives rise to a clone of secondary spermatogonia. After the last round of mitoses, secondary spermatogonia transform into primary spermatocytes, which enter meiosis. First meiotic division results in secondary spermatocytes, and second meiotic division results in spermatids; spermatids transform into spermatozoa.

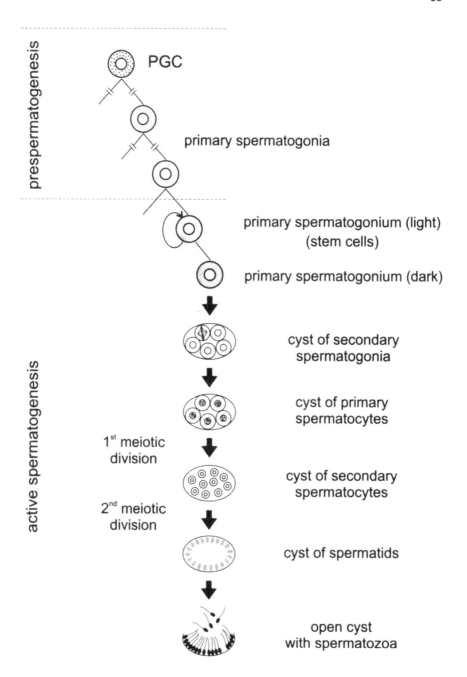

Fig. 7. See caption on page 54.

phase of spermatogenesis the transient premeiotc rRNA amplification takes place. It stops when meiosis starts and spermatogonia transform into primary spermatocytes. Multiple nucleoli are also observed in primary spermatogonia of *Rana lessonae* and *R. ridibunda* (Bartmańska and Ogielska, unpublished) (Fig. 3C).

Two kinds of primary spermatogonia, pale and dark (Fig. 4A), are described in *Pachymedusa dacnicolor* (Rastogi *et al.*, 1988), *Rana esculenta* (Rastogi *et al.*, 1985), *R. lessonae, R. ridibunda, R. esculenta* (Ogielska and Bartmańska, 1999), and *R. cyanophlyctis* (Pancharatna *et al.*, 2000). Pale spermatogonia are bigger with nuclei of mean diameter 12 μm (Ogielska and Bartmańska, 1999), homogenous chromatin, and pale, electron translucent cytoplasm. Spermatogonia belonging to this class are scattered along the whole length of seminiferous tubules close to the tubular wall. They are residual stem cells, which maintain the ability to divide mitotically, being a permanent source of next generations of spermatogonia. Dark spermatogonia are smaller, with nuclei of mean diameter 8.5 μm that contain more hetrochromatin and they have dark electron dense cytoplasm. In pale spermatogonia mitochondria are usually arranged in a group at one side of a nucleus; this arrangement is not so distinct in dark spermatogonia. In both kinds of primary spermatogonia, annulate lamellae, endoplasmic reticulum, ribosomes, and dictyosomes are present (Fig. 8A-C). Rastogi *et al.* (1988), and Ogielska and Bartmańska (1999) suggested that pale spermatogonia are undifferentiated cells, whereas dark ones are differentiated, *i.e.*, they are already devoted to giving rise to secondary spermatogonia.

Secondary spermatogonia. Secondary spermatogonia (Fig. 5A) form a clone of cells deriving from one primary spermatogonium, which divides synchronously. Secondary spermatogonia are connected by cytoplasmic bridges and are enclosed in a cyst formed by Sertoli cells. In *Pachymedusa dacnicolor* a cyst can contain up to 200 secondary spermatogonia, which means that primary spermatogonia underwent not less than 8 mitotic cycles (Rastogi *et al.*, 1988). In two species of *Rana* the number of secondary spermatogonia per cyst ranges from 8 to 127, which is a result of 3-7 mitotic divisions in *R. lessonae*, and 4-7 mitotic cycles in *R. ridibunda* (Bartmańska *et al.*, 2006). In *Xenopus laevis* (Takamune *et al.*, 1995) a cyst contains 213 secondary spermatogonia, *i.e.*, the yield of not less that 7, and in most cases 8 mitotic cycles. The mitotic cycles in secondary spermatogonia in *Xenopus laevis* could be divided into two phases according to morphology of their nuclei: a transition phase from lobulated to round, as well as from bigger to small, and a constancy phase without any change in morphology (Takamune *et al.*, 2001). After the fifth division of the transition phase secondary spermatogonia acquire the ability to enter meiosis and have two possibilities: either enter meiosis or undergo further mitotic divisions, usually 3 more cycles, and then enter meiosis. These findings were additionally confirmed by counting round spermatids displaying expression of *SP4* gene, which is characteristic of mid-pachytene onward. Two morphological types of secondary spermatogonia, clear and dark, were recently described in a hylid frog, *Scinax fuscovarius* (de Oliveira *et al.*, 2003). The two types differ in nuclear morphology. The nucleus is drop-shaped with more condensed chromatin in dark spermatogonia. There are also differences in staining affinity of cytoplasm, which is more translucent in dark spermatogonia. Cysts with clear spermatogonia are localized close to the tubule wall. It may be that clear secondary spermatogonia are in a transition phase, whereas dark secondary spermatogonia are those, which can enter meiosis.

The average diameter of secondary spermatogonia, as well as their nuclei, decreases gradually during proliferation within a cyst. In *Bufo arenarum* the nuclei diameters decrease from about 12 µm to 7.5 µm (Cavicchia and Moviglia, 1983), and in *Rana tigerina* the diameter of nuclei ranges from 6 to 7 µm (Manochantr *et al.*, 2003). The nucleoli become more prominent than in primary spermatogonia, and the nucleus contains increasing number of small heterochromatin blocks situated in the center and close to the nuclear envelope (Manochantr *et al.*, 2003). The number of cysts with secondary spermatogonia, as well as the number of secondary spermatogonia within a cyst, increases with age of sexually active males, as was shown for *Rana cyanophlyctis* (Parachatna *et al.*, 2000).

Primary spermatocytes. After the mitotic multiplication within a cyst is completed, secondary spermatogonia transform into primary spermatocytes, *i.e.*, gain the ability to enter meiosis (Fig. 5B,C). Transformation of secondary spermatogonia into pre-leptotene spermatocytes was studied *in vitro* and *in vivo* in *Xenopus laevis* by Takamune *et al.* (1995). The measurements of nuclear volume revealed that nuclei of pre-leptotene primary spermatocytes are twice as large as nuclei in secondary spermatogonia. Experiments with BrdU labeling indicate that the nuclear volume increase is followed by pre-meiotic DNA replication. The only cytological feature characteristic of primary pre-leptotene spermatocytes is the presence of flattened vesicles immediately after the last mitotic division. The role of the vesicles is unknown.

In *Pachymedusa dacnicolor* (Rastogi *et al.*, 1988), primary spermatocytes increase in diameter about twofold during the first meiotic prophase, as compared to last generation of secondary spermatogonia, *i.e.*, from 9-10 µm to 17 µm. Their cytoplasm contains cisternae of endoplasmic reticulum and the Golgi complex. Primary spermatocytes are sub-staged according to the progress of the prophase of the first meiotic division. Each of these stages has distinctive pattern of chromatin condensation. The degree of chromatin condensation increases constantly, reaching its maximum at metaphase. According to Manochanter *et al.* (2003), chromatin of the leptotene spermatocytes is in form of euchromatin, and there is no heterochromatin facing the nuclear envelope. The nucleolus is present, but not as prominent as in spermatogonia. Chromosomes start to condense and form dense lines (*i.e.*, axial elements). During zygotene synaptonemal complexes began to form between the homologue chromosomes, but some axial elements are still uncoupled. Chromosome ends are attached to the inner surface of nuclear envelope at one side of the nucleus, thereby forming the characteristic bouquet stage. The nucleoli disappear. Fully formed synaptonemal complexes are characteristic of pachytene spermatocytes. This stage is relatively long; the degree of chromatin condensation increases and for this reason pachytene spermatocytes vary in appearance. During diplotene, highly condensed chromosomes are distributed close to the nuclear envelope and the center of the nucleus remains "empty". During diakinesis the nuclear envelope disintegrates, and condensed chromosomes are distributed within the whole nucleus. The next step is metaphase I.

Secondary spermatocytes. As a result of the first meiotic division, small secondary spermatocytes are formed (Fig. 5 B,C). The duration of this stage is relatively short and for this reason cysts containing secondary spermatogonia are rarely observed. The nuclei of secondary spermatocytes are round with some heterochromatin blocks inside (Manochantr *et al.*, 2003). The formation and morphology of secondary spermatocytes were studied *in vitro* in *Xenopus laevis* by Abé *et al.* (1998). In most cases (87.5%) each secondary spermatocyte produces a motile flagellum, which grows up to 6 μm in length during interphase. At the beginning of metaphase II the flagellae disappear, and appear again in resulting spermatids. The number of cysts with spermatocytes, as well as the number of spermatocytes within a cyst, increase with age, being the highest in the oldest sexually active males, as was shown for *Rana cyanophlyctis* (Parachatna *et al.*, 2000).

Spermatids. The second meiotic division results in the formation of spermatids (Fig. 5B,C). Early spermatids differ from secondary spermatocytes by slightly smaller nuclei with less condensed chromatin. Spermatids are connected by cytoplasmic bridges, as were secondary spermatogonia and spermatocytes (Abé, 1988). Morphological changes of spermatids during spermiogenesis were described in detail by Rastogi *et al.* (1988) for *Pachymedusa dacnicolor*, Báo *et al.* (1991) for *Odontophrynus cultripes*, and Manochanter *et al.* (2003) for *Rana tigerina*. In early spermatids (stage I spermatids), nuclei are spherical or slightly oval, chromatin is finely granular and contains membrane-bounded vacuoles, which disappear during late spermiogenesis. Stage I spermatids are still closed inside a cyst.

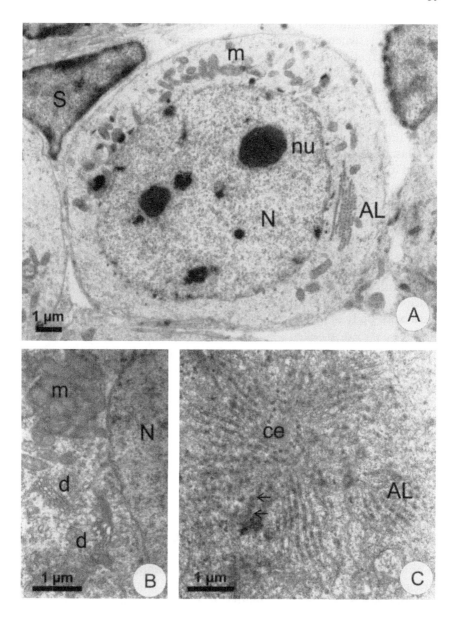

Fig. 8. Ultrastructure of a pale primary spermatogonia in *Rana ridibunda*. Annulate lamellae are arranged around centrioles (ce). and mitochondria are associated with intermitochondrial cement (shown in B). N—nucleus; nu—nucleolus; AL—annulate lamellae; ce—centrioles; d—dictyosomes (Golgi); m—mitochondria; S—Sertoli cell.

Stage II spermatids start to elongate, resulting in bipolar cells with the proximal pole, where the Golgi complex is conspicuous, and the distal pole, where a flagellum is formed. The nuclei of stage II spermatids are eccentrically located and are situated closer to the flagellum. During spermatid differentiation (stage III), cysts open and become cup-shaped with proximal parts of spermatids deeply inserted into processes of the apical part of Sertoli cell. Free distal parts of spermatids are directed to the tubule lumen. Stage IV spermatids have highly condensed chromatin, the cytoplasm has clear appearance, and mitochondria, which start to cluster around the forming flagella, are the predominate organelles.

As was described for *Hyla japonica* (Lee and Kwon, 1992), the cytoplasm of Sertoli cell surrounding the spermatids is strengthen by numerous microtubules lying parallel to the spermatid long axis. In the differentiating sperm cell, centrioles are situated at one side of the nucleus, opposite to the forming acrosome. The one close to the nucleus is a proximal centriole, and is usually situated in a nuclear fosse and can be surrounded by pericentriolar material, whereas the other is a distal centriole. The distal centriole is perpendicular to a long axis of a spermatid and will transform into a basal body of forming microtubular skeleton of a flagellum, *i.e.*, the axoneme.

During fertilization, the proximal centriole together with pericentriolar material will be transmitted to the oocyte, where they will later organize (nucleate) microtubules of the mitotic spindle. This is necessary for further cleavage of a resulting embryo, because oocyte centrioles are lost during late oogenesis.

Spermatozoa. At the end of spermiogenesis, an anuran spermatozoon achieves its characteristic shape, with a head, a midpiece (neck), and a tail (flagellum) (Fig. 9). Detailed comparative morphology of anuran spermatozoa in relation to phylogeny was reviewed recently by Scheltinga and Jamieson (2003).

Head. Nuclei of anuran spermatozoa have various shapes, characteristic of a species. In most cases they are elongated, or helically screwed, with densely packed chromatin. The heterochromatin can be lamellar in an apical part of a spermatozoon head and granular in its distal part (*Xenopus*), or uniformly granular (*Pachymedusa*). The helical shape of the nucleus is supported by microtubules in *Pachymedusa dacnicolor* (Rastogi et al., 1988) and *Odontophrynus cultripes* (Báo et al., 1991). In *Megophrys montana* (Asa and Philips, 1988) and *Xenopus laevis* (Reed and Stanley, 1972), microtubules are absent or not numerous and the shape of nucleus is achieved solely by chromatin condensation. In *Hyla japonica*, microtubules are present around spermatid nuclei and disappear in mature sperm (Lee and Kwon, 1992). Despite the high degree of condensation of chromatin, some apparently empty spaces were found inside the nucleus of *Epidobates flavopictus*, thus forming electron-lucent lacunae (Garda *et al.*, 2002). Such lacunae were also observed in *Scaphiopus couchi*, *Xenopus laevis* and *Bombina variegata*. Scheltinga *et al.* (2001) hypothesize that they might be remnants of the endonuclear canal and axial perforatorium (see below). However, the data supporting this hypothesis are scarce.

The nucleus of spermatozoa of basal genera (*Leiopelma*, *Ascaphus*) flares out abruptly to form distinct "nuclear shoulders", behind which a nucleus forms an elongate cylinder (Scheltinga *et al.*, 2001; Scheltinga and Jamieson, 2003). However, in most anuran species nuclear shoulders are absent.

In many vertebrates, chromatin condensation in spermatid nuclei is accompanied by a change in composition of basic proteins associated with DNA. During this process histones are replaced by protamines. Testis specific basic nuclear proteins of various anuran genera are very diverse. In *Rana pipiens*, protamines were not detected (Zirkin, 1970), but replacement of histones by basic proteins different from those in somatic cells was reported (Bols *et al.*, 1986). In the genus *Rana*, Adler and Gorovsky (1975) described a testis-specific variant of histone H1, and Bols and Kasinsky (1972) reported basic proteins in sperm of *Hyla* and *Bufo* being more arginine-rich than somatic histones. In 11 species and subspecies of the genus *Xenopus*, Mann *et al.* (1982) also reported that basic nuclear proteins of spermatozoa contain more arginine than nuclear proteins of somatic cells. The differences in testis proteins were variable to the extent that these proteins might have been used as molecular markers for species and even subspecies. During the course of spermiogenesis in *X. laevis*, histone H1 is removed from late spermatid, and histones H2A and H2B were reduced in relation to H3 and H4 (Risley and Eckhardt, 1975, 1981).

In the apical part of a spermatid, an acrosomal complex is formed (Scheltinga and Jamieson, 2003). It is a conspicuous organelle composed of the acrosomal vesicle and a subacrosomal substance situated distally, between acrosome and nucleus. The acrosome is formed during spermiogenesis as a result of secretion activity of dictyosomes of the Golgi complex. At the end of spermiogenesis the acrosome changes its shape, and substances accumulated in the acrosomal vesicle become dense and can form granules. In *Xenopus laevis* and *Odontophyrnus cultripes* the acrosomal vesicle collapses at a side adjacent to nucleus, thus forming a thin acrosomal cap, tightly pulled over the nucleus (Báo *et al.*, 1991; Ueda *et al.*, 2002). Acrosomal vesicles contain glycoprotein enzymes necessary for egg envelope digestion during fertilization (Koch and Lambert, 1990; Scheltinga and Jamieson, 2003). Fernandes and Báo (1998) detected various phosphatases (acid phosphatase, thiamine phosphatase, glucose-6-phosphatase) in the acrosome of *Odontophrynus cultripes* (Leptodactylidae).

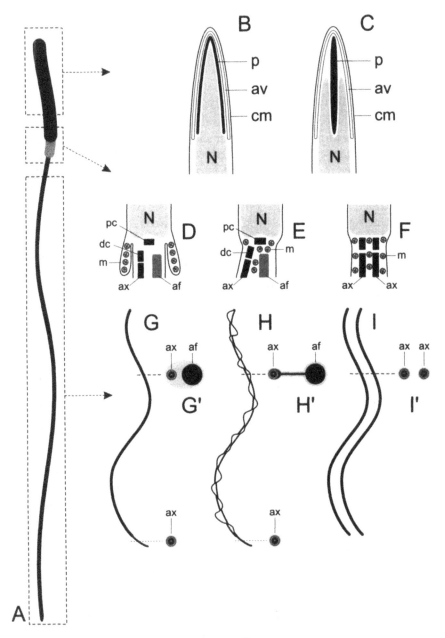

Fig. 9. See caption on the next page.

In *Epidobates flavopictus* (Dendrobatidae), the acrosome is formed by a flat and narrow vesicle situated apically to the nucleus and filled with a homogenous substance, and a subacrosomal substance forming a cone around the nucleus (Garda *et al.*, 2002). The subacrosomal substance can form a rod or a cone. Primarily its role was interpreted as a part of mechanism acting in disruption of egg envelopes during fertilization and for this reason it was also called a perforatorium. As is suggested by Jamieson (1999), Scheltinga *et al.* (2001), and Garda *et al.* (2002), subacrosomal substance and perforatorium are homologous. In most primitive genera (*Ascaphus, Discoglossus, Bombina, Alytes*) the perforatorium is rod-shaped and deeply penetrates nucleus within the endonuclear canal. In Discoglossidea (*Discoglossus pictus, Bombina bombina*) the perforatorium, called the axial perforatorium, is a long rod originating from acrosome and passing distally through acrosomal vesicle and nucleus. In *Ascaphus*, the axial perforatorium coexists with a conical one (Scheltinga and Jamieson, 2003). In another species belonging to primitive Anura, *Leiopelma hochstetteri* (Scheltinga *et al.*, 2001), the acrosome is conical and polygonal on transverse sections, and subacrosomal substance is conical at its distal part and forms a rod in its most apical part.

In evolutionary more advanced Anura (Neobatrachia), the perforatorium is conical (Bufonidae, Dendrobatidae) or, in Ranidae, is lacking

---

Fig. 9. A generalized schematic representation of spermatozoa in Anura based on Scheltinga and Jamieson (2003). A—Model of a spermatozoon with head, midpiece, and tail. B and C represents the apical parts of spermatozoa heads composed mainly of highly condensed nucleus (n) surrounded by cell membrane (cm). Beneath the cell membrane of the apical part of the head, an acrosome vesicle (av) is situated. In most species the perforatorium (p) is conical (as in B) or rod-shaped (as in C). D, E, and F represent various arrangements of midpiece. Two centrioles (proximal, pc, and distal, dc) can be arranged perpendicularly (D), oblique (E) or in parallel (F). Distal centrioles are basal bodies for axonemes (ax) (in most cases one, in some cases two, F). The axial fiber (af) is formed in parallel to the axoneme. Mitochondria can be scattered among centrioles, axoneme, and axial fiber (E, F) or can be gathered in the cytoplasmic collar around the distal part of the head (D). There is usually one tail, without (G) or with (H) the undulating membrane; some anurans have two tails without undulating membranes (I). The tail without undulating membrane is composed of axoneme and axial fiber lying in close distance (G′). In tail with undulating membrane, axoneme is in some distance from axial fiber, and these two axial elements are connected by a thin sheet of the same substance, of which axial fiber is formed (H′). In spermatozoa with two axonemes, neither axial fibers, nor undulating membranes are formed (I′). The terminal portion of the tail is always composed of axoneme, because axial fiber is shorter than accompanying axoneme.

(Jamieson, 1999; Garda *et al.*, 2002; Scheltinga and Jamieson, 2003). Conical perforatoria were also described in the genera *Cyclorana* and *Litoria* (Jamieson *et al.*, 1997). In *Pachymedusa* the perforatorium is composed of filaments (Rastogi *et al.*, 1988). In *Hyla japonica* Lee and Kwon (1992) reported the presence of microtubules. During fertilization, spermatozoon must penetrate glycoprotein layers of jelly coat, as well as the vitelline envelope surrounding the oocyte. This step is achieved as a result of acrosomal reaction, during which the acrosomal cap opens and releases its proteolytic enzymes, sometimes collectively referred to as spermatolysin (Raisman *et al.*, 1980; Cabada *et al.*, 1989). The acrosomal reaction can start in glycoprotein jelly capsule or at the vitelline envelope. When the acrosomal reaction takes place within the jelly coat, the spermatozoon attaches vitelline envelope of an egg with its own cytoplasmic membrane already opened, as is the case in *Discoglossus pictus* (Ueda *et al.*, 2002). In *Xenopus laevis* the acrosome reaction starts after reaching vitelline envelope and is Ca^{+2} dependent, and no formation of an acrosomal process was observed (Ueda *et al.*, 2002). After this contact, the spermatozoon cell membrane fuses with the egg cell membrane, and a male nucleus forms within the ooplasm and soon fuses with a female nucleus.

Midpiece. The midpiece is a region situated between the head and the tail. It contains the nuclear fossa, two centrioles, and the most apical part of axoneme together with its connection with axial fiber (connecting piece). During spermiogenesis, also mitochondria migrate to this part of a spermatid. The mitochondria do not fuse together, which is a case in many animal species (Koch and Lambert, 1990). Mitochondria lie in cytoplasmic mass, as was reported for *Pachymedusa* (Rastogi *et al.*, 1988) and *Xenopus* (Bernardini *et al.*, 1986), or are distributed around the base of flagellum in a fold of cytoplasm called mitochondrial sheath or mitochondrial collar. The mitochondrial collar was described for *Odontophrynus cultripes* (Báo *et al.*, 1991), *Litoria* and *Cyclorana* (Jamieson, 1997), as well as *Leiopelma hochstetteri* (Scheltinga *et al.*, 2001) and others (see the review by Scheltinga and Jamieson, 2003). In *Epidobates flavopictus* (Garda *et al.*, 2002), mitochondria are also present inside the anterior part of undulating membrane of a tail.

Centrioles can be arranged parallel, oblique or perpendicular. The perpendicular orientation is found in most species (Scheltinga and Jamieson, 2003). In a basal species *Leiopelma hochstetteri*, as well as *Scaphiopus* (Scheltinga *et al.*, 2001), two parallel centrioles, orientated in the long axis of the nucleus, lie at the base of nuclear fossa. Two parallel centrioles, each lying in a separate nuclear fossa and constituting basal bodies, are characteristic of spermatozoa with two axonemes, as is the case of *Hyperolius puncticulatus*, *Polypedates leucomystax*, *Telmatobufo australis*, and

3 species of Pelobatidae: *Leptolalax pictus, Spea intermonata, Pelobates syriacus* (Scheltinga and Jamieson, 2003). In *Epidobates flavopictus* (Garda *et al.*, 2002) the proximal centriole lies at an approximate angle of 50° with respect to the longitudinal axis of a spermatozoon. The same situation is reported for *Ascaphus* and *Xenopus*.

Tail. The tail of an anuran spermatozoon is generally composed of the axoneme, the axial fiber (axial rod), and an undulating membrane. The axoneme has the typical arrangement of microtubules, *i.e.*, it is composed of two central microtubules surrounded by nine peripheral doublet microtubules. The axoneme is formed very early during spermiogenesis and was observed in *Xenopus laevis* just after the second meiotic division (Abé *et al.*, 1998). Each newly formed spermatid produce a flagellum containing the axoneme, which achieves 1 μm in lengths within 2-3 hours, and elongates to 20-40 μm during next 2-3 days. The formation of flagella is an intrinsic feature of spermatids and is not induced by Sertoli cells of a cyst.

Concomitantly with the axoneme formation, electron dense material accumulates in the midpiece. It will give rise to an axial fiber running along the axoneme. The axoneme and axial fiber are connected in their most proximal parts forming a connective piece rich in basic proteins (Báo *et al.*, 1991). On transverse sections, axial fibers can be round as in *Odotophrynus cultripes* (Báo *et al.*, 1991) or U-shaped, as is in *Epidobates flavopictus* (Garda *et al.*, 2002).

A thin ribbon-like sheath of cytoplasm stretches between the axoneme and axial filament, forming an undulating membrane. A part of the dense pericentriolar material that is called the juxta-axonemal fiber (Jamieson, 1999) can stay with the axoneme, being attached to one side of axial filament by fine stripes of dense material, as is the case of *Pachymedusa* (Rastogi *et al.*, 1988), or a solid axial strand, as is the case of *Odotophrynus cultripes* (Báo *et al.*, 1991) and *Epidobates flavopictus* (Garda *et al.*, 2002). In *Leiopelma hochstetteri* (Scheltinga *et al.*, 2001), juxta-axonemal fibers are present on both sides of the axoneme and resemble the marginal fibers in Urodela (Koch and Lambert, 1990). The axial filament is shorter than axoneme. For this reason the terminal part of a tail is devoted of undulating membrane and is composed of the axoneme and accompanying juxta-axonemal fiber(s). In some cases the axial filament and juxta-axonemal fiber can fuse and form a paraxonemal rod, parallel to axoneme; in such cases the undulating membrane is reduced, which is the case of *Ascaphus* and *Discoglossus* (Jamieson, 1999). In the genus *Cyclorana* (with exception of *C. cryptotic* and *C. manya*), the undulating membrane is thick due to presence of thick paraxonemal rod (Scheltinga and Jamieson, 2003).

If axial and juxta-axonemal fibers are absent, as is the case in the genera *Rana* and *Xenopus*, the undulating membrane is lacking (Bernardini *et al.*, 1986; Koch and Lambert, 1991; Jamieson, 1999). Lack of undulating membrane is also the case in *Hyla japonica*, although both the axoneme and the axial rod are present (Lee and Kwon, 1992). The same situation was described in other Hylidae, including *Cyclorana manya*, as well as the genera *Acris*, *Plectrohyla*, *Pseudacris* and *Pseudis* (Scheltinga and Jamieson, 2003).

Spermatozoa are released from open cysts into a seminiferous tubule lumen, carrying a lobe of non-rejected electron dense cytoplasm containing numerous vesicles, glycogen, and some mitochondria. The cytoplasmic lobe is rejected later in uro-genital ducts and spermatozoa observed in the Wolffian ducts are already devoid of it. The rejected fragments of cytoplasm, called residual bodies, are phagocytized by epithelial cells of Wolffian ducts, and not by Sertoli cells, which is common in fishes and mammals. Reduction of cytoplasm volume during spermiogenesis in *Rana catesbeiana* reaches more than 95% and, together with 73% reduction of nucleus volume, result in average reduction of the entire cell volume of 87% (Sprando and Russel, 1988).

Kinetics of Spermatogenesis

Primary spermatogonia in an adult male display marked seasonal changes in mitotic activity, being the highest before breeding season, as was shown in *Rana esculenta* (Rastogi *et al.*, 1985). Amphibians are ectothermic animals and the rate of spermatogenesis—as of other life processes—depends on temperature. The mitotic activity is triggered by high temperature (Rastogi *et al.*, 1978) and the increasing level of gonadotropin-releasing hormone (GnRH) by promoting the G1-S transition of cell cycle (Minucci *et al.*, 1996). The presence of spermatids seems to inhibit primary spermatogonia multiplication in adult frogs (Minucci *et al.*, 1992).

Spermatogenesis in temperate zone anuran amphibians is synchronous in most of the seminiferous tubules of a testis (Lofts, 1974). The annual cycle proceeds as a spermatogenic wave in *Rana temporaria*, *R. sylvatica* (Witschi, 1956), and *Bufo bufo* (Jørgensen and Billeter, 1982). Asynchronous continuous spermatogenesis is presumably characteristic of anurans living in constant temperature environments (Jørgensen, 1992).

The duration of spermatogenic wave at optimal temperatures in anurans is 5-6 weeks (reviewed by Jørgensen, 1992). Kalt (1976) studied the process in *Xenopus laevis* at 18°C *in vitro* after H[3] thymidine treatment, from the beginning of meiosis to the end of spermiogenesis. Under these conditions, the premeiotic phase S lasted 6-7 days; leptotene 4 days; zygotene 6 days; pachytene 12 days; diplotene 1 day; first and second meiotic divisions 1 day; spermiogenesis 12 days, which altogether gives 42-43 days, *i.e.*, 6 weeks. Abé and Asakura (1987) and Abé *et al.* (1988) also studied the process for the same species at 22°C. After the completion of the first meiotic division, nuclear envelopes were formed within 10 minutes. The time between telophase I and telophase II lasts 1day, interphase II lasts 18 hours, the time from telophase II to acrosome cap formation in spermatids is about 6 days, and to the end of spermiogenesis almost 8 days. The duration of the spermatogonial mitotic cycle in *Rana esculenta* is approximately 24 hours, and the duration of the sperma- togenesis, beginning with premeiotic S-phase, require 42 days, *i.e.*, 6 weeks (Rastogi *et al.*, 1985, 1990). The progressing wave is correlated with increasing degeneration of cysts. The degeneration and resorption of cysts probably constitute normal final stages of spermatogenic cycle (Jørgensen and Billeter, 1982).

Testis-ova

Testis-ova are diplotene oocytes present sometimes in developing testes of species belonging to the family Ranidae and Rhacophoridae (Fig. 10). They were described in more than a half males *Rana nigromaculata* studied by Iwasawa and Kobayashi (1976), and Kobayashi and Iwasawa (1988). Differentiation of female germ cells inside testes might be caused by dysfunction of hormonal control, *i.e.*, low androgen synthesis or insuffi- cient level of receptors. Testis-ova were also observed in *Rhacophorus arboreus* (Tanimura and Iwasawa, 1989), *Rana catesbeiana* (Hsu *et al.*, 1977, 1979), and occasionally in *Rana lessonae* and *Rana esculenta* (Ogielska and Bartmańska, 1999).

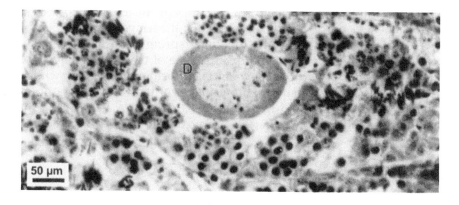

Fig. 10. Testis-ovum (D) in a juvenile testis of *Rana esculenta*.

Sexual Maturity of Males

The age of sexual maturity in amphibians varies among species, populations of the same species, and even individuals of the same population. One of the most important features, dependent on environment, is achievement of proper body size and mass. Sexual maturity is preceded by rapid growth, and then a rate of growth decreases considerably. Males of most of anuran species are not able to reproduce during their first year of life and usually are sexually mature during the second year of life.

In temperate zone amphibians, especially in the family *Ranidae*, so-called precocious spermatogenesis is observed. Iwasawa and Kobayashi (1976), Kobayashi and Iwasawa (1988), and Iwasawa *et al.* (1987) described the phenomenon for *Rana nigromaculata*. During precocious spermatogenesis spermatozoa are formed in testes within a few months after metamorphosis, but they disappear during hibernation. Usually in most males after metamorphosis (July) to first hibernation only primary spermatogonia are observed in seminiferous tubules. In about 30% of males meiosis precociously starts in August, and before first hibernation some spermatozoa are present in a lumen of seminiferous tubules. However, after first hibernation meiocytes and spermatozoa are only rarely observed, and primary spermatogonia are again the main bulk of spermatogenic cells in testes. During the second year of life a "proper" spermatogenesis starts, resulting in production and storage of functional spermatozoa for the next season. Similar observations were obtained for *R. lessonae* and *R. ridibunda* (Ogielska and Bartmańska, 1999). Precocious spermatogenesis was not observed in the genus *Bufo* (Jørgenson and Billeter, 1982).

In a tropical species *Rana cyanophlyctis* the histological composition of testes in one-year-old male (Pancharatna *et al.*, 2000) is similar to that in temperate zone *R. lessonae* and *R. ridibunda* before their first hibernation, *i.e.*, some bundles of spermatozoa are present. In the tropic species spermatogenesis proceeds further, whereas in the temperate zone species it is arrested and continues the following spring. The mechanism of precocious spermatogenesis is not known. One of the reasons may be a temporal increase of androgen level in males before hibernation, which decreases during hibernation below the level that allows retaining spermatozoa (Iwasawa and Kobayashi, 1976; Iwasawa *et al.*, 1987). In tropical species there is no seasonality and no drastic changes in endocrine activity.

Regulation of Spermatogenesis in Anura

The process of spermatogenesis in amphibians is precisely coordinated by the endocrine and nervous systems, generally in a similar way as in other vertebrates. Besides the regulation of testis function, sexual hormones are also involved in differentiation of sexual dimorphism, male behavior, and integration of reproductive processes with the overall metabolism of males. Hormones involved in control of amphibian spermatogenesis represent chemically heterogeneous groups and are synthesized by various cells of the body. According to chemical structure, three types of hormones can be distinguished: lipid, peptide, and monoamine.

Lipid hormones are classified as steroids and eicosanoids. Steroid derivatives of cholesterol consist of four families of hormones: progestagens, androgens, estrogens, and corticosteroids. The first three are regarded as sexual hormones. Steroids are synthesized in smooth endoplasmic reticulum of somatic cells in gonads. They usually act as endocrine hormones, that is, they are released into the bloodstream and circulate throughout the body. Corticosteroids are produced in adrenal glands and are only indirectly involved in reproduction (Licht *et al.*, 1983; Burmaister *et al.*, 2001; Moore and Jessop, 2003; Moore *et al.*, 2005). Eicosanoids, including prostaglandins PGS and leukotrienes, derive from fatty acid substrates of cell membrane phospholipids. They are synthesized in a variety of tissues and diffuse to their targets as locally acting paracrine hormones. Prostaglandins play a relatively important role in reproduction, whereas leukotrienes are of minimal significance.

Peptide hormones include the gonadotropic glycoproteins (follicle-stimulating hormone FSH and luteinizing hormone LH), polypeptides (prolactin PRL and growth hormone GH), and peptides (gonadotropin-releasing hormone GnRH, arginine vasotocin AVT, mesotocin MST, β-endorphins, and peptide growth factors GFs). The peptide hormones are produced mainly in glands of endodermal origin. They may act as endocrine and paracrine hormones, and the secreting cells may also be the target cells (autocrine signaling). GFs are synthesized in a variety of cells and tissues. They control cell proliferation and differentiation acting mainly by paracrine or autocrine ways. In addition, some GFs are bound to the cell membrane and act without diffusion as signal molecules on receptors of neighboring cells (juxtacrine signaling).

Monoamines are represented mainly by melatonin (MEL), which is released from pineal gland under environmental control and modifies the secretory function of hypothalamus and pituitary, thereby influencing reproduction in seasonally breeding amphibians (Kelly and van der Kramer, 1960; Vivien-Roels and Pevet, 1983; Wiechmann *et al.*, 2003) and spermatogonial proliferation (d'Istria *et al.*, 2003). Other monoamines, including neurotransmitters, such as dopamine, epinephrine, and norepinephrine produced in neurosecretory cells in hypothalamus and adrenals, participate in neuroendocrine control of reproduction.

Many hormones regulating male reproductive functions operate on the hypothalamo-pituitary-gonadal (HPG) axis. Information about the level of hormones produced by pituitary and gonads is received by hypothalamus and modulates its activity by negative feedback. The hypothalamus operates in the short loop (between hypothalamus and pituitary) and the long loop (between hypothalamus and gonads). The loops coordinate neuroendocrine functions and enable flow of information concerning physiological state of the animal, up and down the HPG axis.

Hypothalamus

Neurosecretory neurons of hypothalamus synthesize biologically active substances affecting (*via* the pituitary) the function of gonads in response to exogenous signals, or as a result of endogenous rhythm generated by the brain. The substances (GnRHs, arginine vasotocin AVT, and mesotocin MST) are synthesized in different neurosecretory cells.

GnRHs, the main hormones linking the brain with reproductive system, are synthesized in small secretory neurons distributed in hypothalamus and other parts of the brain. Axons of these neurones extend to the median eminence forming the hypothalamo-median eminence tract and release GnRH into capillary blood vessels leading to the *pars distalis* of adenohypophysis (see pituitary) (Jokura and Urano, 1985; Burggren and Just, 1992; Matsumoto and Arai, 1992, Rastogi *et al.*, 1998). GnRH is a decapeptide belonging to the ancient family of signaling molecules involved in regulation of reproduction in animal kingdom (Polzonetti-Magni *et al.*, 1998). To date, several structural variants of GnRHs have been identified in vertebrate nervous tissues. They differ in their amino-acid sequences, but they probably originate from one ancestral molecule (Sherwood *et al.*, 1986, 1997; Cariello *et al.*, 1989; Di Meglio *et al.*, 1991; Muske, 1993; Ja and Millar, 1995; Millar *et al.*, 1996; King and Millar, 1997; Fernald *et al.*, 1999; Yoo *et al.*, 2000; Dubois *et al.*, 2002; Kah *et al.*, 2004, Guilgur *et al.*, 2006). Vertebrate GnRHs fall into three main types: mammalian mGnRH, chicken cGnRH, and salmon sGnRH. At least two of them, mGnRH and cGnRH (cGnRH-II) function in anuran neural tissue (Muskie, 1993; Muske *et al.*, 1994; Licht *et al.*, 1994; Setalo *et al.*, 1994; Collin *et al.*, 1995; Sealfon *et al.*, 1997; Rastogi *et al.*, 1997, 1998; Miranda *et al.*, 1998; Dubois *et al.*, 2002). According to some authors (Cariello *et al.*, 1989; D'Antonio *et al.*, 1992; Fasano *et al.*, 1990; 1993; 1995), sGnRH also appears in anurans; however other studies do not confirm this opinion (Licht *et al.*, 1994; Collin *et al.*, 1995; Dubois *et al.*, 2002).

Immunohistochemical studies in various anuran species show that neurosecretory neurons containing mGnRH and cGnRH-II are localized separately in the brain beginning from the tadpole stages, which suggests different roles of the two forms (King *et al.*, 1994; Muske *et al.*, 1994). In the frog *Rana esculenta* tadpoles, cGnRH-II was detected in mesencephalon at the time when hind limbs begin to develop (Gosner stage 26-27) (D'Aniello *et al.*, 1995; Di Fiore *et al.*, 1996), then mGnRH appears in various areas of telencephalon and diencephalon. After metamorphosis cGnRH-II and mGnRH neurons are recognized also in other parts of the brain (Licht *et al.*, 1994; D'Aniello *et al.*, 1995; Di Fiore *et al.*, 1996; Miranda *et al.*, 1998; Rastogi *et al.*, 1997, 1998; Dubois *et al.*, 2002). Amphibian mGnRHs are species-specific, and are involved in direct regulation of pituitary gonado-tropins (Di Meglio *et al.*, 1991; Fasano *et al.*, 1993; Muske *et al.*, 1994, Licht *et al.*, 1994; King *et al.*, 1994; Collin *et al.*, 1995 Di Matteo *et al.*, 1996; Di Fiore *et al.*, 1996; Dubois *et al.*, 2002; Guilgur *et al.*, 2006). In frog pituitary both forms of GnRHs coexist, however the amount of mGnRH is much higher than that of cGnRH-II (Di Fiore *et al.*, 1996). Daniels and Licht (1980) revealed that the frog pituitary is very sensitive to mGnRH stimulation. After a single injection of GnRH into lymph sac of the bullfrog *Rana catesbeiana* the gonadotropins levels in blood circulation increased several hundred times. However, Troskie *et al.* (2000) described a novel GnRH receptor (XL-GnRH-RI) in *Xenopus laevis* pituitary and midbrain, which has 2000 times higher selectivity to cGnRH than to mGnRH, which is in contrast with the current opinion about the central role of mGnRH as a gonadotropin regulator. Dubois *et al.* (2002) and Guilgur *et al.* (2006) suggested that cGnRH is the evolutionary conserved form. Besides the various parts of brain and spinal cord, it is also widely distributed in peripheral nervous system and sympathetic ganglia, which suggests that it plays a role in neuromodulation and neurotransmition (King *et al.*, 1994; Licht *et al.*, 1994; Muske *et al.*, 1994; Collin *et al.*, 1995; Di Matteo *et al.*, 1996; Troskie *et al.*, 1997, 2000).

GnRH was also detected in gonads. The testis form, which appears to be identical with cGnRH, is known as tGnRH; in *Rana esculenta* tGnRH was detected in cells of somatic and germinal compartment of the gonad (Cariello *et al.*, 1989; Di Matteo *et al.*, 1988, 1990, 1996; Fasano *et al.*, 1988, 1990, 1993, 1995; Minucci *et al.*, 1992). tGnRH stimulates androgen production and proliferation of primary spermatogonia (Di Matteo *et al.*, 1990, 1996, Minucci *et al.*, 1986, 1996). The presence of tGnRH in interstitial and germinal compartments, as well as seasonal changes of its level (Di Matteo *et al.*, 1996) strongly suggest that it is involved in the coordination of local communication between somatic and gametogenic cells, and thereby ensures the optimal conditions for development and differentiation of gametes.

As was mention above, the activity of hypothalamus in synthesis and secretion of GnRHs shows seasonal changes. Jocura and Urano (1985) studied correlations between localization and level of GnRHs in hypo-

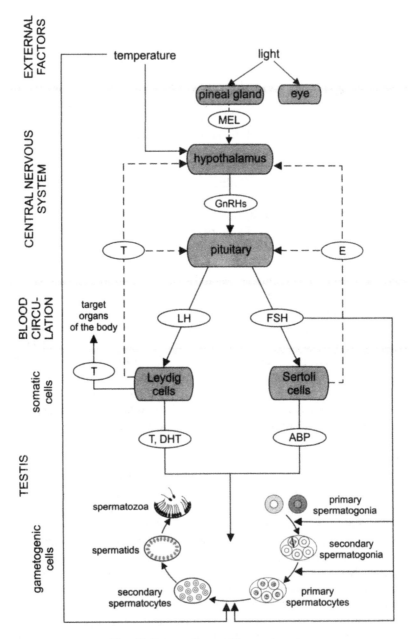

Fig. 11. See caption on the next page.

thalamus with annual cycle in the toad *Bufo japonicus*. They stated that the median eminence in this species displayed low level of GnRH during migration to breeding sites (before spermiation) and its high level when mating in spring. The level of GnRH increased again in autumn, when spermiogenesis started. Rastogi *et al.* (1997) observed the highest GnRHs levels in the brain and pituitary in both sexes of *Rana esculenta* just prior to breeding. Moreover, brain concentration of the GnRHs in males showed significant fluctuations depending on their reproductive status. The authors suggested that seasonal changes of GnRHs concentrations in the brain and pituitary correlated with plasma sex steroids levels, supporting the opinion that steroid hormones affect GnRHs neuronal system.

GnRHs play also an important role in the direct regulation of spermatogenesis, spermiation, and control of steroids biosynthesis during reproductive period (Müller, 1977; Minucci *et al.*, 1986, 1996). GnRHs receptors were demonstrated in anuran gonads by Fasano *et al.* (1990) and Wang *et al.* (2001). It was found that decrease of androgen biosynthesis is caused by strong decrease of enzymatic activity of cytochrome $P450_{C17}$ (Canosa *et al.*, 2002).

Hypothalamus not only sends, but also receives information about hormonal condition of endocrine glands engaged in reproduction. Brain GnRH system is sensitive to concentration of steroids in blood (Tsai *et al.*, 2003, 2005). The amount and mode of GnRH secretion is modulated by the negative feedback exerted by the direct action of androgens on

---

Fig. 11. Summary of main hormonal and environmental factors regulating the activity of the hypothalamo-pituitary-gonadal (HPG) axis during reproductive cycle of Anura. The release of gonadotropine releasing hormones (GnRHs) by hypothalamus stimulates secretion of pituitary gonadotropins (luteinizing hormone LH and follicle stimulating hormone FSH) that regulate the endocrine and spermatogenic activities of the testis. LH acts mainly on Leydig cells and FSH on Sertoli cells, stimulating them to synthesis and secretion of gonadal steroids. FSH may directly regulate the cell cycle of spermatogonia and primary spermatocytes. Leydig cells secrete testosterone (T) and 5 α-dihydrotestosterone (DHT). The main amount of T goes to the blood circulation and stimulates the target organs throughout the body. Some T enters the seminiferous tubules and binds to its receptors in Sertoli cells. In Sertoli cells T may be converted to DHT or to estrogens (E). Steroids (mainly T and DHT) control different stages of spermatogenesis and maintain the function of somatic cells in the testis. Additionally T, DHT, and E operate through the negative feedback mechanism (dashed arrows) on the hypothalamus and pituitary, and cause the significant suppression of further secretion of their hormones. HPG axis is also sensitive to the environmental factors (temperature and photoperiod) that modulate its activity via the pineal gland and its hormone, melatonine (MEL).

hypothalamus. Androgen and estrogen receptors were detected in hypothalamus and other parts of amphibian brain (Callard, 1985; Callard and Callard, 1987; Callard *et al.*, 1978; Paolucci, 2003). Steroid-accumulating cells were found in the frontal part of the *preoptic region* of hypothalamus, in *infundibulum*, as well as in the pituitary in *Rana esculenta* (Di Meglio *et al.*, 1987). Acting directly on the brain, androgens very effectively induce mating behavior (Wada and Gorbman, 1977; Polzonetti-Magni *et al.*, 1984; Moore *et al.*, 2005).

Other two hormones secreted by hypothalamus are arginine vasotocin (AVT) and mesotocin (MST), which are synthesized by large neurons situated in preoptic nuclei of the hypothalamus. AVT and MST are homologues of mammalian vasopressin and oxytocin, respectively. They are stored in neurohypophysis and released when needed. Immuno-cytochemical studies revealed that AVT and MST are synthesized in the brains of *Xenopus laevis*, *Rana ridibunda*, and other amphibian species (Acharjee *et al.*, 2004; Gonzales and Smeets, 1992a,b; Smeets and Gonzales, 2001; Hollis *et al.*, 2005). Both hormones act as neurotransmitters and neuromodulators. The primary function of AVT is regulation of hydro-mineral balance, vascular tone, and glucose metabolism. However, it is also involved in promoting the expression of reproductive behavior, inducing calling in frogs (Burmeister *et al.*, 2001). Opposite to mammals, in which the role of oxytocin is well known, the function of MST in amphibian reproduction is still unclear.

Pituitary

After stimulation of GnRH, the *pars distalis* of adenohypophysis (the glandular part of the pituitary) synthesizes gonadotropins FSH and LH, growth hormone GH, and prolactin PRL. Cells that secrete FSH and LH are distributed mainly in the rostral and ventral regions of *pars distalis*, whereas most of PRL-secreting cells are localized in the rostral region, and GH cells in the caudal region (Mikami, 1992; Polzonetti-Magni *et al.*, 1995). The number, size, and secretory activity of gonadotropic cells in the pituitary change during development and in adults depend on repro-ductive status of a male. The activity is low in tadpoles and increases significantly after metamorphosis under the stimulation of hypothalamic GnRH (Iwasawa and Kera, 1982; Licht *et al.*, 1983; Jokura and Urano, 1985; Delgado *et al.*, 1989). Gonadotropins are stored in gonadotropic cells as granules that are released into the circulation under stimulation of GnRH.

In amphibians, both LH and FSH levels rise during reproductive season (Itoh *et al.*, 1990; Ishi and Itoh, 1992; Kim *et al.*, 1998; Polzonetti-Magni *et al.*, 1998). Both gonadotropins stimulate testis growth, but LH is essential for androgen secretion (Licht, 1986; Herman, 1992). Early stages

of spermatogenesis proceed without or at low levels of circulating gonadotropins, but spermiation needs a "surge" of LH secretion, which increases
blood gonadotropin level by an order of magnitude. The significance of
this phenomenon may be the prevention of premature release of gametes
and synchronization of spermiation with breeding time (Jørgensen, 1992).

Development and differentiation of Leydig cells, steroid synthesis,
spermiogenesis, and spermiation are under control of LH. Action of LH
on Leydig cells of testes shows a high species specificity. In the bullfrog
Rana catesbeiana only conspecific LH was potent in stimulating androgen
secretion, whereas LH of other species of the genus *Rana* had no effect (Lofts,
1961, 1964; Lofts and van Oordt, 1962; Kasinathan and Basu, 1973; Segal
et al., 1979; Jokura and Urano, 1985). The low plasma level of LH was
observed during hibernation; its level increased during mating period
concomitantly with high androgen level (Pierantoni *et al.*, 1984a; Kim *et
al.*, 1998; Polzonetti-Magni *et al.*, 1998).

FSH stimulates mitotic activity of primary and secondary spermatogonia, as well as maintenance of primary spermatocytes (Lofts and
Oordt, 1962; Lofts, 1964; Kasinathan and Basu, 1973; Minucci *et al.*, 1986).
The action of FSH on spermatogonia is mediated by Sertoli cells (Ito and
Abe, 1999; Chattopathyay *et al.*, 2003). FSH also induces synthesis of
androgen binding proteins (ABPs) in Sertoli cells, and the appearance and
activation of LH receptors in Leydig cells (French *et al.*, 1974).

The reproductive cycle in sexually mature males is highly correlated
with the annual activity of adenohypophysis (Rastogi and Chieffi, 1970a,b,
1972a,b; Rastogi *et al.*, 1972a,b, 1976; Callard *et al.*, 1978; Guha and
Jørgensen, 1978b; Lofts, 1984). The influence of adenohypophysis on
spermatogenesis was studied in the temperate zone water frog *Rana
esculenta* (Rastogi *et al.*, 1976). At low temperatures and short photoperiod
(spring), the pituitary secretes mainly LH that stimulates androgen
synthesis in interstitial cells of testes. At that time only primary spermatogonia and open cysts supporting spermatozoa can be detected in
seminiferous tubules. At the beginning of breeding season, a high level of
circulating androgens enables spermiation and inhibits LH secretion. As
the temperature and photoperiod increase, the pituitary starts to secrete
FSH, which stimulates mitotic activity of resting primary spermatogonia.
The first wave of mitoses serves to reestablish the number of primary
spermatogonia, and then the next waves give rise to several generations
of secondary spermatogonia. At the end of breeding season, at high
temperatures and longer photoperiod, FSH secretion increases considerably. As a consequence, a massive transformation of the secondary
spermatogonia into primary spermatocytes, and then into secondary
spermatocytes, are observed. A high level of FSH is maintained until the

end of summer, when cysts containing secondary spermatocytes dominate in testes. In autumn, when temperature falls and days become shorter, the blood level of FSH decreases, whereas the level of LH increases, thereby stimulating the synthesis and secretion of androgens. An increasing androgen level results in inhibition of mitotic activity of primary spermatogonia and transformation of secondary spermatogonia into a new generation of primary spermatocytes. It also allows the completion of meiosis in already existing spermatocytes, resulting in appearance of a new generation of spermatids. During hibernation, androgen level successively increases in gonads and in circulation, reaching its maximum in spring. At this time cysts are filled again with spermatids undergoing spermiohistogenesis to become fully developed spermatozoa for the coming breeding season.

During a typical annual cycle of temperate zone anurans with discontinuous spermatogenesis (*Rana temporaria*) (van Oordt, 1960; van Oordt and Lofts, 1963; Jørgensen, 1992), as well as those with potentially continuous spermatogenesis (*Rana esculenta*) (Lofts, 1964), it is evident that early stages of spermatogenesis proceeds independently of the hormonal activity of interstitial tissue. Development of the interstitium begins in autumn when spermiogenesis starts (Jørgensen et al., 1979).

The role of gonadotropins in amphibian spermatogenesis was described in a number of experiments with intact and hypophysectomized (PDX, pituitary dissected) animals, or in animals with blocked function of pituitary supplied by administration of exogenous gonadotropins. Lack of gonadotropins always results in interruption of spermatogenesis and gradual regression of both spermatogenic cells and interstitial tissue of testes (van Oordt, 1956; Lofts, 1961; Rastogi and Chieffi, 1972a,b; Rastogi et al., 1972a,b, 1976, 1990; Kasinathan and Basu, 1973; Guha and Jørgensen, 1978b; Lofts, 1984; D'Antonio et al., 1992; Minucci et al., 1993, 1994).

It is well documented that gonadoptropins have no effect on maintenance of primary spermatogonia, which are still present in atrophied seminiferous tubules many months after hypophysectomy (Sluiter et al., 1950; Lofts, 1961; Rastogi and Chieffi, 1972a,b; Rastogi et al., 1976; Lofts, 1984). However, gonadotropins, and indirectly also testosterone, are engaged in regulation of proliferation of both primary and secondary spermatogonia (Rastogi et al., 1990). Thus, after hypophysectomy, mitotic activity decreases significantly, especially in secondary spermatogonia (Di Matteo et al., 1988, 1990; Ito and Abe, 1999).

As a result of suppression of mitotic activity of secondary spermatogonia, almost total lack of primary spermatocytes is observed (Lofts, 1961; Bacabella et al., 1963; Rastogi et al., 1976; Guha and Jørgensen, 1978b; Rastogi and Iela, 1980; DiMatteo et al., 1988; Minucci et al., 1992). After the onset of meiosis, further differentiation of spermatogenic cells become independent on gonadotropins, as was shown in the grass frog *Rana temporaria* (van Oordt, 1956). Once the stage of primary spermatocytes is reached, the remaining part of the cycle proceeds without significant degeneration of cells in the absence of gonadotropins (Guha and Jørgensen, 1978b; Jørgensen, 1992).

In contrast to adult amphibians, in which gonads function under gonadotropins guidance, development of larval gonads is independent of the pituitary. In testes of PDX tadpoles of the toad *Alytes obstetricans*, spermatogenetic activity was normal. Even two months after hypophysectomy, testes of juvenile toads contained almost normal numbers of cysts with secondary spermatogonia, whereas in sexually mature toads the process of spermatogenesis practically ceased within one month after surgery (Guha and Jørgensen, 1978b; Jørgensen and Billeter, 1982). Development and differentiation of gonads in both PDX and intact tadpoles is almost the same until late metamorphic climax stages (Pchlemann, 1962; Hsu *et al.*, 1973). This means that the initiation of the first spermatogenic wave in juvenile males is independent of the pituitary (Iwasawa and Kobayashi, 1976; Iwasawa, 1978; Jørgensen and Billeter, 1982; Iwasawa *et al.*, 1987). Although a tadpole pituitary does not synthesize gonadotropins, somatic cells of gonads have gonadotropin receptors and are able to respond to exogenic gonadotropins (Burggren and Just, 1992). The effect of stimulation of juvenile males with exogenous gonadotropins depends on their age and/or physiological state of gonads, therefore indirectly on the season, as was shown in a series of experiments carried out with *Rana nigromaculata*. Males of this species are sexually mature after second hibernation, as most anuran species of the temperate zone. Testes of young males injected with gonadotropins in spring after first hibernation displayed enlargement of seminiferous tubules lumen and release of spermatogonia from cysts. If the experiment was performed in early summer, when spermatogenesis starts in testes of sexually mature animals, multiplication and differentiation of gametogenic cells were observed also in testes of the young males (Iwasawa and Kobayashi, 1976; Iwasawa *et al.*, 1977; Iwasawa, 1978).

Hypophysectomy performed in juvenile males has the same or even more pronounced final effect than in adults. The young males never gained proper weight and body size, and interstitial cells in their gonads were poorly developed. Even if cysts in their testes were filled with primary and secondary spermatogonia, further stages of spermatogenesis did not appear and a male was infertile. Also after hypophysectomy accompanied by one-side castration, juvenile individuals had no compensatory growth of the other testis, as was the case in control adult animals (Guha and Jørgensen, 1978b, 1981; Jørgensen and Billeter, 1982). The effects of hypophysectomy can be reversed by pituitary graft or application of gonadotropins. In PDX male toads of the genus *Bufo* daily injection of 5 IU or about 3.5 µg per 100g^{-1} body weight enabled both the high activity of spermatogenesis and interstitial function, and development of the secondary sex characters (Guha and Jørgensen, 1978a; Jørgensen, 1984).

In testes of frogs after ectopic transplantation of *pars distalis* of the glandular part of pituitary (Rastogi and Chieffi, 1970b, 1972b) spermatogenesis was reduced but interstitial cells functioned; toads after such treatment were able to maintain normal spermatogenesis when interstitial cells regressed (van Dongen *et al.*, 1966). The transplanted part of the pituitary probably secreted too small amounts of gonadotropins, but when transplanted directly to testis, it was able to maintain normal spermatogenesis. Exogenous FSH stimulates the proliferation of spermatogonia, both in PDX and intact males (Rastogi and Chieffi, 1972a,b; Rastogi, 1976; Rastogi *et al.*, 1976; Rastogi and Iela, 1980; Lofts, 1984). LH enables meiosis, differentiation and maturation of spermatids, as well as their release from cysts and differentiation of interstitial tissue of testes. Similar effects were observed after application of mammalian gestation gonadotropins: pregnant mare serum gonadotropin (PMSG) and human chorionic gonadotropin (hCG) to PDX anuran males (Iwasawa, 1978; Rastogi and Iela, 1980), although in normal animals both gonadotropins exerts inhibitory effect on early stages of spermatogenesis (Kasinathan and Basu, 1973). The negative effect of hypophysectomy could be reversed by daily injections of low doses of hCG (Guha and Jørgensen, 1978a). Acting as LH agonist, hCG stimulates multiplications and growth of interstitial cells and restores their specific function, including secretion of androgen and spermiation (Jørgensen, 1992).

As was already mentioned, besides gonadotropins, amphibian adenohypophysis also synthesizes and releases a growth hormone (GH) and prolactin (PRL). GH producing cells are localized mainly in the caudal part of adenohypophysis, and cells that synthesize PRL are often situated close to gonadotropic cells (Mikami, 1992; Polzonetti-Magni *et al.*, 1995). *In vitro* studies revealed that PRL acted as a paracrine factor. PRL together with androgens regulate reproduction and induce sexual behavior (Polzonetti-Magni *et al.*, 1984, 1995; Moore *et al.*, 2005). The blood concentrations of GH and PRL display similar seasonal trends related to cyclic changes of reproductive organs and external dimorphic features. Their lowest levels were found just before and during reproductive phase, whereas the highest ones during hibernation (Mosconi *et al.*, 1994; Polzonetti-Magni *et al.*, 1995).

Testis

Sites of synthesis and secretion of testicular steroids. The relationships between gonads, central nervous system and pituitary are basically similar in all vertebrates (Ball, 1981; Crews and Silver, 1985; Fasolo *et al.*, 1980; Peter, 1983; Jørgensen, 1974, 1992; Pierrantoni *et al.*, 1984; Sealfon *et al.*, 1997). Somatic cells of testes contain receptors for FSH and LH (Ischii and Kubokawa, 1985). In response to pituitary gonadotropins, they synthesize and liberate steroid hormones (mainly androgens and small amounts of estrogens), which control spermatogenesis and sexual behavior of males. Steroids are synthesized mainly in Leydig cells of interstitial tissue, localized in close vicinity to blood vessels. Small amounts of androgens enter Sertoli cells where they are converted to estrogens (Lofts, 1964; Rastogi and Iela, 1980; Lofts, 1984; Minucci *et al.*, 1990).

The main androgens secreted by amphibian testis are testosterone (T), 5 α-dihydrotestosterone (DHT), and androstenedione (A4). Among vertebrates only Anura produce great amounts of testicular DHT that is the most active androgen (Rastogi and Iela, 1980; Licht *et al.*, 1983; Tanaka *et al.*, 1988). DHT and T are usually synthesized in equal amounts, but in some species, for example *Rana catesbeiana*, DHT level is higher (Müller, 1977; Kime and Hews, 1978; Kime, 1980; Siboluet, 1981; Pierantoni *et al.*, 1984a,b; Rastogi *et al.*, 1986; Canosa and Ceballos, 2002). Like other steroids, androgens act on their target tissues through receptors (AR, androgen receptors). ARs of divergent species of vertebrates show high homology in their amino acid sequences, suggesting their conserved nature and function (He *et al.*, 1990; Chattopadhyay *et al.*, 2003). Until now complete frog AR cDNA have been isolated and cloned from *Xenopus laevis* (Ho *et al.*, 1990) and bullfrog *Rana catesbeiana* (Chattopadhyay *et al.*, 2003). Both the bullfrog and other investigated vertebrate species have a single copy of AR gene, whereas *Xenopus laevis* genome contains at least two AR genes (Fischer *et al.*, 1993). In bullfrog gonads ARs are distributed and expressed in Leydig and Sertoli cells (Chattopadhyay *et al.*, 2003).

All precursors and enzymes necessary for synthesis of steroid hormones are synthesized in the testes (Fasano *et al.*, 1989; Gobetti *et al.*, 1991; Polzonetti-Magni *et al.*, 1984). Enzymes involved in steroid production were evidenced histochemically also in early larval gonads of anurans. At least some genera (*Rana, Xenopus*) can produce small amounts of steroid hormones just after hatching and throughout larval life (Chieffi and Botte, 1963). Steroidogenesis increases during development and reaches its maximum at the time of metamorphic climax. This means that the final functional activity of hypothalamus-pituitary-gonadal axis is achieved when the central nervous system is already developed.

It has been shown that gonadal steroids, especially DHT, T, and 17 β estradiol (E2) exert negative feedback on HPG axis. Frogs implanted with DHT or E2 had decreased pituitary and circulatory LH level and marked loss of premeiotic germ cells in their testes, whereas post meiotic stages were not affected (Tsai *et al.*, 2003; 2005).

Influence of androgens on spermatogenesis. Androgens support surviving and development of gametogenic cells acting indirectly through regulation of somatic cells metabolism or directly in germ cell cytoplasm (Sasso-Cerri *et al.*, 2005). Androgens pass across the cell membrane and bind to androgen binding proteins (ABPs) that are cytoplasmic homodimeric glycoproteins with a single steroid-binding site. Then they enter the nucleus and bind to nuclear receptors. In amphibians, ABPs preferentially bind to 17β-estradiol and testosterone (French *et al.*, 1974; for review see Callard, 1987).

Androgens are essential for initiation and maintenance of normal course of spermatogenesis. Their deficiency arrests this process at secondary spermatocyte stage, since spermatids require high level of androgens for their development (Rastogi *et al.*, 1976; Rastogi and Iela, 1980; Minucci *et al.*, 1990, 1992). Developing spermatids are enclosed in cysts, which constitute compartments separated from other testicular tissues by tight junctions between processes of Sertoli cells, thereby forming the blood-testis barrier (Bergmann *et al.*, 1983). It makes possible to achieve a high testosterone concentration inside cysts, suitable for maintenance and maturation of spermatids. After application of anti-androgens, Leydig cells and cysts containing spermatids degenerate (Haider, 1980). Spermatids that mature inside cysts suppress mitotic activity of primary and secondary spermatogonia by paracrine action. Thereby the lack of spermatids, being the result of their physiological release or experimental procedures, triggers mitoses of spermatogonia (Haider, 1980; Rastogi and Iela, 1980; Minucci *et al.*, 1992, 1994).

Intratesticular mechanisms of hormonal regulation. Paracrine and autocrine hormonal interactions are important signaling pathways between somatic and gametogenic cells of the gonads of amphibians (Segal and Adejuwon, 1979; Pierantoni *et al.*, 1984a,b, 2002; Minucci *et al.*, 1986; Risley *et al.*, 1987; Di Matteo *et al.*, 1988; Fasano *et al.*, 1988, 1989; Minucci *et al.*, 1992, 1994). Frog's Leydig cells are able to regenerate after hypophysectomy and pharmacological damage under the paracrine stimulation by neuropeptides (Minucci *et al.*, 1994). Nakayama *et al.* (1999) reported that during *in vitro* experiments also Sertoli cells, stimulated by FSH and androgens, synthesize and secrete factors that control proliferation and differentiation of spermatogonia, acting in a paracrine way. Seroli cells of *Xenopus laevis* supply energy and glutatione metabolism in spermatogenic cells (Risley and Morse-Gaudio, 2005). In Sertoli and Leydig cells, as well as in spermatogonia and spermatocytes, receptors for peptide growth factors IGF I, and IGF II (Insulin-Growth-Like-Factor I and II) were detected. These growth factors stimulate DNA synthesis, spermatogonial proliferation and transformation of secondary spermatogonia into primary spermatocytes.

tGnRH, which is secreted by nonmyelinated neurons innervating the testes, plays an important role in intratesticular control of spermatogenesis. tGnRH modifies metabolism of Leydig, and probably also Sertoli cells, in a paracrine manner. The effect of this stimulation is regulation of spermatogonial cells cycle. The level of tGnRH changes during annual cycle, being high when the androgen level increases and when proliferation of primary spermatogonia starts. tGnRH influences directly proliferation of primary spermatogonia in PDX *Rana esculenta* males, which was also confirmed by *in vitro* experiments (Di Matteo *et al.*, 1988, 1996; Minucci *et al.*, 1986). In intact animals tGnRH is involved in the control of G1/S transition of the cell cycle and the initiation of DNA replication in primary spermatogonia (Minucci *et al.*, 1986, 1992, 1993, 1997). tGnRH regulates the Fos family proteins that stimulate spermatogonial proliferation at the testicular level (Cobellis *et al.*, 2003). tGnRH also regulates testosterone level in both intact and PDX males (Cariello *et al.*, 1989; D'Antonio *et al.*, 1992; Di Matteo *et al.*, 1988, 1990, 1996; Fasano *et al.*, 1989, 1995; Minucci *et al.*, 1990, Segal and Adejuwon, 1979; Pierantoni *et al.*, 1984a,b). It also stimulates a temperature-related synthesis of prostaglandin F_2 by increasing its level prior to, and after the reproduction. The prostaglandins probably promote estradiol synthesis in testes (Gobetti and Zerani, 1992; Gobetti *et al.*, 1991).

Seasonal variation in androgens levels. Seasonal variations in plasma testosterone levels has been monitored in many amphibian species: *Rana catesbeiana* (Licht *et al.*, 1983), *R. esculenta* (Pierantoni *et al.*, 1984a,b; Varriale *et al.*, 1986), *R. perezi* (Delgado *et al.*, 1989), *R. rugosa* (Ko *et al.*, 1998), *Pachymedusa dacnicolor* (Rastogi *et al.*, 1986), and *Bufo japonicus* (Itoh *et al.*, 1990). The level of testosterone in testes increases in autumn and precedes its highest level in the circulation (Pierantoni *et al.*, 1984a,b), then decreases markedly during breeding season, when a new wave of active spermatogenesis occurs (Rastogi *et al.*, 1976; Pierantoni *et al.*, 1984a,b; Delgado *et al.*, 1989; Itoh *et al.*, 1990. In *Rana esculenta* (Rastogi *et al.*, 1976) and *Bufo japonicus* (Itoh *et al.*, 1990) the lowest levels of androgens occur in summer during proliferation of secondary spermatogonia and formation of primary spermatocytes. In contrast, in the Mexican leaf frog *Pachymedusa dacnicolor* that breeds continuously, testosterone level does not fluctuate during annual cycle (Rastogi *et al.*, 1986). It means that the seasonal patterns of plasma testosterone levels are not always associated with spermatogenic activity or breeding season (Ko *et al.*, 1998).

Influence of androgens on sex dimorphism and sexual behavior. Besides gonads ARs are widely distributed in various somatic tissues, which are targets for androgens. AR transcripts have been detected in brain, liver, kidney, heart, larynx, thumb pad, and small intestine; however their expression was much lower than in testes (Sasson and Kelley, 1986; Sasson *et al.*, 1987; Tobias *et al.*, 1991; Variale and Serino, 1994; Tobias and Kelley, 1995). Androgens are engaged in sexual maturation and development of secondary sexual characters (Rastogi *et al.*, 1976; Itoch *et al.*, 1990). They elicit sexual behavior, which confirm that the brain is one of their targets (Wada and Gorbmann, 1977; Polzonetti-Magni *et al.*, 1984; Emerson and Hess, 1996; Chu and Wilczyński, 2001; Paolucci, 2003; Moore *et al.*, 2005). Testosterone implanted into frog's *preoptic nucleus* induced mating behavior very effectively. However, in castrated male *Pachymedusa dacnicolor* and other investigated species, except *Xenopus laevis*, replacement therapy with exogenous testosterone had no effect and failed to stimulate calling behavior (Rastogi *et al.*, 1986; Moore *et al.*, 2005). Estrogen was able to mimic the effect of androgens in manifesting sexual behavior when placed directly to a frog brain (Callard, 1985; Herman, 1992).

Environmental Cues

Fluctuations of the natural environment. Hormonal processes are sensitive to environmental changes influencing the top of the HPG axis (van Oordt, 1960; Rastogi, 1976; Di Meglio and Rastogi *et al.*, 1976, 1978; Siboulet, 1981; Licht *et al.*, 1983; Vivien-Rels and Pevet, 1983; Lofts, 1984; Pierantoni *et al.*, 1984a; Ischii and Kubokawa, 1985; Itoch *et al.*, 1990; Paniagua *et al.*, 1990). Continuous breeding period is characteristic of amphibian species inhabiting tropical or subtropical regions, whereas in temperate zone the temperature and changing photoperiod during the annual cycle result in seasonality of breeding activity. However, there are exceptions to this rule, and some species of the temperate zone have potentially continuous breeding activity, like: *Rana esculenta* (Rastogi *et al.*, 1976, 1978), *R. perezi* (Delgado *et al.*, 1989), *R. porosa brevipoda*, and *R. catesbeiana* (Licht *et al.*, 1983), whereas other members of the family Ranidae: *R. nigromaculata* (Iwasawa and Asai, 1959) and *R. temporaria* (Lofts and van Oordt, 1962) are seasonal breeders. On the other hand, some tropical species have seasonal breeding activity, like *Pachymedusa dacnicolor* (Rastogi *et al.*, 1986).

Higher temperatures usually stimulate gonads indirectly by altering hypothalamus and pituitary secretion; temperature can also act directly by altering gonad response to gonadotropins (Pierantoni *et al.*, 1984a,b, 1985; Rastogi and Iela, 1980). The optimum temperature for active spermatogenesis ranges from 12 to 25°C (Rastogi *et al.*, 1978). An increase of temperature within the optimum range stimulates intensity of spermatogenesis and accelerates sexual maturity in young males; beyond this range of temperatures reproductive processes are inhibited (Easley *et al.*, 1979). For example, experiments with PDX *Rana esculenta* males treated with exogenous gonadotropins under various temperature conditions revealed that in temperature below 4°C primary spermatocytes degenerate.

In species with continuous or potentially continuous breeding activity, the increase or decrease of temperature can respectively stimulate or inhibit spermatogenesis (Iela *et al.*, 1980; Pierantoni *et al.*, 1985). The annual changes in testes weight and size do not reflect their histological structure, because spermatozoa can be present in testes all the year round (Rastogi *et al.*, 1976; Delgado *et al.*, 1989). In seasonal breeders, the spermatogenesis is to high extend independent on the environmental cues, and is regulated by endogenous factors responsible for seasonal level of gonadotropin release and the ability to response by the target interstitial tissue (van Oordt, 1960). *Rana temporaria* does not respond to temperature increase during period of testis inactivity. During winter, testes of this species are several times larger than in summer after breeding season. Changes of the testis size reflect spermatogonial activity, but spermatozoa in seasonal breeders are present only before breeding (van Kamende, 1969; Lofts, 1984; Lofts and van Oordt, 1962; Rastogi *et al.*, 1986; van Oordt, 1956; Ogielska and Bartmańska, 1999; Shalan *et al.*, 2004).

Although the temperature is believed to be the main environmental factor regulating the process of spermatogenesis, some authors emphasize that for gonad activity, both temperature and photoperiod must rise above the minimal critical levels. Thus the light could be a factor determining the response of the gonad to temperature increase (Kelly and van Kamer, 1960; Rastogi, 1976; Rastogi *et al.*, 1976, 1978; Rastogi and Iela, 1980; Delgado *et al.*, 1989; Chieffi and Minucci, 2004). The optimal length of photoperiod ranges from 10 to 16 hours of daylight. Both too short (3 hours) and too long (21 hours) photoperiods inhibit the appearance of primary spermatocytes, and in consequence causes decrease of secondary spermatocytes and spermatid numbers, and increases degeneration of spermatogenic cells. The length of photoperiod is detected by pineal gland, frontal organ, and eyes (Vullings, 1973). The pineal gland and frontal organ have a neural layer resembling that of neural retina of eyes and for this reason are sensitive to light (Kelly and van der Kamer, 1960). Artificially long darkness or removal of the pineal gland and/or eyes inhibit hypothalamic activity (Pierantoni *et al.*, 1985). However Shivakumar (1999) argued against any influence of photoperiod on spermatogenesis in amphibians.

The influence of environmental steroids on the HPG axis. It is well known that a very wide range of chemical compounds, both natural and man-made endocrine-disrupting chemicals (EDCs), mimic the effect of hormones and affect amphibian reproduction. EDCs usually display estrogenic effects and influence sexual differentiation causing feminization in males. Together with estradiol they may interact with the liver of male frogs and newts in a dose depended manner, and induce vitellogenin gene expression. EDCs are known to cause an increase the PRL cells number in pituitary and induces PRL accumulation, possibly due to decrease of hormone release. The inhibitory effects of EDCs on gonadotropins and PRL secretion by pituitary were observed together with an increase of plasma androgens, however they do not cause any significant changes in plasma estradiol levels (Sower et al., 2000; Mosconi et al., 2002; Bögi et al., 2003; Mosconi et al., 2005; Sanderson, 2006).

During tadpole development, the HPG axis is very sensitive to pollution of aquatic systems with EDCs from the agriculture. Depending on the concentration, EDCs disturb or completely destroy the activity of the HPG axis (reviewed by Kloas, 2002). In result the adult frogs Rana catesbeiana and R. clamitans that developed in such reservoirs exhibited serious somatic and gonadal malformations, including sterility. Brain GnRH concentration and the levels of steroid hormones essential for reproduction (androgens and estradiol) were much lower in the malformed frogs than in the normal ones. Since it usually concerns the great number of individuals of a population, it may cause a local amphibian decline.

References

Abé S-I, S Asakura, A Ukeshima. 1988. Formation of flagella during interphase in secondary spermatocytes from Xenopus laevis in vitro. J. Exp. Zool. 246: 65-70.

Abé S-I. 1988. Cell culture of spermatogenic cells from amphibians. Develop. Growth and Differ. 30: 209-218.

Abé S-I, S Asakura. 1987. Meiotic divisions and early mid-spermiogenesis from cultured primary spermatocytes of Xenopus laevis. Zool. Sci. 4: 839-847.

Acharjee S, JL Do-Rego, DY Oh, RS Ahn, K Lee, DG Bai, H Vaudry, HB Kwon, JY Seong. 2004. Molecular cloning, pharmacological characterization, and histochemical distribution of frog vasotocin and mesotocin receptors. J. Mol. Endocrinol. 33: 293-313.

Adler D, MA Gorovsky. 1975. Electrophoretic analysis of liver and testis histones of the frog Rana pipiens. J. Cell Biol. 64: 389-397.

Al-Mukhtar KK, AC Web. 1971. An ultrastructural study of primordial germ cells, oogonia and early oocytes in Xenopus laevis. J. Embryol. Exp. Morph. 26: 195-217.

Asa CS, DM Philips. 1988. Nuclear shaping in spermatids of the Thai leaf frog Megophrys montana. Anat. Rec. 220: 287-290.

Ball JN. 1981. Hypothalamic control of the pars distalis in fishes, amphibians, and reptiles. Gen. Comp. Endocr. 44: 135-170.

Báo SN, GC Dalton, SF de Oliveira. 1991. Spermiogenesis in *Odontophrynus cultripes* (Amphibia, Anura, Leptodactylidae): Ultrastructural and cytochemical studies of proteins using E-PTA. J. Morphol. 207: 303-314.

Bartmańska J, M Ogielska, E Milewska, A Sobik, P Kierzkowski. 2006. Number of primary spermatocytes in testicular cysts of water frogs *Rana ridibunda* and *Rana lessonae* (Amphibia, Anura). Acta Biol. Cracoviensia Series Bot. 48 Suppl 1.

Bergmann M, H Greven, J Schindelmeiser. 1983. Observations of the blood-testis barrier in a frog and salamander. Cell Tissue Res. 323: 189-200.

Bernardini G, R Stipani, G Melone. 1986. The ultrastructure of *Xenopus* spermatozoon. J. Ultra. Mol. Str. Res. 94: 188-194.

Bhaduri JL, SL Basu. 1957. A study of the urogenital system of Salientia. Part I. Ranidae and Hyperolidae of Africa. Ann. Mus. Roy. Congo Bdge. 55: 9-34.

Blüm V. 1985. Vertebrate Reproduction. Springer-Verlag Berlin-New York-Tokyo: 81-85.

Bögi C, J Schwaiger, H Ferling, U Mallow, C Steineck, F Sinowatz, W Kalbfus, RD Negele, I Lutz, W Kloas. 2003. Endocrine effects of environmental pollution on *Xenopus laevis* and *Rana temporaria*. Environ. Res. 93: 195-201.

Bols NC, HE Kasinsky. 1972. Basic protein composition of anuran sperm: a cytochemical study. Can. J. Zool. 50: 171-177.

Bols NC, M Mann, HE Kasinsky. 1986. Detection of sperm histone diversity among vertebrates by alkaline fast green staining. Stain Technol. 61: 111-119.

Burggren WW, JJ Just. 1992. Developmental changes in physiological systems. In: Environmental Physiology of the Amphibians. (Eds) ME Feder, WW Burggreen, pp. 481-492.

Burmaister S, C Somes, W Wilczyñski. 2001. Behavioral and hormonal effects of exogenus vasotocin and corticosterone in the green treefrog. Gen. Comp. Endocrinol. 122: 189-197.

Cabada MO, ME Manes, MI Gomez. 1989. Spermatolysins in *Bufo arenarum*: their activity on oocyte surface. J. Exp. Zool. 249: 229-234.

Callard GV, Z Petro, KJ Ryan. 1978a. Androgen metabolism in the brain and non-neural tissue of the bullfrog *Rana catesbeiana*. Gen. Comp. Endocrinol. 34: 18-25.

Callard GV. 1985. Estrogen synthesis and other androgen converting pathways in the vertebrate brain and pituitary. In: Current Trends in Comparative Endocrinology (Eds) Lofts B and Holmes WN. Hong Kong University Press, 1179-1184.

Callard IP, GV Callard, V Lance, JL Bolaffi, JS Rosset. 1978b. Testicular regulation in nonmammalian Vertebrates. Biol. Reprod. 18: 16-43.

Callard IP, GV Callard. 1987. Sex steroid receptors and non-receptor binding proteins. In: Hormone and Reproduction in Fishes, Amphibians and Reptiles. (Eds) DO Norris, RE Jones. Plenum Press, New York, 355-384.

Cambar R. 1947a. Demonstration experimantale du role inducteur du canal de Wolff dans la morphogenese du mesonephros, chez les Amphibiens anoures. Compt. Rend. Séances Acad. Sci. 225: 823-825.

Cambar R. 1947b. Nouvelle preuve experimentale du role inducteur du canal de Wolff sur la morphogenese du mesonephros des Amphibiens anoures. Compt. Rend. Séances Acad. Sci. 225: 1019-1021.

Cambar R. 1948. La partie anterieure du canal de Wolff peut, experimentalement, induire la morphogenese du mesonephros chez les Amphibiens Anoures. Compt. Rend. Séances Acad. Sci. 226: 1542-1544.

Cambar R. 1949. Donee recentes sur le development du systeme pronephrotique chez les Amphibiens (Anoures en particulier). Ann. Biol. 25: 115-130.

Canosa LF, NR Ceballos. 2002. Seasonal changes in testicular steroidogenesis in the toad *Bufo arenarum* H. Gen. Comp. Endocrinol. 125: 426-434.

Cariello LG, A Romano, L Spagnuolo, S Zanetti, S Fasano, L Minucci, R Di Matteo, R Pierantoni, G Chieffi. 1989. Molecular forms of immunoreactive gonadotropin-releasing hormone in hypothalamus and testis of the frog *Rana esculenta*. Gen. Comp. Endocrinol. 76: 343-348.

Cavicchia JC, GA Moviglia. 1983. The blood-testis barrier in the toad (*Bufo arenarum* Hensel): A freeze-fracture and lanthanum tracer study. Anat. Rec. 205: 387-396.

Chattopathyay S, JH Park, JY Seong, HB Kwon, K Lee. 2003. Cloning and characterization of androgen receptor from bullfrog, *Rana catesbeiana*. Gen. Comp. Endocrinol. 134: 10-17.

Chieffi G, V Botte. 1963. Osservazioni istochimiche sull'attivita della steroide-B-olodeidrogenasi nell' interrenale e nelle gonadi di girihi e adulti di *Rana esculenta*. Riv. Istochim. Norm. Pathol. 9: 172-174.

Chieffi P, S Minucci, G Cobellis, S Fasano, R Pierantoni. 1995. Changes in the proto-oncogene activity in the testis of the frog, *Rana esculenta*, during the annual reproductive cycle. Gen. Comp. Endocrinol. 99: 127-36.

Chieffi P, S Minucci. 2004. Environemtal influence of testicular MAP kinase (ERK1) activity in the frog *Rana esculenta*. J. Exp. Biol. 207: 2209-2213.

Cobellis G, R Mecciarrello, S Minucci, C Palmiero, R Pierantoni, S Fasano. 2003. Cytoplasmic versus nuclear localization of Fos-related proteins in the frog, *Rana esculenta*, testis: In vivo and direct in vitro effect of a gonadotropin-releasing hormone agonist. Biol. Reprod. 68: 954-960.

Collin F, N Chartler, A Fasolo, JM Conlon, F Vandesande, H Vaudry. 1995. Distribution of two molecular forms of gonadotropin releasing hormone (GnRH) in the central nervous system of the frog *Rana ridibunda*. Brain Res. 703: 111-128.

Crews D, R Silver. 1985. Reproductive physiology and behavior interactions in nonmammalian vertebrates. In: Handbook of Behavioral neurobiology (Eds) N Adler, D Pfaff, RW Goy 7: 101-182. New York, Plenum Publishing.

D'Aniello B, C Pinelli, MM Di Fiore, L Tela, JA King, RK Rastogi. 1995. Development and distribution of gonadotropin-releasing hormone neuronal systems in the frog (*Rana esculenta*) brain: Immunohistochemical analysis. Dev. Brain Res. 89: 281-288.

D'Antonio M, S Fasano, R de Leeuw, R Pierantoni. 1992. Effect of gonadotropin releasing hormone variants on plasma and testicular androgen levels in intact and hypophysectomized male frog *Rana esculenta*. J. Exp. Zool. 261: 34-39.

D'Istria M, C Palmiero, I Serino, G Izzo, S Minucci. 2003. Inhibition of the basal and oestradiol-stimulated mitotic activity of primary spermatogonia by melatonin in the testis of the frog, *Rana esculenta*, in vivo and in vitro. Reproduction 126: 83-90.

Daniels E, P Licht. 1980. Effects of gonadotropin-releasing hormone on the levels of plasma gonadotropins (FSH and LH) in the bullfrog *Rana catesbeiana*, Gen. Comp. Endocrinol. 42: 455-463.

de Oliveira C, CA Vicentini, SR Taboga. 2003. Structural characterization of nuclear phenotypes during *Scinax fuscovarius* spermatogenesis (Anura, Hylidae). Caryologia 56: 75-83.

de Oliveira C, C Zanetoni, R Zieri. 2002. Morphological observations of the testes of *Physalemus cuvieri* (Amphibia, Anura). Rev. Chil. Anat. 20: 263-268.

Delgado MJ, P Gutierrez, M Alonso-Bedate. 1989. Seasonal cycles in testicular activity in the frog, *Rana perezi*. Gen. Comp. Endocrinol. 73: 1-11.

Di Fiore MM, JA King, B D'Aniello, RK Rastogi. 1996. Immunoreactive mammalian and chickenII GnRHs In *Rana esculenta* brain during development. Regul. Pept. 62: 119-124.

di Matteo L, S Minucci, S Fasano, M D'Antonio, R Pierantoni, G Chieffi. 1990. Indirect evidence for a physiological role exerted by a "testicular gonado-tropine-releasing hormone" in the frog *Rana esculenta*. Gen. Comp. Endocrinol. 79: 147-153.

di Matteo L, S Minucci, S Fasano, R Pierantoni, B Varriale, G Chieffi. 1988: A gonadotropin releasing hormone (GnRH) antagonist decreases androgen production and spermatogonial multiplication in frog *Rana esculenta*: Indirect evidence for the existence of GnRH-like material receptors in the hypophysis and testis. Endocrinol. 122: 62-67.

di Matteo L, M Vallarino, R Pierantoni. 1996. Localization of GnRH molecular forms in the brain, pituitary and testis of the frog, *Rana esculenta*. J. Exp. Zool. 274: 33-40.

di Meglio M, M Masucci, B D'Aniello, RK Rastogi. 1991. Immunohistochemical localization of multiply forms of gonadotropin releasing hormone in the brain of adult frog. J. Endocrinol. 3: 363-368.

di Meglio M, JI Morrell, DW Pfaff. 1987. Localization of steroid-concentrating cells in the central nervous system of the frog *Rana esculenta*. Gen. Comp. Endocrinol. 67: 149-154.

di Meglio M, RK Rastogi. 1978. Influence of temperature and gonadal steroids on the testis composition in *Rana esculenta*. Experientia 34: 267.

Dubois EA, MA Zandbergen, J Peute, HJ Goos. 2002. Evolutionary development of three gonadotropin-releasing hormone (GnRH) systems in vertebrates. B. Res. Bull. 57: 413-418.

Easley KA, DD Culley, ND Horseman, JE Penkala. 1979. Environmental influences on hormonally induced spermiation in the Bullfrog *Rana catesbeiana*. J. Exp. Zool. 207: 407-450.

Emerson SB, DL Hess. 1996. The role of androgen in oportunistic breeding, tropical frogs. Gen. Comp. Endocrinol. 103: 220-230.

Fasano S, M D'Antonio, P Chieffi, G Cobellis, R Pierantoni. 1995. Chicken GnRH effects on plasma and testicular androgen concentrations in the male frog, *Rana esculenta*, during the annual reproductive cycle. Comp. Biochem. Physiol. Pharmacol. Toxicol. Endocrinol. 112: 79-86.

Fasano S, HJ Goss, C Janssen, R Pierantoni. 1993. Two GnRH fluctuate in correlation with androgen levels in the male frog, *Rana esculenta*. J. Exp. Zool. 266: 277-283.

Fasano S, R Leeuw, R Pierantoni, G Chieffi, PG van Oordt. 1990. Characterization of gonadotropin-releasing hormone (GnRH) binding sites in the pituitary and testis of the frog, *Rana esculenta*. Biochem. Biophys. Res. Commun. 168: 923-932.

Fasano S, S Minucci, L Di Matteo, M D'Antonio, R Pierantoni. 1989. Intratesticular feedback mechanisms in the regulation of steroid profiles in the frog, *Rana esculenta*. Gen. Comp. Endocrinol. 75: 335-342.

Fasano S, S Minucci, R Pierantoni, A Fasolo, L Di Matteo, B Basile, B Varial, G Chieffi. 1988. Hypothalamus-Hypophysis and testicular GnRH control of gonadal activity in the frog, *Rana esculenta*. Seasonal GnRH profiles and annual variation of *in vitro* androgen output by pituitary stimulated testes. Gen. Comp. Endocrinol. 70: 31-40.

Fasolo A, MF Franzoni, V Mazzi. 1980. Evolution of the hypothalamo hypophysial regulation in tetrapods. Boll. Zool. 47, Suppl.: 127-147.

Fernald RD, RB White. 1999. Gonadotropin-releasing hormone genes: Phylogeny, structure, and function. Front. Neuroendocrinol. 20: 224-240.

Fernandes AP, SN Bao. 1998. Cytochemical localization of phosphatases in germ- and Sertoli cells of *Odontophrynus cultripes* (Amphibia, Anura, Leptodactylidae). Biocell. 22: 93-101.

Fischer LM, D Catz, DB Kelley. 1993. An androgen receptor mRNA isoform associated with hormone induced cell proliferation. Propc. Natl. Acad. Sci. 90: 8254-8258.

French FS, SN Nayfeh, EM Ritzen, V Hansson. 1974. FSH and a testicular androgen binding protein in the maintenance of spermatogenesis. Res. Reprod. 6: 2-3.

Garda AA, GR Colli, O Aguiar-Júnior, SM Ricco-Pimentel, SM Báo. 2002. The ultrastructure of the spermatozoa of *Epidobates flavopictus* (Amphibia, Anura, Dendrobatidae), with comments on its evolutionary significance. Tissue & Cell 34: 356-364.

Gobetti A, M Zerani, GF Bolelli, V Botte. 1991. Seasonal changes in plasma prostaglandin F2 alpha and sex hormones in the male water frog, *Rana esculenta*. Gen. Comp. Endocrinol. 82: 331-336.

Gobetti A, M Zerani, LB Cardellini. 1992. A possible role of prostaglandin E2 in reproduction of the male water frog, *Rana esculenta*. *In vivo* and *in vitro* studies. Prostaglandins 44: 277-289.

Gobetti A, M Zerani. 1992. Mammalian GnRH involvement in prostaglandin F2 alpha and sex steroid hormones testicular release in two amphibian species: the anuran water frog *Rana esculenta*, and the Urodele crested newt, *Triturus carnifex*. Gen. Comp. Endocrinol. 87: 240-248.

Golmann G, LJ Borkin, P Roth. 1993. Genic and morphological variation in the fire-bellied toad, *Bombina bombina* (Anura, Discoglossidae). Zool. Jb. Syst. 120: 129-136.

Gonzales A, WJ Smeets. 1992a. Comparative analysis of the vasotocinergic and mesotocynergic cells and fibers in the brain of two amphibians, the anuran *Rana ridibunda* and the urodele *Pleurodeles waltlii*. J. Comp. Neurol. 315: 53-73.

Gonzales A, WJ Smeets. 1992b. Distribution of vasotocin- and mesotocin-like immunoreactivities in the brain of the South African claved frog *Xenopus laevis*. J. Chem. Neuroanat. 5: 465-479

Guha KK, CB Jørgensen. 1978a. Effect of human chorionic gonadotropin and salmon gonadotropin on testis in hypophysectomized toads (*Bufo Bufo Bufo* L.) Gen. Comp. Endocrinol. 36: 371-379.

Guha KK, CB Jørgensen. 1978b. Effect of hypophysectomy on structure and function of testis in adult toads *Bufo bufo bufo* (L.) Gen. Comp. Endocrinol. 34, 201-210.

Guha KK, CB Jørgensen. 1981. Growth response of testis tissue to partial castration in toads, *Bufo bufo bufo* (L.) J. Zool. (London) 193: 171-181.

Guilgur LG, Moncaut NP, Canario AVM, Somoza G. 2006. Evolution of GnRH ligands and receptors in gnathostomata. Comp. Biochem. Physiol. 144: 272-283.

Guraya SS. 2001. Comparative cellular and molecular biology of testis in vertebrates. Trends in Endocrine, Paracrine and Autocrine Regulation of Structure and Functions, Science Publishers, p. 91.

Haider S.G. 1980. Histophysiological effects of an antiandrogen (cyproterone acetate) on the testis of the frog *Rana temporaria*. Acta Anat. (*Basel*). 106: 387-391.

He WW, LM Fischer, S Sun, DL Bilhartz, XP Zhu, CY Young, DB Kelley, DJ Tindall. 1990. Molecular cloning of androgen receptors from divergent species with a polymerase chain reaction technique: Complete cDNA sequence of the mouse androgen receptor and isolation of androgen receptor cDNA probes from dog, guinea pig and clawed frog. Biochem. Biophys. Res. Commun. 171: 697-704.

Herman, CA. 1992. Endocrinology. In: Environmental Physiology of the Amphibians, (Eds) ME Feder and WW Burggren. pp. 40-54.

Hiragond NC, SK Saidapur. 2000. The excurrent duct system of sperm transport in *Rana cyanophlyctis*, *Rana limnocharis*, *Polypedates limnocharis*, *Microhyla rubra*, *Bufo melanosticus* and *Bufo fergusonii*. Zool. Sci. 17: 453-458.

Hollis MD, J Chu, EA Walthers, BL Heppner, BT Searcy, FL Moore. 2005. Neuroanatomical distribution of vasotocin and mesotocin in two urodele amphibians (*Plethodon shermani and Taricha granulosa*) based on in situ hybridization histochemistry. Brain Res. 1035: 1-12.

Hsu ChY, HJ Wang. 1981. Production of sterile gonads in tadpoles by 17β-ureide steroid. Proc. Nat. Sci. Council China, part B: Biol. Sci. 5: 322-327.

Hsu CY, CH Chiang, HM Liang. 1973. Effect of hypophysectomy on sex transformation in frog tadpoles. Endocrinology (Japan) 20: 391-396.

Iela L, R Pierantoni, RK Rastogi. 1980. Effect of temperature and light on the production of androgens in the male *Rana esculenta*. Experientia. 36: 256.

Ischii S, K Kubokawa. 1985. Adaptation of vertebrate gonadotropin receptors to environmental temperature. In: Current Trends in Comparative Endocrinology, (Eds) B. Lofts and WN Holmes. Hong Kong University Press, pp. 751-754.

Ishi S, M Itoh. 1992. Amplexus induces surge of luteinizing hormone in male toads, *Bufo japonicus*. Gen. Comp. Endocrinol. 86: 34-41.

Ito R, SI Abe. 1999. FSH-initiated differentiation of newt spermatogonia to primary spermatocytes in germ-somatic cell reagregates cultured within a collagen matrix. Int. J. Dev. Biol. 43: 111-116.

Itoh M, M Inoue, S Ischii. 1990. Annual cycle of pituitary and plasma gonadotropins and sex steroids in a wild population of the toad *Bufo japonicus*. Gen. Comp. Endocrinol. 109: 13-23.

Iwasava H, M Kobayashi. 1976. Development of the testis in the frog *Rana nigromaculata* with special reference to germ cells maturation. Copeia: 461-467.

Iwasawa H, H Michibata. 1972. Comparative morphology of sperm storage portion of Wolffian duct in Japanese Anurans. Annot. Zool. Jap. 45: 218-233.

Iwasawa H, M Yamada, M Kobayashi. 1977. Response of immature testis to exogenous gonadotropins in young *Rana nigromaculata*. Sci. Rep. Niigata Univ. Ser. D. 14: 15-20.

Iwasawa H, K Yamaguchi. 1984. Ultrastructural study of gonadal development in *Xenopus laevis*. Zool. Sci. 1: 591-600.

Iwasawa H. 1978. Spermatogonial responsiveness to mammalian gonadotropins in subadult *Rana nigromaculata*. Gen. Comp. Endocrinol. 34: 1-5.

Iwasawa H, T Nakazawa, T Kobayashi. 1987. Histological observations on the reproductive organs of growing *Rana nigromaculata* frogs. Sci. Rep. Niigata Univ. Ser. D (Biology) 24: 1-13.

Iwasawa H, O Asai. 1959. Histological observation on the seasonal change of the testis and the thumb pad in the frog *Rana nigromaniculata*. Yor. Fac. Sci., Niigata Univ. 2: 213-218.

Iwasawa H, Y Kera. 1982. Structural changes in gonadotropin producing cells of male frogs *Rana nigromaculata* in the process of sexual maturation. Gen. Comp. Endocrinol. 46: 236-245.

Ja KG, RP Millar. 1995. Evolutionary aspects of gonadotropine releasing hormone and its receptor. Cell. Mol. Neurobiol. 15: 5-23.

Jamieson BGM, E Meyer, DM Scheltinga. 1997. Sperm ultrastructure of 6 Australian hylid frogs from 2 genera (*Litoria* and *Cyclorana*)—phylogenetic implications. J. Submicr. Cytol. Pathol. 29: 443-451.

Jamieson BGM. 1999. Spermatozoal phylogeny of the Vertebrata. In: The Male Gamete: From Basic Science to Clinical Applications, (Ed.) Gagnon C. Vienna (USA). Cache River Press, pp. 303-331.

Jokura Y, A Urano. 1985. An immunochistochemical study of seasonal changes in luteinizing hormone-releasing hormone and vasotocin in the forebrain and the neurohypophysis of the toad, *Bufo japonicus*. Gen. Comp. Endocrinol. 59: 238-245.

Jørgensen CB, E Billeter. 1982. Growth, differentiation and function of the testes in the toad *Bufo bufo bufo* (L.), with special reference to regulatory capacities: effects of unilateral castration, hypophysectomy, and excision of Bidder's organ. J. Exp. Zool. 221: 225-236.

Jørgensen CB, LO Larsen, B Lofts. 1979. Annual cycle of fat bodies and gonads in the toad *Bufo bufo* bufo (L.) compared with cycles in other temperate zone anurans. Det. Kong. Danske Vidensk. Selskab. Biol. Skr. 22: 1-37.

Jørgensen CB. 1984. Testis function in the toad Bufo bufo (L.) (Amphibia, Anura) at organ and subunit levels Vidensk. Medd. Dan. Nat. hist. Foren. 145: 117-130.

Jørgensen CB. 1992. Growth and reproduction In: Environmental Physiology of the Amphibians (Eds) ME Feder and WW Burggren. pp. 439-466.

Kah O, C Lethimonier, JJ Lareyre. 2004. Gonadotrophin-releasing hormone (GnRH) in the animal kingdom. J. Soc. Biol. 198: 53-60.

Kalt M, JG Gall. 1974. Observations on early germ cell development and premeiotic ribosomal DNA amplification in *Xenopus laevis*. J. Cell. Biol. 62: 460-472.

Kalt MR. 1976. Morphology and kinetics of spermatogenesis in *Xenopus laevis*. J. Exp. Zool. 195: 393-408.

Kasinathan S, SL Basu. 1973. Effect of hormones on spermatogenesis in hypophysectomised *Rana hexodactyla* (Lesson) Acta Morph. Acad. Sci. Hung. 21: 249-259.

Kelly DE, JC van de Kamer. 1960. Cytological and histochemical investigation on the pineal organ of the adult frog *(Rana esculenta)*. Z. Zelforsch. 52: 618-630.

Kim KH, WB Im, H Choi, S Ischii, HB Kwon. 1998. Seasonal fluctuation in pituitary gland and plasma level of gonadotropic hormones in *Rana*. Gen. Comp. Endocrinol. 109: 13-23.

Kime DE, EA Hews. 1978. Androgen biosynthesis *in vitro* by testes from amphibia. Gen. Comp. Endocrinol. 35: 280-288.

Kime DE. 1980. Comparative aspects of testicular androgen biosynthesis in nonmammalian vertebrates. In: Steroids and Their Mechanism of Actions in Non-mammalian Vertebrates. (Eds) G Delrio and J Brachet. Raven Press, New York, pp. 17-31.

King JA, RP Millar. 1997. Co-ordinated evolution of GnRH and its receptors. In: GnRH Neurone: Gene to Behaviour, (Eds) Parhar IS, Sakuma Y. Brian Shuppan Tokyo, pp. 51-57.

King JA, AA Stenewald, JD Curlewis, EF Rissman, RP Millar. 1994. Differential regional distribution of gonadotropin releasing hormones in brains and plasma of ranid frogs. Gen. Comp. Endocrinol. 94: 186-198.

Kloas W. 2002. Amphibians as a model for the study of endocrine disruptors. Int. Rev. Cytol. 216: 1-57.

Ko SK, HM Kang, WB Im, HB Kwon. 1998. Testicular cycles in three species of korean frogs: *Rana nigromaculata, Rana rugosa, Rana dybowskii*. Gen. Comp. Endocrinol. 111: 347-358.

Kobayashi T, H Iwasawa. 1988. Dynamics of precocious spermatogenesis in *Rana nigromaculata*. Copeia 4: 1076-1071.

Koch RA, CC Lambert. 1990: Ultrastructure of sperm, spermiogenesis, and sperm-egg interaction in selected invertebrates and lower vertebrates which use external fertilization. J. Electron. Microsc. Technique 16: 115-154.

Lee YH, AS Kwon. 1992. Ultrastructure of spermiogenesis in *Hyla japonica* (Asnura, Amphibia). Acta Zool. (Stockholm) 73: 49-55.

Licht P, BR McCreery, R Barnes, R Pang. 1983. Seasonal and stress related changes in plasma gonadotropins, sex steroids, and corticosterone in the bullfrog, *Rana catesbaiana*. Gen. Comp. Endocrinol. 50: 124-145.

Licht P, PS Tsai, J Sotowska-Brochocka. 1994. The nature and distribution of gonadotropin-releasing hormones in brains and plasma of ranid frogs. Gen. Comp. Endocrinol. 94: 186-198.

Licht P. 1986. Suitability of the mammalian model in comparative reproductive endocrinology. In: Comparative Endocrinology, Development and Directions, (Ed.) C.L. Ralph, Liss, New York: 95-114.

Lofts B, PGWJ Van Oordt. 1962. Some effects of high temperature upon the pituitary and testis in the common frog, *Rana temporaria*. Gen. Comp. Endocrinol. 2: 614-625.

Lofts B. 1961. The effect of follicle stimulating hormone and luteinizing hormone on the testis of hypophysectomized frogs (*Rana temporaria*). Gen. Comp. Endocrinol. 1: 179-213.

Lofts B. 1964. Seasonal changes in the functional activity of the interstitial and spermatogenic tissues of the green frog, *Rana esculenta*. Gen. Comp. Endocrinol. 4: 550-562.

Lofts B. 1974. Reproduction. In: Physiology of the Amphibia, (Ed.) B Lofts. Academic press, New York, Vol. 2, pp. 107-218.

Lofts B. 1984. Amphibians. In: Marshall's Physiology of Reproduction, (Ed.) GE Lamming. Vol. 1. Churchill Livingstone, Edinburgh.

Madej Z. 1964. Studies on the fire bellied toad (*Bombina bombina* (Linnaeus, 1761)) and yellow bellied toad (*Bombina variegata* (Linnaeus, 1758)) of Upper Silesia and Moravian Gate. Acta Zool. Cracov. IX: 292-334.

Mann M, RA Risley, RA Eckhardt, HE Kasinsky. 1982. Characterization of spermatid/sperm basic chromosomal proteins in the genus *Xenopus* (Anura, Pipidae). J. Exp. Zool. 222: 173-186.

Manochantr S, P Sretarugsa, C Wanichanon, J Chavadej, P Sobhon. 2003. Classification of spermatogenic cells in *Rana tigerina* based on ultrastructure. Science Asia 29: 241-254.

Matsumoto A, Y Arai. 1992. Hypothalamus. In: Atlas of Endocrine Organs, (Eds) Matsumoto Akira, Ishii S. Springer-Verlag, pp. 25-37.

Mikami SI. 1992. Hypophysis, In: Atlas of Endocrine Organs. Vertebrates and Invertebrates. (Eds) A Matsumoto and S Ischii. Springer-Verlag, pp. 39-60.

Minucci S, L Di Matteo, S Fasano, G Chieffi Baccari, R Pierantoni. 1992. Intratesticular control of spermatogenesis in the frog *Rana esculenta*. J. Exp. Zool. 264: 113-118.

Minucci S, L Di Matteo, S Fasano, G Chieffi-Baccari, R Pierantoni. 1994. Regeneration of the testicular interstitial compartment after ethane dimethane sulfonate treatment in the hypophysectomized frog *Rana esculenta*: independence of the pituitary control. Gen. Comp. Endocrinol. 95: 84-91.

Minucci S, L Di Matteo, R Pierantoni, B Variale, RK Rastogi, G Chieffi. 1986. *In vivo* and *in vitro* stimulatory effect of a gonadotropin-releasing hormone analog (HOE766) on spermatogonal multiplication in the frog *Rana esculenta*. Endocrinology 119: 731-736.

Minucci S, S Fasano, M D'Antonio, R Pierantoni. 1993. Dopamine regulation of testicular activity in intact and hypophysectomized frogs *Rana esculenta*. Experientia 49(1): 65-67.

Minucci S, S Fasano, L Di Matteo, G Chieffi-Baccari, R Pierantoni. 1990. Morphological and hormonal changes in the frog, *Rana esculenta* testis after administration of ethane dimethane sulfonate. Gen. Comp. Endocrinol. 79: 335-345.

Minucci S, S Fasano, R Pierantoni. 1996. Induction of S-phase entry by a gonadotropin releasing hormone agonist (busurelin) in the frog, *Rana esculenta* primary spermatogonia. Comp. Biochem. Physiol. Pharmacol. Toxicol. Endocrinol. 113: 99-102.

Minucci S, L Di Matteo, P Chieffi, R Pierantoni, S Fasano. 1997. 17 β –estradiol effects on mast cell number and spermatogonial mitotic index in the testis of the frog, *Rana esculenta*. J. Exp. Zool. 1; 278: 93-100.

Miranda LA, DA Paz, JM Affanni, GM Somoza. 1998. Identification and neuro-anatomical distribution of immunoreactivity for mammalian gonadotropin-releasing hormone (mGnRH) in the brain and neural hypophyseal lobe of the toad *Bufo arenarum*. Cell Tissue Res. 293: 419-425.

Moore FL, SK Boyd, DB Kelley. 2005. Historical perspective: Hormonal regulation of behaviors in amphibians. Horm. Behav. 48: 373-383.

Moore IT, TS Jessop. 2003. Stress, reproduction, and adrenocortical modulation in amphibians and reptiles. Horm. Behav. 43: 39-47.

Moriguchi Y, A Tanimura, H Iwasawa. 1991. Annual changes in the Bidder's organ of the toad *Bufo japonicus formosus*: Histological observation. Sci. Rep. Niigata Univ. Ser. D Biol. 28: 11-17.

Mosconi G, O Carnevali, MF Franzoni, E Cottone, I Lutz, W Kloas, K Yamamoto, S Kikuyama, AM Polzonetti-Magni. 2002. Environmental Estrogens and Reproductive Biology in Amphibians. Gen. Comp. Endocrinol. 126: 125-129.

Mosconi G, I Di Rosa, S Bucci, L Morosi, MF Franzoni, AM Polzonetti-Magni, R Pascolini. 2005. Plasma sex steroid and thyroid hormones profile in male water frogs of the *Rana esculenta complex* from agricultural and pristine areas. Gen. Comp. Endocrinol. 142: 318-324.

Mosconi G, K Yamamoto, O Carnevali, M Nabissi, A Polzonetti-Magni, S Kikujama. 1994. Gen. Comp. Endocrinol. 93: 380-387.

Müller CH. 1977. In vitro stimulation of 5 alpha-dihydrotestosterone and testorone secretion from bullfrog testis by nonmammalian and mammalian gonadotropins. Gen. Comp. Endocrinol. 33. 109-121.

Muske LE, JA King, FL Moore, RP Millar. 1994. Gonadotropin releasing hormones in microdissected brain regions of an amphibian: Concentration and anatomical distribution of immunoreactive mammalian GnRH and chicken GnRH II. Regul. Pept. 50: 277-289.

Muske LE. 1993. Evolution of gonadotropin releasing hormone (GnRH) neuronal systems. Brain Behav. Evol. 42: 215-230.

Nakayama Y, T Yamamoto, S Abe. 1999. IGF-I, IGF-II and insulin promote differentiation of spermatogonia to primary spermatocytes in organ culture of newt testes. Dev. Biol. 43: 343-347.

Nielson M, K Lohman, J Sullivan. 2001. Phylogeography of the tailed frog (*Ascaphus truei*): implications for the biogeography of the Pacific Northwest. Evolution 55: 147-160.

Obert HJ. 1976. Die Spermatogenese bei der Gelbbauchunke (*Bombina vareigata variegata* L.) im Verlauf der jarhlichen Aktivitatsperiode und die Korrelation zur Paarungsfufaktivitat (Discoglossodae, Anura). Z. Mikrosk. Anat. Forsch. Leipzig. 90: 908-924.

Ogielska M, J Bartmańska. 1999. Development of testes and differentiation of germ cells in water frogs of the *Rana esculenta*—complex (Amphibia, Anura). Amphibia-Reptilia 20: 251-263.

Pancharatna K, S Chandran, S Kumbar. 2000. Phalangeal growth marks related to testis development in the frog *Rana cyanophlyctis*. Amphibia-Reptilia 21: 371-379.

Paniagua R, B Fraile, FJ Sàez. 1990. Effects of photoperiod and temperature on testicular function in amphibians. Histol. Histopathol. 5: 365-378.

Paolucci M. 2003. An androgen receptor in the brain of the green frog *Rana esculenta*. Life Sci. 73: 265-274.

Pchlemann FW. 1962. Experimentale Untersuchungen zur determination und differenzierung der hypophyse bei Anuren (*Pelobates fuscus, Rana esculenta*) Roux' Arch. Entw. Mech. Org. 153: 551-602.

Pecio A, R Piprek, M Ogielska, J Bartmańska, JM Szymura. 2007 Evidence for Sertoli cells in the testes of *Bombina variegata* (Anura: Bombinatoridae). Abstr. Int. Congress Vert. Morph., Paris 2007: 95.

Peter RE. 1983. Evolution of neurohormonal regulation of reproduction in lower vertebrates. Amer. Zool. 23: 685-695.

Pierantoni R, G Cobellis, R Meccariello, C Palmiero, S Minucci, S Fasano. 2002. The amphibian testis as model to study germ cell progression during spermatogenesis. Comp. Bioch. Physiol. B 132: 131-139.

Pierantoni R, S Fasano, L Di Matteo, S Minucci, B Variale, G Chieffi. 1984b. Stimulatory effect of a GnRH agonist (busurelin) in *in vitro* and *in vivo* testosterone production by the frog (*Rana esculenta*) testis. Mol. Cell. Endocrinol. 38: 215-219.

Pierantoni R, L Iela, M d'Istria, S Fasano, RK Rastogi, G Delrio. 1984a. Seasonal testosterone profile and testicular responsiveness to pituitary factors and gonadotropin releasing hormone during two different phases of the sexual cycle of the frog (*Rana esculenta*). J. Endocrinol. 102: 387-392.

Pierantoni R, S Minucci, L Di Matteo, S Fasano, B Variale, G Chieffi. 1985. Effect of temperature and darkness on testosterone concentration in the testes of intact frogs (*Rana esculenta*) treated with gonadotropin releasing hormone analog (HOE 766). Gen. Comp. Endocrinol. 58: 128-130.

Polzonetti-Magni A, V Botte, L Bellini-Cardellini, A Gobetti, A Crasto. 1984. Plasma sex hormones and post-reproductive period in the green frog, *Rana esculenta* complex. Gen. Comp. Endocrinol. 54: 372-377.

Polzonetti-Magni A, O Carnevali, K Yamamoto, S Kirujama. 1995. Growth hormone and Prolactin in Amphibian Reproduction. Zool. Sci. 12: 683-694.

Polzonetti-Magni AM, G Mosconi, O Carnevali, K Yamamoto, Y Hanaoka, S Kikuyama. 1998. Gonadotropin and reproductive function in the anuran amphibian, *Rana esculenta*. Biol. Reprod. 58: 88-93

Poska-Teiss L. 1933. Spermatogonien von *Bufo vulgaris* Laur. und ihr Vergleich mit larvalen somatischen Zellen desselben Tieres. Z. Zellfor. Mikr. Anat. 17: 347-419.

Pudney J. 1993. Comparative cytology of the non-mammalian vertebrate Sertoli cells. In: The Sertoli Cell, (Eds) LD Russell and MD Griswold, Cache River Press, Clearwater F2.

Pudney J. 1995. Spermatogenesis in nonmammalian vertebrates. Microsc. Res. Tech. 32: 459-497.

Raisman JS, RW de Cunio, MO Cabada, EJ del Pino, MI Mariano. 1980. Acrosome breakdown in *Leptodactylus chaquensis* (Amphibia, Anura) spermatozoa. Dev. Growth Diff. 22: 289-297.

Rastogi RK, JT Bagnara, L Iela, MA Krasovich. 1988. Reproduction in the Mexican leaf frog, *Pachymedusa danicolor*. IV. Spermatogenesis: a light and ultrasonic study. J. Moprhol. 197: 277-302.

Rastogi RK, G Chieffi, C Mormorino. 1972. Effect of Methalibure (IC33.828) on the pars distalis of of pituitary, testis and thumb pad of the green frog, *Rana esculenta* L. Z. Zellforsch. 123: 430-440.

Rastogi RK, G Chieffi. 1970a. Cytological changes in the pars distalis of pituitary of the green frog *Rana esculenta* L., during the reproductive cycle. Z. Zellforsch. 111: 505-518.

Rastogi RK, G Chieffi. 1970b. A cytological study of the pars distalis of the pituitary gland of normal, gonadectomized and gonadectomized, steroid hormone treated green frog, *Rana esculenta* L. Gen. Comp. Endocrinol. 15: 247-263.

Rastogi RK, G Chieffi. 1972a. Hypothalamic control of the hypophyseal gonadotropic function in the adult male green frog *Rana esculenta* L. J. Exp. Zool. 181: 263-270.

Rastogi RK, G Chieffi. 1972b. Inhibition of pituitary gonadotropic effects in the pars distalis-ectomized *Rana esculenta* by Methalibure (ICI 33,828). Z. Zellforsch. 129: 51-55.

Rastogi RK, L Di Matteo, S Minucci, M Di Meglio, L Iela. 1990. Regulation of spermatogonial proliferation in the frog (*Rana esculenta*): an experimental analysis. J. Zool. Lond. 1990: 201-211.

Rastogi RK, M Di Meglio, L Di Matteo, S Minucci, L Iela. 1985. Morphology and cell population kinetics of primary spermatogonia in the frog (*Rana esculenta*) (Amphibia: Anura). J. Zool. Lond. 207: 319-330.

Rastogi RK, L Iela, G Delrio, T Bagnara. 1986. Reproduction in the Mexican leaf frog *Pachymedusa dacnicolor*: 2. The male. Gen. Comp. Endocrinol. 62: 23-35.

Rastogi RK, L Iela, G Delrio, M Di Meglio, A Russo, G Chieffi. 1978. Environmental influence on testicular activity in the frog, *Rana esculenta*. J. Exp. Zool. 206: 49-64.

Rastogi RK, L Iela, K Saxena, G Chieffi. 1976. The control of spermatogenesis in the green frog *Rana esculenta* J. Exp. Zool. 196: 151-166.

Rastogi RK, L Iela. 1980. Steroidogenesis and spermatogenesis in Anuran Amphibia: A brief survey. In: Steroids and Their Mechanism of Action in Non-mammalian Vertebrates, (Eds) G Delrio and J Brachet. Raven Press, New York, pp. 131-146.

Rastogi RK, JA King, MM Di Fiore, C Pinelli. 1997. Sex and reproductive status related brain content of mammalian and chicken-II GnRHs in *Rana esculenta*. J. Neurocrinol. 9: 519-522

Rastogi RK, DL Meyer, C Pinelli, M Fiorentino, B D'Aniello. 1998. Comparative analysis of GnRH neuronal system in the amphibian brain. Gen. Comp. Endocrinol. 112: 330-345.

Rastogi RK. 1976. Seasonal cycle in Anuran (Amphibia) testis: The endocrine and environmental controls. Boll. Zool. Agrar. Bachic. 43: 151-172.

Reed SC, HP Stanley. 1972. Fine structure of spermatogenesis in the South African clawed toad *Xenopus laevis* Daudin. J. Ultrastruct. Res. 41: 277-295.

Risley MS, RA Eckhadrdt. 1975. Basic protein changes during spermatogenesis in *Xenopus laevis*. J. Cell. Biol. 67: 362a.

Risley MS, RA Eckhadrdt. 1981. H1 histones of *Xenopus laevis*. Dev. Biol. 84: 79-87.

Risley MS, A Miller, DA Bumcrot. 1987. In vitro maintenance of spermatogenesis in *Xenopus laevis* testis explants cultured in serum-free media. Biol. Reprod. 36: 985-997.

Risley MS, M Morse-Gaudio. 2005. Comparative aspects of spermatogenic cell metabolism and Sertoli cell function in *Xenopus laevis* and mammals. J. Exp. Zool. 261: 185-193.

Roosen-Runge EC. 1977. The process of spermatogenesis in animals. Devel. Biol. and Cell. Biol. Series 5. Cambridge Univ. Press Cambridge, New York, Melbourne: 115-122.

Sanderson JT. 2006. The steroid hormone biosynthesis pathway as a target for endocrine-disrupting chemicals. Toxicol. Sci. 94: 3-21.

Sasso-Cerri E, E Freymüller, SM Miraglia. 2005. Testosterone-immunopositive primordial germ cells in the testis of the bullfrog, *Rana catesbeiana*. J. Anat. 206: 519-523.

Sasson D, DB Kelley. 1986. The sexually dimorphic larynx of *Xenopus laevis*; development and androgen regulation. Am. J. Anat. 177: 457-472.

Sasson DA, GE Gray, DB Kelley. 1987. Androgen regulation of muscle fiber type in the sexually dimorphic larynx o9f *Xenopus laevis*. J. Neurosci. 7: 3198-3206.

Scheltinga DM, BGM Jamieson. 2003. Spermatogenesis and the mature spermatozoon: Form, function and phylogenetic implications. In: Reproductive Biology and Phylogeny of Anura, (Ed.) BGM Jamieson, Vol. 2, pp. 119-251.

Scheltinga DM, BGM Jamieson, KE Eggers, DM Green. 2001. Ultrastructure of the spermatozoon of *Leiopelma hochstetteri* (Amphibia, Anura, Leiopelmatidae). Zoosystema 23: 157-171.

Schlaghecke R, V Blum. 1978. Seasonal variations in fat body metabolism of the green frog *Rana esculenta* (L.). Experientia 34: 1019-20.

Sealfon SC, H Weinstein, RM Millar. 1997. Molecular mechanisms of ligand interaction with the Gonadotropin-Releasing Hormone Receptor. Endocrine Rev. 182: 180-205.

Segal SJ, CA Adejuwon. 1979. Direct effect of LHRH on testicular steroidogenesis in Rana pipiens. Biol. Bull. 157: 393-394.

Setalo G, G Lazar, T Kozicz. 1994. The gonadotropin-releasing hormone (GnRH) neuron system of the clawed toad *Xenopus laevis*. Acta Biol. Hung. 45: 427-440.

Sever DM, WC Hamlett, R Slabach, B Stephenson, PA Verrell. 2003. Internal fertilization in the Anura with special reference to mating and female sperm storage in *Ascaphus*. In: Reproductive Biology and Phylogeny of Anura, (Ed.) BGM Jamieson, Vol. 2, pp. 319-341.

Sever DM, EC Moriarty, LC Rania, WC Hamlett. 2001. Sperm storage in the oviduct of the internal fertilizing frog, *Ascaphus truei*. J. Morphol. 238: 143-155.

Shalan AG, SD Bradshaw, PC Withers, G Thompson, MFF Bayomy, FJ Bradshaw T Stewart. 2004. Spermatogenesis and plasma testosterone levels in *Cyclorana maini* and *Neobatrachus sutor* during aestivation. Gen. Comp. Endocrinol. 136: 90-100.

Sherwood NM, K von Schalburg, DW Lescheid. 1997. Origin and evolution of

GnRH in vertebrates and invertebrates. In: GnRH Neurone: Gene to Behaviour, (Eds) IS Parhar and Y Sakuma. Brain Shuppan Tokyo, 3-25.

Sherwood NM, RT Zoeller, FL Moore. 1986. Multiple forms of gonadotropin-releasing hormone in amphibian brains. Gen. Comp. Endocrinol. 61: 313-322.

Shivakumar GS. 1999. Light has no role in spermatogenesis in the frog *Rana cyanophlyctis* (Schneider). Indian J. Exp. Biol. 37 (3): 319-321

Siboulet R. 1981. Variations sansonnierens de la teneur plasmatique en testosterone et dihydrotestosterone chez le crapaud de Mauritanie *(Bufo mauritanicus)*. Gen. Comp. Endocrinol. 43: 71-75.

Sluiter JW, GJ van Oordt, M Grasvelt. 1950. Spermatogenesis in normal and hypophysectomized frogs *(Rana temporaria)* following gonadotropin administration. Acta Endocrinol. 4: 1-15.

Smeets WJ, A Gonzales. 2001. Vasotocin and mesotocin in the brains of amphibians. Microsc. Res. Tech. 54: 125-136.

Sower AS, KL Reed, KJ Babbitt. 2000. Limb malformations and abnormal sex hormone concentrations in frogs. Environ. Health Prospect. 108: 1-7.

Sprando RL, LD Russell. 1988. Spermiogenesis in the bullfrog *(Rana catesbeiana)*: A study of cytoplasmic events including cell volume changes and cytoplasmic elimination. J. Morphol. 198: 303-319.

Takamune K, T Kawasaki, S Ukon. 2001.The first and the second mitotic phases of spermatogonial stage in *Xenopus laevis*: secondary spermatogonia which have differentiated after completion of the first mitotic phase acquire an ability of mitosis to meiosis conversion. Zool. Sci. 18: 577-583.

Takamune K, K Teshima, M Maeda, SI Abé. 1995. Characteristic features of preleptotene spermatocytes in *Xenopus laevis*: Increase in the nuclear volume and first appearance of flattened vesicles in these cells. J. Exp. Zool. 273: 264-270.

Tanaka S, H Iwasawa, K Wakabayashi. 1988. Plasma levels of androgens in growing frogs of *Rana nigromaculata*. Zool. Sci. 5: 1007-1012.

Tanimura A, H Iwasawa. 1988. Ultrastructural observations on the origin and differentiation of somatic cells during gonadal development in the frog, *Rana nigromaculata*. Develop. Growth Differ. 30: 681-691.

Tanimura A, H Iwasawa. 1989. Origin of somatic cells and histogenesis in the primordial gonad of the Japanese tree frog, *Rhacophorus arboreus*. Anat. Embryol. 180: 165-173.

Tanimura A, H Iwasawa. 1992. Ultrastructural observations of the ovary and Bidder's organ in young toad, *Bufo japonicus formosus*. Sci. Rep. Niigata Univ. D Biol. 29: 27-33.

Tobias ML, DB Kelley. 1995. Sexual differentiation and hormonal regulation of the laryngeal synapse in *Xenopus laevis*. J. Neurobiol. 28: 515-526.

Tobias ML, ML Marin, DB Kelley. 1991. Temporal constraints on androgen directed laryngeal masculinization in *Xenopus laevis*. Dev. Biol. 147: 260-270.

Troskie BE, JP Hapgood, RP Millar, N Illing. 2000. Complementary deoxyribo-nucleic acid cloning, gene expression, and ligand selectivity of a novel gonadotropin releasing hormone receptor expressed in the pituitary and midbrain of *Xenopus laevis*. Endocrinology 141: 1764-1771.

Troskie BE, JA King, RP Millar, Y Peng, J Kim, H Figueras, N Illing. 1997. GnRH

II in amphibian sympathetic ganglia. Neuroendocrinol 6: 396-402.

Tsai PS, AE Kessler, JT Jones, KB Wahr. 2005. Alternation of the hypothalamic-pituitary-gonadal axis and androgen treated adult male leopard frog *Rana pipiens*. Reprod. Biol. Endocrinol. 3: 2-15.

Tsai PS, JB Lunden, JT Jones. 2003. Effect of steroid hormones on spermatogenesis and GnRH release in male leopard frogs *Rana pipiens*. Gen. Comp. Endocrinol. 134: 330-338.

Ueda Y, N Yoshizaki, Y Iwao. 2002. Acrosome reaction in sperm of the frog, *Xenopus laevis*: Its detection and induction by oviductal pars recta secretion. Dev. Biol. 243: 55-64.

van Kaemende JAM. 1969. Effect of a rise in ambient temperature on the pars distalis of the pituitary, the internal gland the interstitial tissue of the testis in the common frog *Rana temporaria*, during hibernation. Z. Zellforsch. 95: 620-630.

van Oordt PGWJ, B Lofts. 1963. The effects of high temperature on gonadotropin secretion in the male common frog *Rana temporaria* during autumn. J. Endocrinol. 27: 136-137.

van Oordt PGWJ. 1956. Regulation of the spermatogenetic cycle in the common frog (*Rana temporaria*), Acta Physiol. Pharmacol. Neerl. 5(2): 223-224.

van Oordt PGWJ. 1960. The influence of internal and external factors in the regulations of the spermatogenesis cycle in Amphibia. Symp. Zool. Soc. London 2: 29-52.

Variale B, I Serino. 1994. The androgen receptor mRNA is upregulated by testosterone in both the Harderian gland and thumb pad of the frog *Rana esculenta*. J. Steroid. Biochem. Mol. 51: 259-265.

Varriale B, R Pierantoni, M Di Matteo, S Minucci, S Fasano, M D'Antonio, G Chieffi. 1986. Plasma and testicular estradiol and plasma androgen profile in the male frog *Rana esculenta* during annual cycle. Gen. Comp. Endocrinol. 64(3): 401-404.

Viertel B, S Richter. 1992. Anatomy: viscera and endocrines. In: Tadpoles. The Biology of Anuran Larvae, (Eds) RW McDiarmid, R Altig. The University of Chicago Press, pp. 92-148.

Vivien–Roels B, P Pevet. 1983. The pineal gland and the synchronization of reproductive cycles with variations of the environmental climatic conditions, with special reference to temperature. Pineal Res. Rev. 1: 91-143.

Vullings HGB. 1973. Influence of light and darkness on the hypothalamo-hypophyseal system of blinded and unilaterally blinded frogs *Rana temporaria*. Z. Zellforsch. 136: 355-364.

Wada M, A Gorbman. 1977. Relation of mode of administration of testosterone to evocation of male sex behaviour in frogs. Horm. Beh. 8: 310-319.

Wake MH. 1980. The reproductive biology of *Nectophrynoides malcolmi* (Amphibia: Bufonidae) with comments on the evaluation of reproductive modes in the genus *Nectophrynoides*. Copeia 2: 193-209.

Wang L, J Bogerd, HS Choi, JY Seong, JM Soh, SY Chun, M Blomenröhr, BE Troskie, RP Millar, WH Yu, SM McCann, HB Kwon. 2001. Three disting types of GnRH receptor characterized in the bullfrog. Proc. Natl. Acad. Sci. USA 98: 361-366.

Wiechmann AF, MJ Vrieze, CR Wirsig-Wiechmann. 2003. Differential distribution of melatonin receptors in the pituitary gland of *Xenopus laevis*. Anat. Embryol. 206 (4): 291-299.

Witschi E. 1929. Studies on sex differentiation and sex determination in amphibians. I. Development and sexual differentiation of the gonads of *Rana sylvatica*. J. Exp. Zool. 52: 235-265.

Witschi E. 1956. Development of Vertebrates. Saunders Company, pp. 14-23.

Yoo MS, HM Kang, HS Choi, JW Kim, BE Troskie, RP Millar, HB Kwon. 2000. Molecular cloning, distribution and pharmacological characterization of a novel gonadotropin releasing hormone ([Trp8)] GnRH) in frog brain. Mol. Cell. Endocrinol. 164: 197-204.

Zancanaro C, F Merigo, M Digito, G Pelosi. 1996. Fat body of the frog *Rana esculenta*: an ultrastructural study. J. Morphol. 227: 321-334.

Zirkin BR. 1970. The protein composition of nuclei during spermiogenesis in the leopard frog, *Rana pipiens*. Chromosoma (Berl.) 31: 231-240.

Zuber-Vogeli M, F Xavier. 1965. La spermatogenese de *Nectophrynoides occidentalis* au cours du cycle annuel. Bull. Soc. Zool. France 90: 261-267.

3

Spermatogenesis and Male Reproductive System in Amphibia—Urodela

Mari Carmen Uribe Aranzábal

Structure of Testis in Urodela

The male reproductive system of urodeles is composed of a pair of elongated testes, fat bodies, a system of efferent ducts, and a cloacal gland complex. The testes are attached to the dorsal side of body cavity by the mesorchium. The testes are parallel to the mesonephros, Wolffian (or mesonephric) ducts, and rudimentary Müllerian (or paramesonephric) ducts (Kingsbury, 1901; Branca, 1904; Adams, 1940; Baker and Taylor, 1964; Baker, 1965; Lofts, 1984, 1987; Brizzi *et al.*, 1985; Norris, 1997; Uribe, 2001, 2003). Each testis is surrounded by *tunica albuginea* formed by a thin layer of connective tissue and superficial squamous mesothelium.

Each testis is composed of one or several lobes. Humphrey (1922) concluded that sexually immature males have a simple testis composed of only one lobe, whereas mature males have multiple testes composed of several lobes, and the multiple teses must have arisen by addition of new lobes after sexual maturity. Adams (1940) observed that bigger individuals of *Triturus viridescens* tend to have heavier multilobed testes than small ones with simple testes. Lofts (1984, 1987) and Bergmann (1994) confirmed that the number of testicular lobes depends on age of the animal. Lofts (1984) observed in *Salamandra salamandra* that males passes through four annual cycles before the second lobe develops, then additional lobes may be added in two years intervals. Therefore, age of animal must be considered as determining factor in development of multiple lobes, at least in some species of Urodela. Testis with multiple lobes are described by several authors in various species: Humphrey (1922) in *Desmognathus fuscus*,

Diemyctylus viridescens, Diemyctylus torosus, and *Salamandra atra;* Adams (1940) in *Triturus viridescens;* Baker (1965) in *Taricha granulosa, Diemyctylus viridescens* and *Cynops pyrrhogaster;* Lofts (1987) and Tso and Lofts (1977a) in *Trituroides hongkongenesis;* Fraile *et al.* (1989b) in *Triturus marmoratus;* Lofts (1984), Schindelmeiser *et al.* (1983, 1985), and Bergmann (1994) in *Salamandra salamandra.* On the other hand, some species have simple testes with one lobe, regardless of their age. The simple testes were described in *Eurycea quadridigitata* (Trauth, 1983), *Salamandrina terdigitata* (Brizzi *et al.*, 1985), *Ambystoma dumerilii* (Uribe *et al.*, 1994)

The lobes of the testis are joined in a linear chain by narrow bridges of connective tissue (Humphrey, 1922; Lofts, 1987; Bergmann, 1994) containing germ cells and covered by the peritoneal epithelium, also containing germ cells (Baker, 1965; Tso and Lofts, 1977a; Tanaka *et al.*, 1980; Lofts, 1984; Bergmann, 1994). Each lobe is a basic testicular unit, and is morphologically and functionally similar to other lobes. The lobes have the structural organization of miniatures testes, containing the same distribution and types of germ and somatic cells (Lofts, 1987).

Testicular Lobules (Seminiferous Tubules)

Several authors described the structure of testis in various species of urodeles such as *Necturus maculosus* (Humphrey, 1921; Pudney *et al.*, 1983; Lofts, 1987; Pudney, 1995), *Desmognatus fuscus, Diemyctylus viridescens, Plethodon erithronotus, P. glutinosus, Spelerpes bilineatus, Ambystoma punctatum, Salamandra atra* (Humphrey, 1921), *Ambystoma tigrinum* (Carrick, 1934; Moore, 1975; Norris *et al.*, 1985; Lofts, 1987), *Trituroides hongkongensis* (Tso and Lofts, 1977a,b), *Taricha granulosa* (Moore *et al.*, 1979; Lofts, 1987), *Cynops pyrrhogaster* (Tanaka *et al.*, 1980), *Triturus cristatus carnifex* (Franchi *et al.*, 1982; Lofts, 1987), *Salamandra salamandra* (Bergmann *et al.*, 1982, 1983; Schindelmeiser *et al.*, 1983, 1985; Lofts, 1984; Lecouteux *et al.*, 1985, Bergmann, 1994), *Ambystoma mexicanum* (Miltner and Armstrong, 1983, Armstrong, 1989, Uribe, 2001, 2003), *Salamandrina terdigitata* (Brizzi *et al.*, 1985), *Triturus alpestris* (Lofts, 1987), *Triturus marmoratus* (Fraile *et al.*, 1989a, 1990), *Ambystoma dumerilii* (Uribe *et al.*, 1994; Uribe, 2001, 2003). The internal structure of the testicular lobes consists of abundant longitudinal lobules (being the equivalent of seminiferous tubules in Anura), which communicate by the system of intratesticular tubules (*rete testis*) with efferent ducts (Baker, 1965; Williams *et al.*, 1984; Bergmann, 1994). The testicular lobules are separated by trabecules of thin connective tissue similar to that of the tunica albuginea.

102

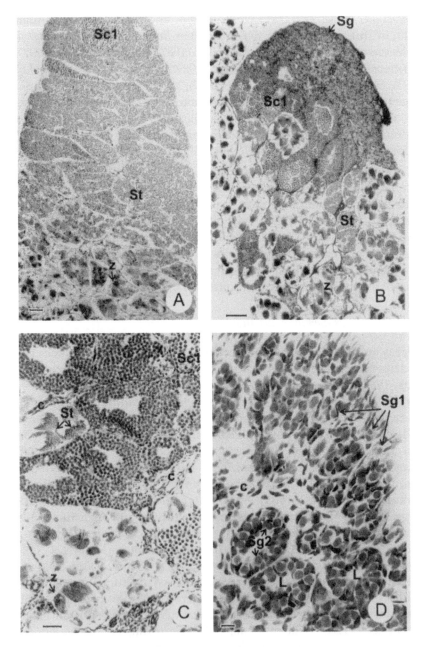

Fig. 1. See caption on the next page.

Sertoli Cells and Cysts Formation

Each lobule contains a great number of cysts with synchronously developing germ cells at various stages of spermatogenesis. The distribution of cysts displays the cephalo-caudal gradient within a lobule in respect to the stage of spermatogenesis (Fig. 1A,B). The distribution of cysts within lobules is reflected in overall morphology of a testis. Lobules containing cysts with spermatogenic cells at successive stages of spermatogenesis form zones (Fig. 1B,C) of spermatogonia, primary spermatocytes, secondary spermatocytes, spermatids, and spermatozoa (Moore, 1975; Tso and Lofts, 1977a; Trauth, 1983; Lecouteux *et al.*, 1985; Bergmann, 1994; Uribe *et al.*, 1994; Pudney, 1995). The cephalic region of a testis contains spermatogonia, either as individual cells or in small groups (Fig. 1D). The caudal region of a testis contains cysts with spermatozoa (Fig. 2A). Pudney (1995) suggested that the continuous formation of cysts at cephalic end of the lobule causes their migration along the lobule (Fig. 1A,B) and, consequently, their displacement toward the system of intratesticular efferent ducts (*rete testis*).

A cyst is formed when a primary spermatogonium becomes surrounded by a Sertoli cell (also called a follicular cell), which forms the wall of a cyst (Schindelmeiser *et al.*, 1985; Fraile *et al.*, 1989b, 1990; Grier, 1993; Bergmann, 1994). During spermiation, when the cysts open and spermatozoa leave the testis, the acidophilic Sertoli cells (Fig. 2D) remain inside the lobule (Fig. 2D) and undergo strong morphological changes (Fig. 3C,D), acquiring a pattern of steroid-secreting cells. In some empty cysts abnormal spermatozoa can be observed (Fig. 3A), which are later phagocytized by Sertoli cells (Schindelmeiser *et al.*, 1985). Thereafter, the Sertoli cells differentiate into secretory cells and then degenerate (Carrik, 1934; Pudney *et al.*, 1983; Bergmann, 1994). At that time, the interstitial Leydig cells differentiate into glandular cells.

Fig. 1. Testes of *Ambystoma dumerilii*. A, B—During (Spring) (A), (Autumn) (B) in longitudinal sections. Spermatogenesis procceding in sucessives zones along the cephalo-caudal axis. Spermatogonia (Sg), primary spermatocytes (Sc1), spermatids (St), spermatozoa (z). Compare the abundant primary spermatocytes during April and scarce during October. A—Stained with hematoxylin-eosin (H-E). Bar = 0.3 mm B—Stained with Masson's trichrome. Bar = 0.3 mm. C—Longitudinal section. Lobules containing primary spermatocytes (Sc1), spermatids (St), during the activation of spermatogenesis and lobules containing spermatozoa (z) from the previous reproductive cycle. Interlobular connective tissue (c). Masson's trichrome. Bar = 0.1 mm. D—Cephalic region of the testis containing primary spermatogonia (Sg1), and secondary spermatogonia (Sg2). Note the secondary spermatogonia forming clusters of cells, containing a central lumen (L) and surrounded by connective tissue (c). H-E. Bar = 30 μm.

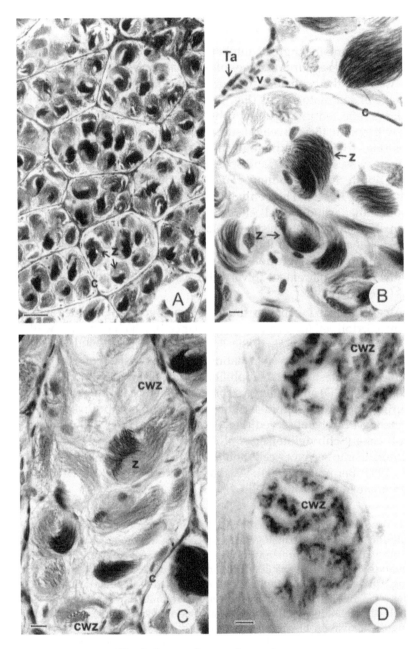

Fig. 2. See caption on the next page.

The synchronous development of germ cells inside a cyst throughout spermatogenesis is a result of the presence of intercellular bridges between them, but also suggests the possible interactions between the germ cells and Sertoli cells. These interactions include maintenance of permeability barrier to germ cells during spermatogenesis, endocrine activity, phagocytosis of degenerating germ cells and residual bodies, and formation of specific antigens (Lazard, 1979; Franchi et al., 1982; Schindelmeiser et al., 1985; Grier, 1993; Pudney, 1995).

Interstitial Tissue

The organization of the interstitial tissue varies according to the stage of spermatogenesis. The Leydig cells stay undifferentiated during spermatogenesis and hypertrophy during spermiation (Fig. 3B) when Sertoli cells degenerate (Humphrey, 1921; Tso and Lofts, 1977b; Pudney and Callard, 1984; Fraile et al., 1989b). During spermiation, cells of interstitial tissue localized around lobule walls (pericystic cells) (Imai and Tanaka, 1978) differentiate and hypertrophy, thereby transforming into Leydig cells (Fig. 3C,D), which form glandular tissue (Humphrey, 1921; Lazard, 1976, 1979; Pudney et al., 1983; Pudney and Callard, 1984; Lofts, 1984; Lecouteux et al., 1985; Bergmann, 1994). Leydig cells, before their transformation into glandular tissue, are structurally similar to fibroblasts (Imai and Tanaka, 1978; Fraile et al., 1990). Pudney et al. (1983) revealed that spermiation, followed by regression of the mature lobules, is an event that signals hypertrophy of the interstitial tissue. Fraile et al. (1989b) suggested that Sertoli cells control the differentiation of Leydig cells by a local diffusible agent.

Leydig cells differentiate into steroidogenic cells (Humphrey, 1921; Tso and Lofts, 1977b; Lazard, 1979; Bergmann et al., 1983; Pudney et al., 1983; Pudney and Callard, 1984; Lofts, 1984; Norris et al., 1985; Schindelmeiser et al., 1985; Fraile et al., 1989b, 1990; Bergmann, 1994). Moore et al. (1979)

Fig. 2. Spermiation in the testes of *Ambystoma dumerilii*. A—Before spermiation. Several lobules containing abundant cysts with spermatozoa (z). Thin connective tissue (c) around the lobules. Masson's trichrome. Bar = 0.1 mm. B, C—During spermiation. Lobules containing cysts with spermatozoa (z). Note the reduction of the number of cysts (Fig. 2C) compared with those seen in Figure. 2A. There are cysts without spermatozoa (cwz), containing only Sertoli cells. Connective tissue (c) around the lobules. The tunica algubinea (Ta) contains blood vessels (v). B—H-E. Bar = 30 µm. C—Alcian-blue. Bar = 30 µm. D—After spermiation. The cysts without spermatozoa (cwz) contains rests of Sertoli cells (S). H-E. Bar = 10 µm.

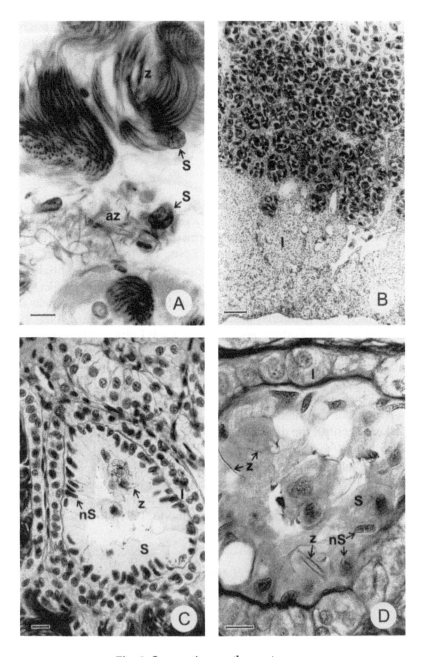

Fig. 3. See caption on the next page.

and Lecouteux *et al.* (1985) suggested that in *Taricha granulosa* and *Salamandra salamandra* these cells are the sites of androgen synthesis (testosterone and 5 α-dihydrotestosterone). The active steroidogenesis in Leydig cells were confirmed by histochemical studies in *Ambystoma tigrinum* (Moore, 1975), *A. mexicanum* (Lazard, 1976, 1979), *Trituroides hongkongensis* (Tso and Lofts, 1977a), *Cynops pyrrhogaster pyrrhogaster* (Imai and Tanaka, 1978), *Necturus maculosus* (Pudney *et al.*, 1983), and *A. tigrinum* (Norris *et al.*, 1985).

Cytological features found in steroid-producing cells, such as well-developed abundant smooth endoplasmic reticulum, abundant lipid droplets, prominent Golgi complexes and mitochondria with dense granular matrix, were observed in Leydig cells in *Trituroides hongkongensis* (Tso and Lofts, 1977b), *Cynops pyrrhogaster pyrrhogaster* (Imai and Tanaka, 1978), and *Triturus marmoratus* (Fraile *et al.*, 1990).

The interstitial tissue of *Ambystoma dumerilii* (Fig. 3C,D), and *A. mexicanum* (Uribe and Mejia-Roa, unpublished data) that forms a distinct ample region of a testis hypertrophies during autumn and winter, when abundant spermatozoa are present in testicular lobules (Fig. 3B). This tissue, formed by Leydig cells, is evidently reduced during spring, when spermatozoa are scarce, and the next generation of germ cysts appears. Initially, the Leydig cells are small and cuboidal in shape. During spermiation, interstitial cells hypertrophy and attain 35-55 μm in diameter (Fig. 3C,D) (Uribe, 2001). At the end of spermiation, this tissue regresses constantly and consists of smaller and irregular cells with pycnotic nuclei and amorphous and lightly stained cytoplasm. During the regression period, Leydig cells suffer rapid involution. Maturation of Leydig cells occurs when testosterone levels increases, whereas involution of Leydig cells is noticed, when the testosterone level decreases (Fraile *et al.*, 1990). The morphological changes of the interstitial tissue are closely correlated with the seasonal cycle of spermatogenesis (Williams *et al.*, 1984; Norris *et al.*, 1985; Armstrong, 1989).

Fig. 3. Spermiation in the testes of *Ambystoma dumerilii*. A—During spermiation lobules contain cysts with spermatozoa (z), and a cyst with some abnormal spermatozoa (az). Sertoli cell's nuclei (S) are seen at the periphery of the cysts. H-E. Bar = 30 μm. B—Cross section of a testis of during spermiation. Lobules contain cysts with spermatozoa (z), and interstitial tissue (I) that forms an ample region of the testis. H-E. Bar = 0.3 mm. C, D—The lobules contain rests of spermatozoa (z) among hypertrophied Sertoli cells (S) with ovoidal nuclei situated at the basal end of the cells. Borders of a lobule are indicated by arrow heads. Interstitial tissue (I) transforms into Leydig cells (l) that surround the lobule. C—H-E. Bar = 30 μm. D—PAS. Bar = 30 μm.

Urogenital Connections

Sperm-collecting ducts constitute a complex system of intratesticular efferent ducts (*rete testis*) (Fig. 4A), transverse tubules (*vasa efferentia*) (Fig. 4B,C,D), Wolffian ducts (Fig. 4B), rudimentary Müllerian ducts (Fig. 5A), and cloaca (Baker and Taylor, 1964; Baker, 1965; Trauth, 1983; Williams *et al.*, 1984; Norris *et al.*, 1985; Norris, 1987, 1997; Uribe, 2001, 2003). The morphology of efferent ducts changes along with seasonal cycle of testes. The diameter of the ducts increases during spermiation, when spermatozoa are abundant and epithelial cells are larger (Norris *et al.*, 1985).

Male reproductive ducts of urodeles was reviewed by Norris (1987), who consider that efferent ducts of more primitive families (Crytobranchidae and Sirenidae) function for both the urine and spermatozoa transportation to the cloaca. In more advanced families (Ambystomatidae, Plethodontidae, Salamandridae), the efferent ducts are separated from excretory ducts and transport only spermatozoa.

Baker and Taylor (1964) described urogenital system in several species of the family Ambystomatidae, i.e., *Ambystoma maculatum, A. opacum, A. texanum, A. taipoldeum,* and *A. tigrinum,* with emphasis on development between connections of testis and the Wolffian duct. Wolffian ducts originate on the lateral edge of kidney and each of them enters the cloaca separately (Duellman and Trueb, 1986).

Trauth (1983) described the histology of efferent ducts in *Eurycea quadridigitata,* Williams *et al.* (1984) in *Eurycea lucifuga* and *E. longicauda,* Norris *et al.* (1985) and Norris (1987) in *Ambystoma tigrinum,* Zalisko and Larsen (1988) in *Rhyacotriton olympicus,* and Uribe (2001, 2003) in *Ambystoma mexicanum* and *A. dumerilii.* The walls of the system of efferent ducts are composed of an outer *serosa*, a muscle layer, a connective tissue layer, and a simple epithelium lining the lumen. A squamous peritoneal epithelium and a thin sheath of connective tissue form the *serosa*. The muscle layer is composed of smooth muscle fibers arranged circularly around the ducts. Connective tissue is composed of extracellular matrix with collagen fibers, fibroblasts, melanocytes, blood vessels and nerve fibers. The inner epithelium is squamous, cubic or columnar with various apical specializations, such as cilia or stereocilia, depending on a region of efferent ducts (Williams *et al.*, 1984).

Intratesticular ducts (*rete testis*) are embedded in the interlobular connective tissue of the testis (Fig. 4A). Their lumen is lined with cuboidal epithelium (Williams *et al.*, 1984).

Transverse tubules (*vasa efferentia*) run from a testis to the cephalic region of kidney and join the Wolffian duct. The junction of transverse tubules and nephric tubules in kidney forms an epididymal complex

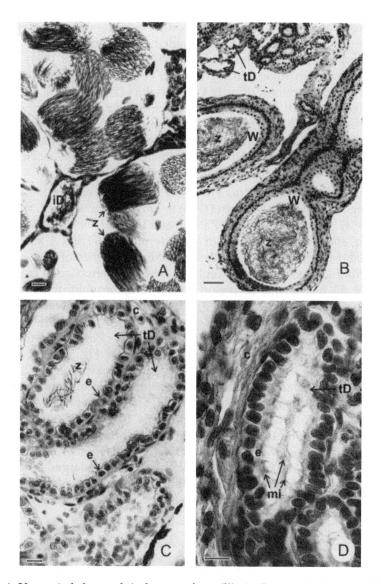

Fig. 4. Urogenital ducts of *Ambystoma dumerilii*. A—Intratesticular duct (iD) in the connective tissue (c) around the lobules of the testis. The lobules contain cysts with spermatozoa (z). H-E. Bar = 30 μm. B—Transverse tubules (tD) and Wolffian duct (W) containing abundant spermatozoa (z). H-E. Bar = 0.1 mm. C—Transverse tubules (tD) with cubic epithelium (e) and spermatozoa (z) in the lumen. Connective tissue (c) around the tubules. Alcian-blue. Bar = 30 μm. D— Transverse tubules (tD) showing columnar epithelium (e) with long microvilli (mi) or stereocilia on the apical end. Connective tissue (c) around the tubules. H-E. Bar = 20 μm.

(Williams *et al.*, 1984). Therefore, the kidney consists of two regions: the cephalic genital (or sexual) kidney called epididymis, and the caudal excretory kidney (Baker and Taylor, 1964; Williams *et al.*, 1984). In the family Plethodontidae, kidney lacks a cephalic extension and the transversal ducts connect directly Wolffian duct (Williams *et al.*, 1984). In transverse tubules of *Ambystoma dumerilii* and *A. mexicanum* (Uribe, 2001, 2003) there are two different types of epithelium: one is composed of cuboidal cells (Fig. 4C), whereas the other is composed of cells with long microvilli, stereocilia and cilia at the apical end, and their cytoplasm contains vacuoles suggesting the secretory activity (Fig. 4D).

The Wolffian duct (*ductus Wolffii*) is the largest among the sperm-collecting ducts (Fig. 4B). Trauth (1983) in *Eurycea quadridigitata*, Williams *et al.* (1984) in *Eurycea bislineata*, Zalisko and Larsen (1988) in *Rhyacotriton olimpicus*, and Uribe (2001, 2003) in *Ambystoma dumerilii* and *A. mexicanum* described the morphology of epithelium lining the lumen of various parts of the Wolffian duct. The most cephalic region is composed of columnar cells (Fig. 5A,B). The middle region, constituting the majority of the duct, has cuboidal epithelium (Fig. 5C,D). The caudal region has tall columnar secretory cells. In addition to the columnar or cuboidal epithelial cells, spherical or ovoidal cells with round nuclei and clear cytoplasm occur apically in the epithelium of the entire Wolffian duct (Fig. 5B,C,D). Zalisko and Larsen (1988) suggested that the apical cells, in addition to cilia, may protect the cuboidal epithelial cells from possible abrasion caused by circulation of sperm masses inside the duct. These authors also suggested that cilia are continually replaced, possibly because of the abrasion. Therefore, the epithelium may have two important functions: partial protection of the duct wall by the apical cells and secretion by the columnar cells. The most caudal portion of the Wolffian duct is enlarged for storage of spermatozoa prior to breeding season (Williams *et al.*, 1984; Norris *et al.*, 1985; Norris, 1987; Zalisko and Larsen, 1988) and become secretory, as

---

Fig. 5. Urogenital ducts of *Ambystoma mexicanum*. A—End cephalic region of the Wolffian duct (W) and the vestigial Mullerian duct (M). The connective (c) tissue around the ducts contains melanocytes (me). H-E. Bar = 30 μm. B—Cephalic region of the Wolffian duct (W), containing spermatozoa (z) in the lumen (L) and lined by columnar epithelium (e). Connective tissue (c) around the duct. Masson's trichrome. Bar = 30 μm. C—Middle region of the Wolffian duct (W), containing spermatozoa (z) in the lumen (L) and lined by columnar epithelium (e), but shorter than those seen of the cephalic region in the Figure 5B. H-E. Bar = 30 μm. D—Detail of the Fig. 5C. Middle region of the Wolffian duct (W), containing spermatozoa (z) in the lumen and lined by columnar epithelium (e) with epithelial apical cells (ea). The connective tissue (c) contains melanocytes (me). H-E. Bar = 30 μm.

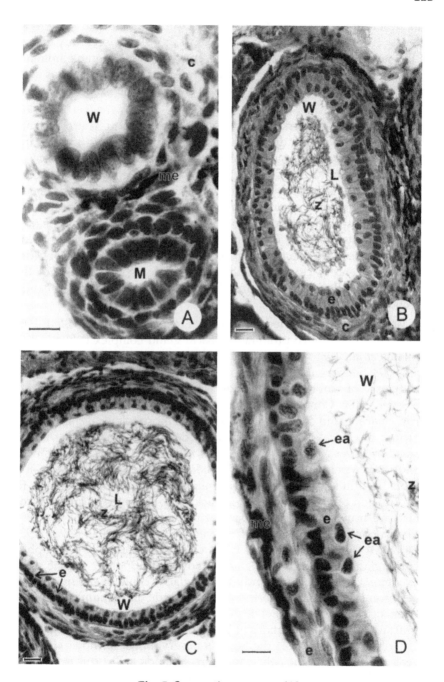

Fig. 5. See caption on page 110.

was described for *Notophtalmus viridescens, Taricha torosa, A. tigrinum* (Norris, 1987), and *A. dumerilii* (Uribe, 2001, 2003). The secretory activity of the caudal portion of Wolffian duct may provide favorable milieu to maintain stored spermatozoa until the breeding season (Williams *et al.*, 1984). The Wolffian duct opens into the cloaca on the urogenital papilla (Adams, 1940).

Rudimentary Müllerian ducts (Fig. 5A) persist in male urodeles and run parallel to Wolffian ducts. Müllerian ducts develop in females into oviducts, but in males it was suggested that they secrete part of the seminal fluid necessary for the transfer of spermatozoa to the female cloaca (Norris, 1987). Adams (1940) described Müllerian ducts in *Triturus viridescens* as thin threads extending anteriorly beyond the Wolffian ducts along the mid-dorsal line to the anterior end of body cavity. Müllerian ducts in *Triturus viridescens* (Adams, 1940), *Taricha granulosa* and *Cynops pyrrhogaster* (Baker, 1965) arise near the pectoral girdle, enter Wolffian duct surrounding stroma, and end in cloaca near the Wolffian ducts terminates. For this reason the Müllerian ducts cannot be observed externally in caudal direction of this region.

Male Urodela, except for the species of family Sirenidae (Willett, 1965; Sever *et al.*, 1996), have prominent cloacal gland complex (Willett, 1965; Williams *et al.*, 1985; Norris *et al.*, 1985; Norris, 1987; Sever, 1981, 1991, 1994, 2003; Sever *et al.*, 1996; Romo *et al.*, 1996; Uribe, 2001). The complex is composed of simple tubular glands, which open to the cloaca (for review see Sever, 2003). Concomitantly with seasonal cycle of testis and efferent ducts, cloacal gland complex shows seasonal morphological changes. The structure of cloacal glands of several urodelan species has been widely studied by Sever and his colleagues (for review, see Sever, 2003). Cloacal glands are hypertrophied during breeding season. They produce substances used in sex and species recognition and secretions necessary for the formation of spermatophores - a packages of bundles of spermatozoa, which are transferred to the cloaca of the females (Zalisko *et al.*, 1984; Williams *et al.*, 1985; Norris, 1987; Sever, 1975, 1991, 2003). Sever (1981, 1994, 2003) classified four basic types of cloacal glands: two basophilic (ventral gland and Kingsbury gland) and two acidophilic (pelvic gland and pheromone-producing gland). Ventral glands could be subdivided into anterior and posterior. Anterior ventral glands are usually the largest clusters of cloacal glands. Pelvic glands could be subdivided into dorsal, lateral and caudal. Observations of Williams *et al.* (1985), Romo *et al.* (1996), and Uribe (2001), support the Sever's observations. Romo *et al.* (1996) revealed that acidophilic pelvic and dorsal glands in *Triturus marmoratus marmoratus* secrete neutral mucins, whereas basophilic ventral and Kingsbury glands secrete acid mucins. Licht and Sever (1990) compare the cloacal glands of metamorphosed and neotenic (i.e., retaining larval features in sexually mature animals) individuals of the species *Ambystoma gracile,* concluding that no differences were evident in the cloacal anatomy between metamorphosed and neotenic animals.

Spermatogenesis in Urodela

Spermatogenesis occurs in testes under the control of neuroendocrine system, showing cyclic changes influenced by environmental factors, such as temperature, rainy season and photoperiod. In most urodele species, spermatogenesis shows a circannual reproductive cycle, closely correlated with the seasons of the year (Moore, 1975; Sever, 1975; Tso and Lofts, 1977a; Miltner and Armstrong, 1983; Lofts, 1984; Norris *et al.*, 1985; Uribe *et al.*, 1991; Fraile *et al.*, 1990; Norris, 1997; Ricote *et al.*, 2002). Testes, efferent ducts, and cloacal glands increase in size and weight during breeding season, and decrease after spermiation and thereby histological structure of these organs changes throughout the year. The renewal of germ cells has frequently been observed in spring, when proliferation of spermatogonia and early differentiation of spermatogenic stages occurs, whereas later stages of spermatogenesis predominate in summer. Spermiation occurs in autumn and winter, when spermatozoa are abundant and fill the most of the lobules. Such a pattern of spermatogenesis is common in urodeles and is characteristic of post-nuptial spermatogenesis.

According to these observations, the reproductive cycle of male urodeles may be divided into two main phases: a period of spermatogenesis during spring and summer, and a period of spermiation and regression during autumn and winter. During spermatogenesis, spermatozoa are formed by mitotic multiplication of spermatogonia followed by meiosis. During spermiation, spermatozoa are progressively released from testes to efferent ducts (Tso and Lofts, 1977a; Lazard, 1979; Lofts, 1984; Williams *et al.*, 1984; Norris *et al.*, 1985; Schindelmeiser *et al.*, 1985; Armstrong, 1989; Fraile *et al.*, 1989a; Bergmann, 1994; Norris, 1997).

Another classification of male reproductive cycle was recently provided by Ricote *et al.* (2002), who considered that testicular cycle of *Triturus marmoratus marmoratus* comprises three periods, according to seasonal changes: (a) proliferative period, characterized by proliferation of spermatogonia until the formation of round spermatids (April-June), (b) spermiogenic period, in which round spermatids transform into spermatozoa (July-September), and (c) regression period, when Sertoli cells and interstitial Leydig cells become glandular (October-April). The glandular tissue degenerates at the beginning of the spermiogenesis. Analyzing the annual cycle of *Triturus marmoratus marmoratus*, Ricote *et al.* (2002) suggested that products of genes controlling cell cycle (*p53, p21, Rb* and *phosph-Rb*) seem to be involved in control of various periods of testicular cycle. During the proliferation period, the expression of *p53* and *p21* decreases, and that of *phosph-Rb* increases and enhances cell proliferation, in contrast to spermiation and regression period when increase in *p53* expression activates *p21* expression, which inhibits *Rb* phosphorylation arresting cell cycle in G1. *Phosph-Rb* induces the entry of cell in S phase of cell-cycle.

Morphology of Germ Cells

Spermatogenesis in urodeles was studied by several authors who described all stages of germ cell maturation in a variety of species: Kingsbury (1901) in *Desmognathus fuscus,* Humphrey (1921); Pudney *et al.* (1983), and Pudney (1995) in *Necturus maculosus;* Carrick (1934), Moore (1975), and Norris *et al.* (1985) in *Ambystoma tigrinum;* Tso and Lofts (1977a) in *Tritutoides hongkongensis;* Miltner and Armstrong (1983), and Armstrong (1989) in *A. mexicanum;* Brizzi *et al.* (1985) in *Salamandrina terdigitata;* Fraile *et al.* (1989a), and Ricote *et al.* (2002) in *Triturus marmoratus;* Bergman *et al.* (1982), Schindelmeisner *et al.* (1983, 1985), Bergman (1994) in *Salamandra salamandra;* Uribe *et al.* (1994) in *A. dumerilii.* Because their observations are consistent with those made for *Ambystoma mexicanum* and *A. dumerilii* (Uribe, 2003), the following description of germ cells morphology (Figs. 6 and 7) will concern the latter species.

Primary Spermatogonia

Primary spermatogonia (Figs. 1D, 6A) are the biggest germ cells (40-50 μm in diameter). They are localized in the cephalic region of a testis, or in connective tissue bridges between lobes in multilobular testis. Primary spermatogonia proliferate mitotically (Fig. 6A), thereby renewing the pool

---

Fig. 6. Spermatogenesis in *Ambystoma dumerilii.* A—Cephalic region of the testis*i.* Primary spermatogonia (Sg1) with spheric nucleus and lightly stained cytoplasm, surrounded by connective tissue (c). Cluster of secondary spermatogonia (Sg2) with spheric nucleus, the nuclei of Sertoli cells (S) are irregular in shape and situated around he spermatogonia. Central lumen (L) of the cluster. One spermatogonia shows mitotic process during metaphase (m). H-E. Bar = 30 μm. B—Periphery of the testis. Tunica albuginea (Ta) with blood vessels (v). Lobules containing cysts of primary spermatocytes in meiosis during zygotene stage (Scz) pachytene stage (Scp). H-E. Bar = 30 μm. C—Primary spermatocytes during pachytene stage (Scp). Note the thick fibrillar chromatine. Scarce connective tissue (c) of the interlobular region. Masson's trichrome. Bar = 30 μm. D—Primary spermatocytes during diplotene stage (Scd) with pairs of chromosomes showing chiasms. H-E. Bar = 10 μm.

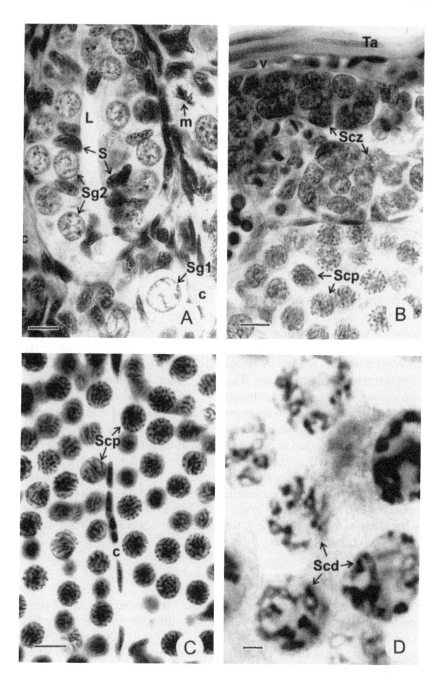

Fig. 6. See caption on page 114.

of germ cells and giving rise to secondary spermatogonia. They are spherical cells with light granular cytoplasm and irregular or spherical nuclei containing granular chromatin and one or two nucleoli. Primary spermatogonia are single cells or are arranged in small groups. They are surrounded by a Sertoli (follicle) cell, which forms a cyst wall at later stages of spermatogenesis.

Secondary Spermatogonia

Secondary spermatogonia attain 35-45 µm in diameter and contain spherical nuclei similar to that observed in primary spermatogonia (Figs. 1D, 6A). They form clusters of cells adjoining processes of the Sertoli cell. A central lumen appears in the clusters. Secondary spermatogonia enter meiosis and transform into primary spermatocytes.

Primary and Secondary Spermatocytes

Primary spermatocytes are spherical cells, similar in size to secondary spermatogonia (35-45 µm in diameter). These cells contain duplicated chromosomes at different stages of meiotic prophase I: leptotene (fine reticular chromatin), zygotene (fine fibrillar pattern of duplicated homologous chromosomes in synapsis) (Fig. 6B), pachytene (duplicated chromosomes in crossing-over) (Fig. 6B,C), and diplotene (separation of homologous duplicated chromosomes, with chiasms) (Fig. 6D). The primary spermatocytes at pachytene stage are relatively abundant, whereas those in leptotene, zygotene, and especially in diplotene are rarely seen. This reflects the duration of meiotic stages, among which pachytene is the longest, leptotene and zygotene are shorter, and diplotene is the shortest. The primary spermatocytes enter metaphase I, anaphase I, and telophase I, resulting in two secondary spermatocytes. Secondary spermatocytes are spherical cells and have in average 18-20 µm in diameter. They are seen less frequent, since they divide mitotically after short interphase, rapidly giving rise to two spermatids.

Spermatids and Spermatozoa

Early spermatids are haploid cells spherical in shape, and 14-17 µm in diameter. They transform into spermatozoa throughout spermiogenesis, and become progressively elongated cells (Fig. 7A,B). As spermiogenesis proceeds, the nuclei of spermatids become smaller and the chromatin shows increasing degree of condensation.

Spermatozoa are arranged in a swirl inside a cyst, with their heads oriented in the same direction (Fig. 7C,D). The total length of spermatozoa of urodeles is usually longer than those of other amphibians, and other vertebrates (for review, see Scheltinga and Jamieson, 2003). The shortest spermatozoa were reported for *Hynobius nebulosus* with a length of 156 μm, whereas the longest (about 1000 μm) was observed in *Necturus maculosus*. The lengths of spermatozoa were also measured in *Ambystoma mexicanum* (444 μm) and *A. dumerilii* (451 μm) (Brandon *et al.*, 1974). The lengths of spermatozoa were reported in *E. bislineata* (459 m), and *E. lucifuga* (523 μm), *Desmognathus wrighti* (504 μm), *D. aeneus* (388 μm); *Plethodon dorsalis* (535 m), *P. cinereus* (507 μm), and *P. dunni* (626 μm) (Wortham *et al.*, 1977; Scheltinga and Jamieson, 2003). The biological significance of differences in the lengths of spermatozoa is unknown.

The morphology of urodelan spermatozoa, with the exception of sirens, is relatively uniform.

A wide review of the structure of mature spermatozoa of urodeles was recently presented by Scheltinga and Jamieson (2003). The spermatozoa consist of an elongated large head (Fig. 7C,D). The head of the spermatozoa contains the acrosome and nucleus, connecting piece or neck, and tail containing axoneme, undulating membrane and axial fiber (Wortham *et al.*, 1977, 1982; Pudney, 1995). The acrosomal complex, situated in the apical part of the spermatozoa, can have various shapes: a terminal sharp point, as in *Necturus viridescens*, or a terminal knob as in *Pleurodeles waltlii* (Fawcett, 1970). The acrosomal complex is composed of acrosome vesicle and *perforatorium*. The acrosome vesicle is composed of homogenous electron-dense material of relatively constant diameter, between 4.6 and 32 μm. The *perforatorium* is situated in subacrosome position. It is an electron-dense arrow-shaped structure with the apical axial part and the distal conical part. The axial part of the *perforatorium* is surrounded by the acrosomal vesicle in its apical end and extends caudally into the endonuclear canal, i.e., a tubular invagination of the nuclear envelope in the anterior part of the nucleus. The perforatorium can extend to one third of the nucleus length, with an exception of plethodontids which have no endonuclear canal. The conical part of the *perforatorium* unsheathes the anterior part (*rostrum*) of the nucleus. The acrosome vesicle and *perforatorium* contain proteolytic enzymes enabling the passage of spermatozoa during fertilization across the jelly layers and vitelline envelope which surround the egg. The acrosome usually has an extension called the acrosomal barb, seen as a lateral projection which curves, giving appearance of a harpoon (Fawcett, 1970; Wortham *et al.*, 1982). The barb is observed in ambystomatids, plethodontids, rhyacotritonids, and salamandrids, but is absent in amphiumids, cryptobranchids, dicamptodontids, and hynobiids. The functional significance of the barb is unknown (Scheltinga and Jamieson, 2003), but Wortham *et al.* (1982) suggested that the harpoon morphology could be efficient in egg activation.

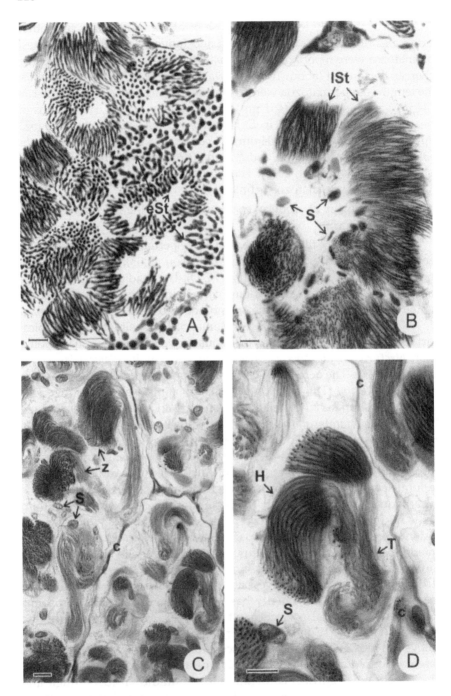

Fig. 7. See caption on the next page.

The nucleus has an elongated shape with the narrower anterior part (*rostrum*). Postreriorly, the nucleus has a fossa, or, as in plethodontids and amphiumids, is flat. The fossa is a concavity that contains the connecting piece, which is a long cylinder that fits into the fossa. The connecting piece is composed of the basic proteins and contains two centrioles. The proximal centriole lies close to the nucleus and the distal one is a part of the basal body of the axonema. The basal body is formed by a dense ring around the distal centriole (Wortham *et al.*, 1977; Uribe *et al.*, 1994; Pudney, 1995). The posterior end of the connecting piece is continuous with the axial fiber which is the principal supporting structure of the tail. The axial fiber is parallel to the axoneme and the undulating membrane (Fawcett, 1970; Brandon *et al.*, 1974; Scheltinga and Jamieson, 2003).

A cytoplasmic droplets of diverse shapes, attached anywhere from the head to the connecting piece, are frequently observed in urodele spermatozoa. These droplets contain some vesicles, dense material and mitochondria. Wortham *et al.* (1977) examined the morphology of the spermatozoa of 28 species of plethodontid salamanders and observed that the cytoplasmic droplet was present in all of them. These authors suggest that the droplet may play some roles during the prolonged storage of spermatozoa in the female's spermatheca. The surface topography of the cytoplasmic droplet varies and probably reflects different physiological stages. Wortham *et al.* (1982) observed rough surface of the droplets in ambystomatid spermatozoa and smooth in plethodontid spermatozoa.

The tail is connected to the caudal end of the connecting piece and consists of a supporting axial fiber attached to the lateral axoneme by the undulating membrane (Brandon *et al.*, 1974). In the anterior-posterior direction, the tail is composed of three regions: midpiece, principal piece, and endpiece or terminal piece. The midpiece contains mitochondria, axoneme, undulating membrane and axial fiber. The principal piece contains the same structures with the exception of the mitochondria, the limit of the axial fiber and the undulating membrane is in the caudal end

---

Fig. 7. Spermiogenesis in *Ambystoma dumerilii*. A—Lobules containing several cysts with early spermatids (eSt). Note the synchrony of development of spermatids within each cyst, and the progressive elongation of the spermatids. H-E. Bar = 30 µm. B—Lobules containing several cysts with late spermatids (lSt). Sertoli cell's nuclei (S) are seen around the cysts. H-E. Bar = 30 µm. C—Lobules containing several cysts with spermatozoa (z). Sertoli cell's nuclei (S) are seen around the cysts. Scarce connective tissue (c) of the interlobular region. H-E. Bar = 30 µm. D—Detail of the Figure 7C. Cysts with spermatozoa showing the head (H) and the tail (T). Sertoli cell's nuclei (S). Scarce connective tissue (c) of the interlobular region. H-E. Bar = 30 µm.

of the principal piece. Then, the short end piece contains only the axoneme (Scheltinga and Jamieson, 2003).

Data on sperm morphology are essential for the studies on the biology of reproduction and may contribute to clarification of evolutionary and taxonomic relationship among populations in a variety of species, as it is suggested by Brandon *et al.* (1974), Wortham *et al.* (1977), Sheltinga and Jamieson (2003).

The spermatophore is a structure characteristic of most male urodeles, with the exception of species belonging to the families Hynobiidae and Cryptobranchidae, which have external fertilization, and some species of the family Sirenidae, where spermatozoa are discharge without formation of spermatophores (Lofts, 1984). Spermatophores are formed inside the cloaca. They are composed of a package of spermatozoa released from the Wolffian duct, and a gelatinous capsule formed by the secretions of the cloacal gland complex (Lofts, 1984). The spermatophores are formed during the courtship, deposited by a male, and picked up by cloacal labia of a female. The spermatozoa are stored in the roof of the female cloaca until ovulation. Russell *et al.* (1981) and Zalisko *et al.* (1984) described the structure and histochemistry of spermatophores of several species of the families Ambystomatidae, Salamandridae, and Plethodontidae. Spermatophores consist of a gelatinous base and a stalk surmounted by a sperm-containing gelatinous cap. In *Necturus*, the spermatophores are deposited to a female directly by cloacal apposition, and the stalk may be absent (Lofts, 1984). At the junction of the stalk and a cap, a filamentous material was described, apparently serving as an adhesive which maintains together the cap and the stalk (Russell *et al.*, 1981).

Acknowledgements

I thank Gerardo Gómez-Ríos for processing some histological specimens, and valuable discussions during the early progress of this work; Marcela Esperanza Aguilar Morales who assisted with tissue processing (alcian blue and PAS techniques) with excellent results; José Antonio Hernández Gómez who kindly assisted with the digital preparation of figures. Many thanks to Maria Ogielska for critical reading and valuable comments that significantly improved this manuscript.

References

Adams EA. 1940. Sexual conditions in *Triturus viridescens*. III The reproductive cycle of the adult aquatic form of both sexes. Amer. J. Anat. 66: 235-276.

Armstrong JB. 1989. Spermatogenesis, pp 36-41. In: Developmental Biology of the Axolotl, (Eds) J.B. Armstrong and G.M. Malacinski Columbia University Press, New York.

Baker CL. 1965. The male urogenital system of the Salamandridae. J. Tennessee Acad. Sci. 40: 1-5.

Baker CL, WW Jr Taylor. 1964. The urogenital system of the male *Ambystoma*. J. Tennessee Acad. Sci. 39: 1-10.

Bergmann M. 1994. The morphology of the testis in *Salamandra salamandra* (L.), In: Mertensiella. Supplement zu Salamandra, (Eds) H Greven and B Thiesmeier. Proc. Symp. Biology of *Salamandra* and *Mertensiella* 4. Bonn, pp. 75-82.

Bergmann M, J Schindelmeiser, H Greven. 1982. The zone of mature spermatozoa in the testis of *Salamandra salamandra* (L.) (Amphibia, Urodela). Z. Mikrosk. Anat. Forsch. 96: 221-234.

Bergmann M, J Schindelmeiser, H Greven. 1983. The glandular tissue in the testis of *Salamandra salamandra* (L.) (Amphibia, Urodela). Acta Zoologica. 64: 123-130.

Branca A. 1904. Les premier stades de la formation du spermatozoide chez l'axolotl. Arch, Zool. Exp. Gen. 4 Ser: 105-113.

Brandon RA, J Martan, JWE Wortham, DCC Englert. 1974. The influence of interspecific hybridization on the morphology of the spermatozoa of *Ambystoma* (Caudata, Ambystomatidae). J. Reprod. Fert. 41: 275-284.

Brizzi R, C Calloni, S Vanni. 1985. Spermatogenetic cycle in *Salamandrina terdigitata* (Lacepede, 1788) (Amphibia: Salamandridae). Z. Mikrosk.-anat. Forsh, Leipzig. 99: 271-292.

Carrick R. 1934. The spermatogenesis of the axolotl (*Ambystoma tigrinum*). Trans. Royal Society, Edinburgh. 58: 63-76.

Duellman WE, L Trueb. 1986. Biology of Amphibians. McGraw-Hill. New York, pp. 670.

Fawcett DW. 1970. A comparative view of sperm ultrastructure. Biol. Reprod. Suppl. 2: 90-127.

Fraile B, R Paniagua, MC Rodríguez, FJ Saez. 1989a. Effect of photoperiod and temperature on spermiogenesis in the marbled newt, *Triturus marmoratus marmoratus*. Copeia 1989: 357-363.

Fraile B, R Paniagua, MC Rodríguez, FJ Saez A Jiménez. 1989b. Annual changes in the number, testosterone content and ultrastructure of glandular tissue cells of the testis in the marbled newt, *Triturus marmoratus*. J. Anat. 167: 85-94.

Fraile B, R Paniagua, FJ Saez, R Paniagua. 1990. The cycle of follicular and interstitial cells (Leydig cells) in the testis of the marbled newt, *Triturus marmoratus*. J. Morphol. 204: 89-101.

Franchi E, M Camatini, I de Curtis. 1982. Morphological evidence of a permeability barrier in urodele testis. J. Ultrastructure Research. 80: 253-263.

Grier HJ. 1993. Comparative organization of Sertoli cells including the Sertoli cell barrier, In: The Sertoli Cell, (Eds) LD Russell and MD Griswold. Cache River Press, Clearwater, Florida, pp. 704-739.

Humphrey RR. 1921. The interstitial cells of the urodele testis. Am. J. Anat. 29: 213-279.

Humphrey RR. 1922. The multiple testes in urodeles. Biol. Bull. 43: 45-67.

Imai K, S Tanaka. 1978. Histochemical and electron microscopic observations on the steroid hormone-secreting cells in the testis of the Japanese red-bellied newt, *Cynops pyrrhogaster pyrrhogaster*. Develop. Growth and Differ. Vol. 20 No. 2: 151-167.

Kingsbury BF. 1901. The spermatogenesis of *Desmognathus fusca*. Am. J. Anat. 1: 99-135.

Lazard L. 1976. Spermatogenesis and 3β-HSDH activity in the testis of the axolotl. Nature (London) 264: 796-797.

Lazard L. 1979. Steroidogenesis in axolotl testis. Histochemistry of two major enzymes related to cell type, spermatogenesis, and substrate. Gen. Comp. Endocrinol. 39: 381-387.

Lecouteux A, DH Garnier, T Bassez, J Joly. 1985. Seasonal variations of androgens, estrogens, and progesterone in the different lobules of the testis and in the plasma of *Salamandra salamandra*. Gen. Comp. Endocrinol. 58: 211-221.

Licht LE, Sever DM. 1990. Cloacal anatomy of metamorphosed and neotenic salamanders. Can. J. Zool. 69: 2230-2233.

Lofts, B. 1984. Amphibians. In: Marshall's Physiology of Reproduction, (Ed.) GE Lamming. Vol. 1. Reproductive Cycles of Vertebrates. Churchill Livingstone, Edinburgh, pp. 127-205.

Lofts B. 1987. Testicular function. In: Hormones and Reproduction in Fishes, Amphibians and Reptiles (Eds) DO Norris, RE Jones Plenum Press, New York, pp. 288-298.

Miltner MJ, JB Armstrong. 1983. Spermatogenesis in the Mexican axolotl, *Ambystoma mexicanum*. J. Exp. Zool. 227: 255-263.

Moore FL. 1975. Spermatogenesis in larval *Ambystoma tigrinum:* Positive and negative interactions of FSH and testosterone. Gen. Comp. Endocrinol. 26: 525-533.

Moore FL, CH Muller, JL Specker. 1979. Origin and regulation of plasma dihydrotestosterone and testosterone in the rough-skinned newt, *Taricha granulosa*. Gen. Comp. Endocrinol. 38: 451-456.

Norris DO. 1987. Regulation of male gonaducts and sex accesory structures, In: Hormones and Reproduction in Fishes, Amphibians and Reptiles, (Eds) DO Norris, RE Jones, Plenum Press, New York, pp. 327-354.

Norris DO. 1997. Vertebrate Endocrinology, Academic Press, New York.

Norris DO, MF Norman, MK Pankak, D Duvall. 1985. Seasonal variation in spermatogenesis, testicular weights, vasa deferentia and androgen levels in neotenic tiger salamander, *Ambystoma tigrinum*. Gen. Comp. Endocrinol. 60: 51-57.

Pudney J. 1995. Spermatogenesis in nonmammalian vertebrates. Microsc. Res. Tech. 32: 459-497.

Pudney J, GV Callard. 1984. Organization of interstitial tissue in the testis of the salamander *Necturus maculosus* (Caudata: Proteidae). J. Morphol. 181: 87-95.

Pudney J, JA Canick, P Mak, GV Callard. 1983. The differentiation of Leydig cells, steroidogenesis, and the spermatogenetic wave in the testis of *Necturus maculosus*. Gen. Comp. Endocrinol. 50: 43-66.

Ricote M, JM Alfaro, I Garcia-Tuñon, MI Arenas, B Fraile, R Paniagua, M Royuela. 2002. Control of the annual testicular cycle of the marbled-newt by p53, p21, and Rb gene products. Mol. Reprod. Develop. 63: 201-209.

Romo E, MP De Miguel, MI Arenas, L Frago, B Fraile, R Paniagua. 1996. Histochemical and quantitative study of the cloacal glands of *Triturus marmoratus marmoratus* (Amphibia: Salamandridae). J. Zool., London. 239: 177-186.

Russell LD, RA Brandon, EJ Zalisko, J Martan. 1981. Spermatophores of the salamander *Ambystoma texanum*. Tissue and Cell. 13(3): 609-621.

Scheltinga DM, BGM Jamieson. 2003. The mature spermatozoa, In: Reproductive Biology and Phylogeny of Urodela, (volume Ed.) D Sever and BGM Jamieson (Series Ed.). Science Publishers, Enfield (NH), USA, pp. 203-274.

Schindelmeiser J, H Greven, M Bergmann. 1983. The immature part of the testis in *Salamandra salamandra*, (L.) (Amphibia, Urodela). Arch. Histol. Jap. 46(2): 159-172.

Schindelmeiser J, M Bergmann, H Greven. 1985. Cellular differentiation in the urodele testis, In: Functional Morphology in Vertebrates (Eds) HR Duncker, G Fleischer, Proceedings of the 1st International Symposium on Vertebrate Morphology Giessen, 1983 Fortschritte der Zoologie, Band 30. Gustav Fischer Verlag, Stuttgart, New York, pp. 445-447.

Sever DM. 1975. Morphology and seasonal variation of the mental hedonic glands of the dwarf salamander (*Eurycea quadridigitata*), (Holbrook). Herpetologica 31: 241-251.

Sever DM. 1981. Cloacal anatomy of male salamanders in the families Ambystomatidae, Salamandridae and Plethodontidae. Herpetologica 37: 142-155.

Sever DM. 1991. Comparative anatomy and phylogeny of the cloacae of salamanders (Amphibia: Caudata). I. Evolution at the family level. Herpetologica 47(2): 165-193.

Sever DM. 1994. Comparative anatomy and phylogeny of the cloacae of salamanders (Amphibia: Caudata). VII. Plethodontidae, Herpetological Monographs No. 8. The Herpetologists' League, Inc., pp. 276-337.

Sever DM. 2003. Courtship and mating glands, In: Reproductive Biology and Phylogeny of Urodela, D Sever (volume Ed.), and BGM Jamieson (Series Ed.). Science Publishers, Enfield (NH), USA, pp. 323-381.

Sever DM, LC Rania, JD Krenz. 1996. Reproduction of the salamander *Siren intermedia* Le Conte with especial reference to oviducal anatomy and evidence for sperm storage and internal fertilization. J. Morphol. 227: 335-348.

Tanaka S, H Iwasawa, K Imai. 1980. Formation of germ cell-like cells in the peritoneal epithelium of the testes of estrogen-treated adult males of the Japanese red-bellied newt, *Cynops pyrrhogaster pyrrhogaster*. Develop. Growth and Differ. 22(4): 611-626.

Trauth SE. 1983. Reproductive biology and spermathecal anatomy of the dwarf salamander (*Eurycea quadridigitata*), in Alabama. Herpetologica. 39: 9-15.

Tso ECF, B Lofts. 1977a. Seasonal changes in the newt, *Trituroides hongkongenesis*

124

testis. I. A histological and histochemical study. Acta Zoologica, Stockholm. 58: 1-8.

Tso ECF, B Lofts. 1977b. Seasonal changes in the newt, *Trituroides hongkongenesis* testis. II. An ultrastructural study on the lobule boundary cell. Acta Zoologica, Stockholm. 58: 9-15.

Uribe MCA. 2001. Reproductive systems of caudata amphibians, In: Vertabrate Functional Morphology. Horizon of Research in the 21st Century, (Eds) HM Dutta, JS Datta Munshi, Science Publishers, Enfield, USA, pp. 267-293.

Uribe MCA. 2003. The testes, spermatogenesis and male reproductive ducts, In: Reproductive Biology and Phylogeny of Urodela, (volume ed.) D Sever and (Series ed.) BGM Jamieson. Science Publishers, Enfield (NH), USA, pp. 183-202.

Uribe MCA, G GómezRíos, RA Brandon. 1994. Spermatogenesis in the urodele *Ambystoma dumerilii*. J. Morphol. 222: 287-299.

Uribe MCA, G Gómez Ríos, C López Arriaga. 1991. Cambios morfológicos del testículo de *Ambystoma dumerilii* durante un ciclo anual. Bol. Soc. Herpetol. Mex. 3: 13-18.

Willet JA. 1965. The male urogenital system in the Sirenidae. J. Tennessee Acad. Sci. 40: 9-17.

Williams AA, RA Brandon, J Martan. 1984. Male genital ducts in the salamanders *Eurycea lucifuga* and *Eurycea longicauda*. Herpetologica 40(3): 322-330.

Williams AA, J Martan, RA Brandon. 1985. Male cloacal gland complex of *Eurycea lucifuga* and *Eurycea longicauda* (Amphibia: Plethodontidae). Herpetologica 41: 272-281.

Wortham JWE Jr, RA Brandon, J Martan. 1977. Comparative morphology of some Plethodontid salamander spermatozoa. Copeia 1977(4): 666-680.

Wortham JWE Jr, JA Murphy, J Martan, RA Brandon. 1982. Scanning electron microscopy of some salamander spermatozoa. Copeia 1982(1): 52-60.

Zalisko EJ, RA Brandon, J Martan. 1984. Microstructure and histochemistry of salamander spermatophores (Ambystomatidae, Salamandridae and Plethodontidae). Copeia 1984(3): 739-747.

Zalisko EJ, JH Larsen Jr. 1988. Ultrastructure and histochemistry of the vas deferens of the salamander *Rhyacotriton olympicus:* Adaptations for sperm storage. Scanning Microscopy 2: 1089-1095.

4

Spermatogenesis and Male Reproductive System in Amphibia—Gymnophiona

Jean-Marie Exbrayat

Introduction

The anatomy of male reproductive system in Gymnophiona has been described in the following species: *Typhlonectes compressicauda* (Exbrayat and Sentis, 1982; Exbrayat and Delsol, 1985; Exbrayat, 1985, 1986a,b, 1993, 2000; Exbrayat *et al.*, 1986; Exbrayat and Dansard, 1992, 1994; Anjubault and Exbrayat, 1998; Pujol and Exbrayat, 2000); *Uraeotyphlus menoni* (Chatterjee, 1936); *Ichthyophis glutinosus* (Spengel, 1876; Sarasin and Sarasin, 1887-1890; Semon, 1892; Seshachar, 1936, 1937, 1939, 1942a,b,c, 1943, 1945); *Hypogeophis rostratus* (Tonutti, 1931); *Ichthyophis beddomei* (Bhatta, 1987). The following species were studied more recently by Wake (1968a,b, 1970a,b, 1972, 1981, 1995): *Schistometopum gregorii, Herpele squalostoma, Boulengerula boulengeri, Uraeotyphlus oxyurus, Geotrypetes seraphinii, Siphonops brasiliensis, Chthonerpeton viviparum, Scolecomorphus kirkii, Scolecomorphus uluguruensis, Scolecomorphus vittatus, Caecilia (Oscaecilia) ochrocephala, Idiocranium russeli, Dermophis mexicanus*. In very recent works, Smita *et al.* (2003, 2004a,b, 2006) using light and electron microscopy, give a deep study of Sertoli and germ cells in two asiatic species, *Ichthyophis tricolor* and *Uraeotyphlus narayani*. George *et al.* (2004), Akbarsha *et al.* (2006) give the histological and ultrastructural study of Müllerian ducts in *Uraeotyphlus narayani*. In a recent paper Measey *et al.* (2008) described a structure of testes and spermatogenic activity in *Boulengerula taitanus*.

The testes are paired elongated organs situated in the posterior part of the body, parallel to gut and kidneys (Fig. 1). They are composed of 1-22 lobes, which size and number vary among species (Wake, 1968b). In some species of the genus *Scolecomorphus* lobes can fuse, whereas in others their

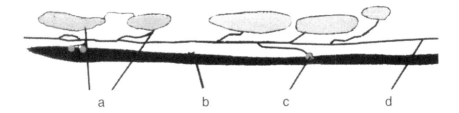

Fig. 1. Schematic representation of the structure of testis lobes in *Typhlonectes compressicauda*. a—lobes; b—kidneys; c—glomerula; d—*rete testis*. After Exbrayat 1986a, modified.

number is originally low. In some individuals of *Dermophis mexicanus* testis is composed of a single elongated lobe with constrictions (Wake, 1968b). Each lobe is composed of several ovoid or spherical lobules (or locules, see Spengel, 1876, and Seshachar, 1936), which are equivalent to seminiferous tubules of Anura and other vertebrates.

The Müllerian ducts persist in males as paired structures observed in the third posterior part of the body. They are functional glands (Wake, 1981; Exbrayat, 1985) producing fluid component of semen. Kidney ducts and Wolffian ducts evacuate the sperm to cloaca. An erective phallodeum, which is extruded during internal fertilization, is part of the male cloaca (Tonutti, 1931; Wake, 1972; Exbrayat, 1991, 1996). A pair of fat bodies is parallel to the testes (Sarasin and Sarasin, 1887-1890; Taylor, 1968; Wake, 1968a,b; Exbrayat, 1988).

Structure of the Testis in Gymnophiona

Seminiferous Lobules (Seminiferous Tubules)

In *Typhlonectes compressicauda*, each lobule is 1 to 2 mm³ in volume (Exbrayat, 1986a; Exbrayat and Estabel, 2006). Connective and interstitial tissues separate lobules. Each lobule is connected to longitudinal duct by a single sperm-collecting canal, being a part of *rete testis* (Fig. 1). In each lobule germ cells are enclosed in cysts (Fig. 2A,B). Number and stages of germ cells per lobule is the same in all lobes of the same testis in a given period of spermatogenic cycle (Exbrayat, 1986a,b; Pujol and Exbrayat, 2000).

Sertoli Cells and Cyst Formation

Each lobule contains cysts composed of big Sertoli cells enclosing germ cells. In earlier studies, lobules of Gymnophiona were described as units

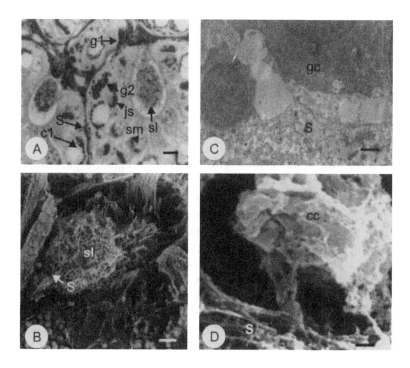

Fig. 2. Sertoli cells in testes of. *Typhlonectes compressicauda* (after Exbrayat 1986a). A—Cross section of a testis (Januarybreeding period). c1—primary spermatocytes; g1—primary spermatogonium; g2—secondary sperma-togonium; sm—mature spermatozoa; sl—free spermatozoa; js—young spermatozoa; S—Sertoli cell. Scale bar = 90 μm. B—SEM view of a testis (January, breeding period). S—Sertoli cell; sl—free spermatozoa. Scale bar = 20 μm. C—TEM view of a testis showing the contact between germ cell and Sertoli cells. gc—germ cell (primary spermatocyte); S—Sertoli cell. Scale bar = 1.5 μm. D—SEM of a Sertoli cell. cc—body of the cell containing the nucleus; S—Sertoli cell. Scale bar = 1 μm.

containing filamentous "matrix" with fat droplets, in which isogenic groups of germ cells were floating (Sarasin and Sarasin, 1887-1890; Seshachar, 1936, 1942a; Wake, 1968b; de Sa and Berois, 1986). Isolated ovoid bodies, described as Sertoli cells by Seshachar (1942a,b), were observed in the matrix close to a lobule wall, whereas round bodies, intensively stained with nuclear dyes, were scattered in the matrix. Another kind of round cells was observed among spermatids and spermatozoa. These cells were described as hypertrophied degenerated spermatocytes or spermatogonia (Seshachar, 1936). Recent studies with transmission (TEM) and scanning (SEM) microscopes (Exbrayat, 1986a,b, 2000; Exbrayat and Dansard, 1994; Exbrayat and Estabel, 2006; Smita *et al.*, 2006) resulted in new interpretation of lobule composition. Structures described by Seshachar (1936) as degenerated germ cells are in fact Sertoli cells, whereas round bodies stained with nuclear dyes are nuclei of the Sertoli cells. Sertoli cells are big with spongy cytoplasm, which resembles filamentous matrix, as analyzed in light microscope after fixation in Bouin's fluid or formaldehyde. After fixation in iso-osmotic buffered glutaraldehyde and paraformaldehyde solutions, the cytoplasm is homogenous, as was shown for *Ichthyophis beddomei* (Bhatta *et al.*, 2001). Sertoli cells are polarized, with a bulk of cytoplasm around nuclei (Fig. 2D). "Matrix" observed at the TEM level is in fact the cytoplasm of Sertoli cells with regular organelles, large vacuoles and fat droplets (Fig. 2C).

Sertoli cells differentiate from follicle cells (Lofts, 1974) surrounding primary spermatogonia. When spermatogonia multiply, follicle cells increase in size and their nuclei become lobulated. Differentiated Sertoli cells engulf spermatogonia which multiply, giving rise to a group of secondary spermatogonia. Cysts containing secondary spermatogonia detach from a lobule wall to the center of a lobule, being pushed by new generation of forming cysts. At these times, the nuclei of Sertoli cells decrease in size and become spherical. Their cytoplasm becomes filled with fat droplets. When a cyst opens, Sertoli cells start to degenerate, their size decreases and they finally reach sperm-collecting duct, where they can be recognized among spermatozoa (Fig. 2A,B). This probably explains the presence of some round isolated cells among groups of spermatids or spermatozoa, as was originally described by Seshachar (1936), although Smita *et al.* (2003, 2004a,b, 2006) give another interpretation of the cells, suggesting that they are amoeboid cells.

Smita *et al.* (2004a,b, 2006) give a TEM study of Sertoli cells in *Ichthyophis tricolor* and *Uraeotyphlus narayani*. Their interpretation resembles that given by Exbrayat and Dansard (1994). The results provided by the contemporary studies can definitively give the real structure of gymnophionan testis.

During the breeding cycle of adult *Typhlonectes compressicauda* cytoplasm of Sertoli cells displays histochemical variations. During breeding period (from January to May-June), the cytoplasm is acidic glycoproteinous. At the end of breeding period, it looses acidic character, and during last phase of spermatogenesis (from July to October), the Sertoli cells become again acidic.

Interstitial Tissue

Connective tissue, which separates lobules inside a lobe, contains blood vessels, nerves and Leydig cells. Size variations affect this tissue according to the phase of sexual cycle. At time when primary spermatogonia are the dominant class of germ cells in testes (December and January) interstitial tissue is well developed, as was shown in *Gymnopis multiplicata* (Wake, 1968b). The amount of interstitial tissue decreases during active spermatogenesis.

Interstitial tissue was studied in details in *Typhlonectes compressicauda* (Anjubault and Exbrayat, 1998, 2000b, 2004b, 2006). In this species, interstitial tissue increases immediately before breeding period (from November to February) when it reaches its maximal volume (Fig. 3A). Leydig cells are big (25 µm in diameter) and polygonal with centrally situated nuclei. During breeding period (February to April), Leydig cells reach their maximal size, display δ5 3β hydroxysteroid dehydrogenase (δ5 3β HSDH) activity, and a substantial presence of testosterone (Anjubault and Exbrayat, 1998, 2004b). After breeding season (May), Leydig cells decrease in size and during active spermatogenesis (June), they are flat with a small nucleus and a small amount of cytoplasm poor in lipids (Fig. 3B). At this period, testosterone is also cytochemically detected in follicle (Sertoli) cells surrounding spermatogonia. In November, the labeling is weak. The interstitial tissue decreases in volume until October.

Differentiation and Development of Testis

Development of gonads in Gymnophiona has been known for some species for more than one hundred years (Spengel, 1876; Sarasin and Sarasin, 1887-1890; Brauer, 1902; Tonutti, 1931; Marcus, 1939). Seshachar (1936) provided detailed description of testis differentiation for *Ichthyophis glutinosus*. More recently, general data on development of gonads in Gymnophiona have been published by Wake (1968b), and studied in details in *Typhlonectes compressicauda* by Anjubault and Exbrayat (2000a, 2004a, 2006).

At birth (stage 34) gonads are sexually differentiated. At that time the primordia of lobules are formed. They are composed of groups of cells situated at the end of the *rete testis*. Each lobule contains several primary spermatogonia, some of which being enveloped by somatic follicle (future Sertoli) cells (Lofts, 1974) (Fig. 4A). In several individuals follicles composed of cells resembling Sertoli cells of adults envelop secondary spermatogonia (Fig. 4B). Interstitial tissue is scarce and Leydig cells are absent.

In animals 12- to 15-month old (Fig. 4C), the lobules increase in size. Interstitial tissue appears and Leydig cells differentiate. More secondary spermatogonia appear and follicle cells differentiate into Sertoli cells, which become higher and their nuclei move to their basal parts, close to a lobule wall. At light microscopy level, their cytoplasm is filled with filamentous structures and fat droplets can be visualized on frozen sections. During the first year after birth (July to July), the germ cells present in testes are

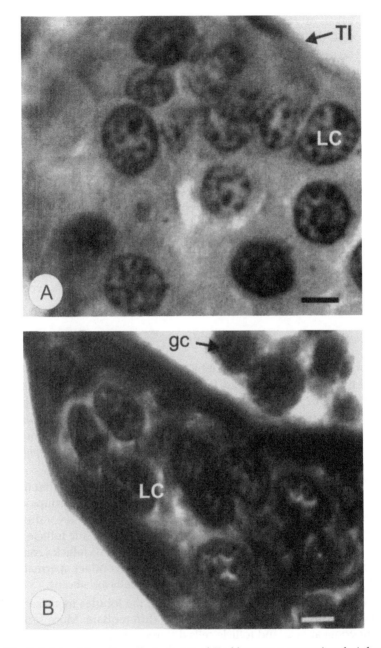

Fig. 3. Inlets of Leydig cells in testes of *Typhlonectes compressicauda* (after Exbrayat, 1986a). A—Before breeding (April); B—During period of sexual quiescence (June). gc—germ cells; TI—interstitial tissue; LC—Leydig cells. Scale bar = 10 µm.

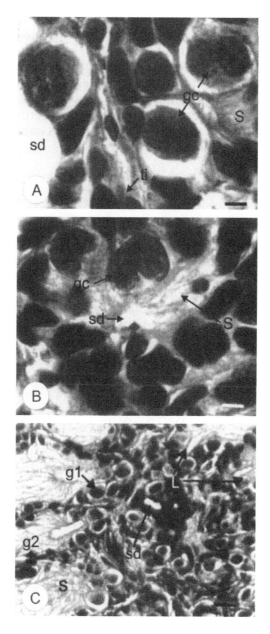

Fig. 4. Development of testes in *Typhlonectes compressicauda* (after Exbrayat, 1986a).
A—Testis of a newborn; B—Testis of a young, less than one year old; C—Testis
of a young, one year old. sd—sperm-collecting ductule; gc—germ cells; S—Sertoli
cell; ti—interstitial tissue, g1—primary spermatogonia; g2—secondary spermatogonia; L—lobule (locule). A, B—Scale bar = 10 μm, C—Scale bar = 30 μm.

exclusively primary and secondary spermatogonia. During the second year of life (January/February), meiosis starts and various stages of prophase I can be observed. In 18-month old animals (May/June), first spermatids and spermatozoa are observed. However, they degenerate and are eliminated. This phenomenon is reminiscent to the precocious spermatogenesis observed in some Anura in which spermatozoa are observed after meta-morphosis, before they degenerate. Proliferation of spermatogonia is still observed. In August, the composition of germ cells, as well as size of testes, resembles those of adult animals.

Urogenital Connections

Cloaca

In her important work, Wake (1972) provides comparisons of the morpho-logy of cloaca in several genera and species of Gymnophiona. Some studies were devoted to description of cloaca in *Hypogeophis rostratus* (Tonutti, 1931, 1932, 1933; Taylor, 1968; Wake, 1972, 1977, 1981). Cloacal zones were des-cribed in *Hypogeophis rostratus, Boulengerula taitana, Gymnopis multiplicata, Typhlonectes compressicauda* (Tonutti, 1931, 1932, 1933; Wake, 1972; Exbrayat, 1991, 1996). Morphological changes during growth and breeding cycle of cloaca in *Typhlonectes compressicauda* was described by Bons (1986) and Exbrayat (1991, 1996).

The cloaca is attached to dorsal region of body by connective tissue of dorsal mesentery, which also wraps the urinary bladder. The cloaca in male Gymnophiona can be divided into three regions. The most anterior region, cloacal ampulla, receives the intestine, bladder, and common trunk of fused Wolffian and Müllerian ducts. The narrow median region forms central duct and a pair of blind sacs, although not in all species. The posterior region consists of phallodeum—a unique structure among amphibians, but common in all Gymnophiona. The phallodeum is a copulatory organ used in internal fertilization. A ventral muscle, *musculus retractor cloacae*, links cloaca to the ventral median region of muscular belt of the body permitting return of phallodeum to its resting position after sexual act.

The internal wall of cloaca is covered by some folds and bulbs arranged in a more or less complex manner, depending on species. A pair of blind sacs is present in some species (Wake, 1972). They derive from folds of the dorsal cloacal wall. It seems that the most primitive species have the most complex cloacal structures (Wake, 1972). In the primitive genera and species (Ichthyophidae and Rhinatrematidae) the sacs are big, whereas in more advanced species, such as *Typhlonectes compressicauda*, the sacs are small. In some species (*Boulengerula boulengeri, Microcaecilia unicolor, Geotrypetes seraphinii*) they are reduced, whereas in others they are widely open to the central duct of cloaca. In *Scolecomorphus*, several types of lateral blind sacs have been observed (Wake, 1972). In *Scolecomorphus uluguruensis* two blind sacs were found.

At birth, the cloaca in male *Typhlonectes compressicauda* is an undifferentiated duct. Blind sacs are not yet developed (Exbrayat, 1996). Epithelial cells lining the anterior region produce acidic mucous, whereas epithelium of the median region is composed of mucous cells and cells with large vacuoles. The same type of epithelium, with less mucous cells and squamous cells covers the region of phallodeum. In 18-month old animals the median region of cloaca is lined by stratified epithelium, which is covered by acidic mucous provided by goblet cells of the blind sacs.

Phallodeum

The phallodeum is an extension of the cloaca, with lumen inside. The lumen has epithelial lining composed of various types of epithelium, depending on a zone of the organ and its physiological state. The phallodeum is covered by a desquamating cornified and stratified epithelium with several mucous cells. The epithelium has local thickenings, giving rise to spines, which are extruded during copulation (Fig. 5). The wall is composed of connective tissue rich in blood vessels, numerous longitudinal folds, longitudinal layer of striated muscles, and a peripheral connective sheath. In Scolecomorphidae, some spicules originate from plates of cartilaginous tissue are inserted in cloacal wall (Noble, 1931; Nussbaum, 1985; Taylor, 1968, 1969; Wake, 1998). The number, size and arrangement of spicules vary according to the species, and these variations do not seem to be linked to size of animal and sexual maturity (Wake, 1998). It is still not known whether these structures are involved in reproduction (Wake, 1998). External morphology of phallodeum is also useful as taxonomic feature. Spicules of phallodeum in several species of the genus *Scolecomorphus* were described by Wake (1998). The phallodeum is enveloped by a connective tissue case, from which it is separated by periphallodeal space. Phallodeum slides into the connective capsule during its eversion after sexual act.

The mechanism of extrusion of erective phallodeum was studied in details in *Hypogeophis rostratus* (Tonutti, 1931). In this species phallodeum is a tube covered by spinous structures that are more or less developed according to the period of sexual cycle. At breeding, the muscular wall contracts and pushes the cloaca, which slides into the connective tissue case, permitting extrusion of its posterior end. The cloacal tube is reversed and the spinous internal epithelium constitutes now the external surface of phallodeum (Fig. 6). After breeding, the *musculus retractor cloacae* is used for the return of the cloaca into its connective case.

134

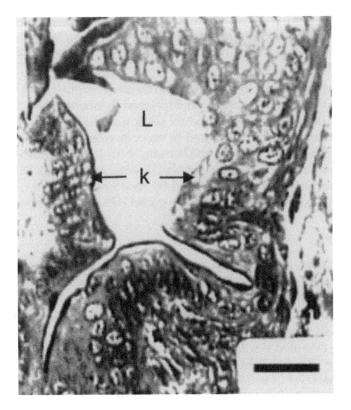

Fig. 5. Posterior part of the cloaca in male *Typhlonectes compressicauda* during breeding period. k—keratinized cells; L—central lumen of the cloaca. Scale bar = 30 μm. After Exbrayat, 1991.

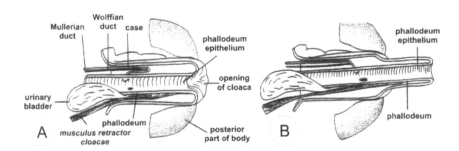

Fig. 6. Schematic representation of a cloaca in male *Typhlonectes compressicauda*. A—During sexual quiescence; B—During sexual act, showing extrusion of the phallodeum. Modified from Exbrayat, 1991.

Each Wolffian duct is an elongated tube that lays laterally, parallel to kidney. Several studies on Wolffian ducts have been published by Wake (1970a) in several species, Sakai *et al.* (1986, 1988a,b) in *Typhlonectes compressicauda*, Cavalho and Junqueira (1999) in *Siphonops annulatus*. In *Typhlonectes compressicauda*, each Wolffian duct is lined with a pseudo-stratified columnar epithelium, surrounded by a dense connective tissue. Three cell types have been described (Sakai *et al.*, 1986, 1988a,b). The Wolffian cells proper are columnar with the apical part limiting the lumen of duct, some intercalated cells are also columnar with a dark cytoplasm containing numerous mitochondria, and basal cells are found against the basal lamina and are simple in structure. Sperm and urine leave the kidney using the unmodified Wolffian duct that is a more primitive condition than that of other amphibians (Wake, 1970a).

Spermatogenesis in Gymnophiona

Morphology of Germ Cells

Spengel (1876), Semon (1892), Tonutti (1931) and Seshachar (1936, 1937, 1942a, 1943, 1945) described spermatogenic cells differentiation in *Caecila gracilis, Ichthyophis glutinosus* and *Siphonops annulatus*. More recently male germ cells were studied in *Typhlonectes compressicauda* (Exbrayat and Sentis, 1982; Exbrayat, 1986a,b, Exbrayat and Estabel, 2006), in *Chthonerpeton indictinctum* (De Sa and Berois, 1986), and *Boulangerula taitanus* (Measey *et al.* (2008). Smita *et al.* (2003, 2004a,b, 2006) gave a detailed review about the spermatogenesis and spermiogenesis in Caecilians. The ultrastructure of germ cells has also been related to the phylogeny (Jamieson, 1999; Scheltinga and Jamieson, 2006).

In all species of Gymnophiona studied so far, differentiating germ cells are enclosed in cysts formed by big Sertoli cells.

Primary Spermatogonia

Primary spermatogonia are grouped around the sperm-collecting duct, often being embedded in epithelial cells of a lobule wall (Fig. 2A). They are spherical cells and in *Typhlonectes compressicauda* their diameter ranges from 20 to 50 μm. They are surrounded by 2-4 Sertoli cells. Similar observations were also provided for other Gymnophiona (Stieve, 1920, quoted from Roosen-Runge, 1977). Cytoplasm of primary spermatogonia is not intensively stained; nucleus is big, round or irregular in outline, depending on the cell cycle. Seshachar (1936) provided a detailed study of primary spermatogonia in *Ichtyophis glutinosus*. In *Typhlonectes compressi-cauda*, primary spermatogonia between breeding seasons (from August to

October) are not as big as those described in *Ichtyophis glutinosus*. As revealed in TEM and SEM, these cells and accompanying follicle (further Sertoli) cells form a plate attached to a lobule wall. Primary spermatogonia multiply and give rise to secondary spermatogonia.

Secondary Spermatogonia

At the multiplication phase, arising secondary spermatogonia migrate against the lobule wall. As was described by Seshachar (1937) for *Ichthyophis glutinosus*, one primary spermatogonium gives rise to 128 cells after 7 mitotic cycles. Stieve (1920, quoted from Roosen-Runge, 1977), indicate that their number within one cyst is 64-128, which means that they undergo 6-7 mitotic cycles. Secondary spermatogonia are smaller than primary spermatogonia, and in *Typhlonectes compressicauda* have a diameter of 13 μm. They have big nuclei and small amount of cytoplasm.

Primary and Secondary Spermatocytes

Secondary spermatogonia increase in volume and differentiate into primary spermatocytes, still enclosed in cysts at the periphery of lobules, close to the wall. Primary spermatocytes are arranged in a layer lining a cyst wall, which is seen on sections as a "crown" surrounding a central, optically empty space (Fig. 2A). In SEM, primary spermatocytes appear spongy with some filamentous extensions protruding to the center of a cyst (Fig. 7A). Cytoplasmic bridges connect primary spermatocytes and primary spermatocytes contact Sertoli cells by intercellular bridges.

Secondary spermatocytes are smaller than primary spermatocytes, and are rarely found on sections because of short duration of this meiotic stage. They are still enclosed in cysts migrating towards the center of a lobule. Their morphology revealed in SEM is similar to that of primary spermatocytes.

Spermatids

Spermiogenesis was studied in *Ichthyophis glutinosus*, *Uraeotyphlus narayani* and *Gegenophis carnosus* (Seshachar, 1936, 1943, 1945), *Typhlonectes compressicauda* (Exbrayat, 1986a,b; Exbrayat and Sentis, 1982; Exbrayat and Estabel, 2006), *Chthonerpeton indistinctum* (De Sa and Berois, 1986), *Uraeotyphlus narayani*, *Ichthyophis tricolor* (Smita *et al.*, 2003, 2004a,b). Spermiogenesis is similar in all the species studied so far. Early spermatids are spherical. The acrosome is secreted by dictyosomes and is more or less conspicuous in the anterior pole of the cell. Mitochondria are scattered throughout the cytoplasm and form an aggregation near acrosome. Detailed

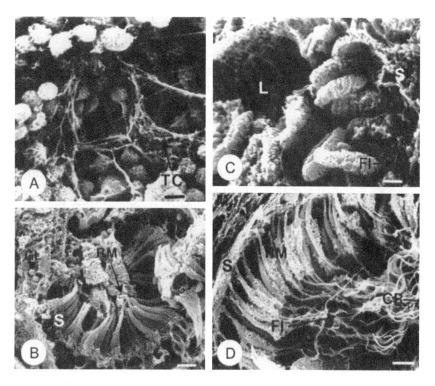

Fig. 7. Spermatogenesis in *Typhlonectes compressicauda* (after Exbrayat, 1986a). A—SEM view of a primary spermatocyte in contact with filamentous extensions of primary spermatocytes protruding to the center of a cyst; B—TEM view of young spermatozoa; C—SEM view of spermatids; D—SEM view of mature spermatozoa. TC—filamentous extensions of primary spermatocytes; S—Sertoli cell; Fl—flagellum of a spermatid; L—cyst lumen; GL—fat vacuole; RM—contact zone between germ cells and Sertoli cells; CR—center of the cyst. A—Scale bar = 7 µm; B—Scale bar = 15 µm; C and D—Scale bar = 3 µm.

cytological description of changes during spermiogenesis was given by Seshachar (1943, 1945), Jamieson (1999), and Smita *et al.* (2005).

When viewed by SEM, spermatids are single cells with no cytoplasmic bridges and have a spongy appearance (Fig. 7C). Several round cells with spongy external membrane are always scattered among flagellum. At light microscopy, these cells have intensively stained nucleus surrounded by a narrow layer of cytoplasm. Early spermatids are arranged inside a cyst in a similar way to spermatocytes, forming a "crown" when observed on sections. When spermatids elongate, they are closely packed and contact each other, and become cone-shaped. Flagellae are oriented to the center of cyst. The head of the spermatids is always in contact with Sertoli cells (Fig. 7B).

The centrioles, originally located in the acrosome area, migrate to the opposite (further distal) pole of a cell where they differentiate into distal and proximal centrioles. The proximal centriole is situated close to the nuclear envelope. The distal centriole, parallel to further long axis of spermatid, provides a basal body of a flagellum. During spermiogenesis, spermatids and their nuclei elongate. The acrosome is narrowly linked to the nucleus by an anterior nuclear fosse (Jamieson, 1999; Scheltinga and Jamieson, 2006).

The flagellum elongates at the distal pole of the cell and becomes increasingly narrow and hardly visible after cytological staining, until it reaches its maximal size (4 × 10 µm in *Ichthyophis kohtaoensis*, 3 × 25 µm in *Typhlonectes compressicauda*). At this time dictyosomes from the Golgi complex migrate along the side of the nucleus toward the posterior pole of a cell, where they will be detached and rejected, together with a thin layer of surrounding cytoplasm. Mitochondria also migrate to the posterior pole of a cell and become disposed around the flagellum in a spiral manner. Nucleus then condenses and elongates. At this time, the midpiece is well seen. The cytoplasm excess is eliminated.

Spermatozoa

At the end of spermiogenesis the spermatozoa are formed. The shape of acrosome is species specific (ampulla in *Typhlonectes compressicauda*, hook in other species), the nucleus is condensed and elongated, and cytoplasm is scarce with proximal and distal centrioles, a midpiece and flagellum (Scheltinga *et al.*, 2003; Scheltinga and Jamieson, 2006).

In *Typhlonectes compressicauda*, as in other species, the formation of spermatozoa was divided into three stages (Exbrayat and Sentis, 1982; Exbrayat, 1986a,b). The youngest spermatozoa are closely packed and have residual eosinophilic cytoplasm. On a section of a cyst they display a characteristic fan shape. The products of their cytoplasm rejection are observed near the midpiece. In SEM, the proximal pole of spermatozoon appears smooth but the distal part is still spongy. The heads of spermatozoa are in contact with folded membrane of the Sertoli cells (Fig. 7B). At the next stage spermatozoa lay parallel to each other and form bundles situated in the center of a cyst (Fig. 7D). No residual cytoplasm has been observed. In SEM, the cytoplasm extensions of round cells scattered among the flagella and interpreted as degenerative Sertoli cells, look like a network. At the last stage cysts open, spermatozoa are released from cysts, and are seen free in the funnel-shaped area close to sperm-collecting duct of a lobule; at various seasons of the year they are also observed in *vasa efferentia*.

Age of Sexual Maturity

The age of sexual maturity has been poorly studied in Gymnophiona. Development of male genital tract from birth to sexual maturity has been studied in *Dermophis mexicanus* (Wake, 1980) and *Typhlonectes compressicauda* (Exbrayat, 1986a,b; Exbrayat and Dansard, 1994, Anjubault and Exbrayat, 2004b, 2006). In these species, each testis resembles a thread on which lobes develop at birth (like beads on a thread). After one year the beginning of spermatogenesis was observed in 90% of male *Dermophis mexicanus*. After two years active spermatogenesis is observed in only few lobules in each testis. Maturation of *Dermophis mexicanus* seems to occur during the third year of life, when males seemed to be ready for breeding. These data show that the sexual maturity was acquired in two years old animal, e.g. during their third year of life, like in *Dermophis mexicanus*. It seems that during this period the hormonal control of breeding begins to be efficient, although detailed studies are lacking.

Hormonal Control of Spermatogenesis

Sexual Cycle in Males

Gymnophiona are seasonal breeders, which is reflected in discontinuous spermatogenetic cycle in males. The studies were provided for *Ichthyophis* (Bhatta, 1987; Sarasin and Sarasin, 1887-1890; Seschachar, 1936, 1937, 1943); *Gymnopis multiplicata* (Wake, 1968b); *Dermophis mexicanus*, (Wake, 1980, 1995); *Typhlonectes compressicauda* (Exbrayat and Sentis, 1982; Exbrayat and Delsol, 1985; Exbrayat, 1986a,b, 2006).

Spermatogenesis in Gymnophiona is postnuptial. In the Indian species *Ichthyophis beddomei* breeding season occurs in January-February (Bhatta, 1987). In *Ichthyophis glutinosus* from Sri Lanka, germ cell number in testes during winter is scarce, with exception of primary spermatogonia. In spring, they multiply, give rise to secondary spermatogonia, which enter meiosis and differentiate into spermatozoa for the next breeding season. In *Ichthyophis* sp. from Mysore (India), spermatogenesis begins in March, spermatozoa are released from lobules in December; and are not detectable until March, when a new spermatogenetic cycle begins (Seshachar, 1936). In *Gymnopis multiplicata*, from Costa Rica breeding season occurs in December-January (Wake, 1968b). From December until March, testes contain only primary spermatogonia. Spermatogenesis starts in June and is completed in October, when all germ cell stages are observed in the lobules. In *Dermophis mexicanus* from Guatemala, the spermatogenesis remains active throughout a long period of the year (June until March) and breeding occurs from March until June. During this period, testes decrease

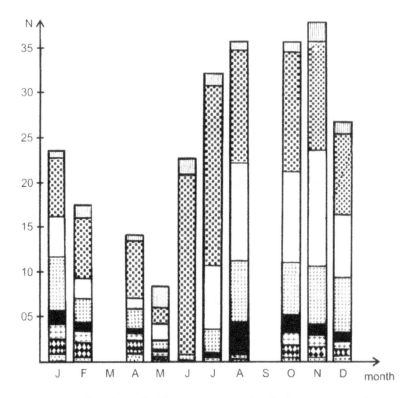

Fig. 8. Male sexual cycle in *Typhlonectes compressicauda*; spermatogenesis occurs from June to August; period of sexual quiescence is observed from August until December at which evacuation of germ cells begins; from December to May-June, a spermatogenesis is also observed to reconstitute the stock of post-meiotic germ cells. ▥ primary spermatogonia; ▤ secondary spermatogonia; ☐ primary spermatocytes; ▨ secondary spermatocytes and young spermatids; ■ old spermatids; ▥ young spermatozoa; ▥ mature spermatozoa; ▥ free spermatozoa. N—number of germ cell/μm² of section. After Exbrayat, 1986a.

in size due to release of spermatozoa (Wake, 1980, 1995). In *Typhlonectes compressicauda* from French Guyana, cyclic changes were also observed (Fig. 8). The cycle is closely linked to the seasonal variations, breeding occurs during the rainy season and the sexual quiescence during the dry season (Exbrayat and Sentis, 1982; Exbrayat, 1986a,b, 2006). Testes have maximal size in December-January, just at the beginning of the breeding season in May-June. From February until May, they decrease in size as a result of spermatozoa release. In April-May, the last spermatozoa are released and the lowest stock of germ cells is observed. A new wave of spermatogenesis starts again in June, in July spermatids and the first spermatozoa appear. In August, a new stock of germ cells is established in testes and remains constant until November, and gradual increase of testis size is observed until December. From December until May, numerous spermatozoa are seen in efferent ducts (*vasa efferentia*), and the total number of germ cells progressively decreases. However, the number of spermatids and spermatozoa remains constant from August to April, suggesting that new generations of spermatozoa are formed and replace spermatozoa already evacuated.

The annual reproductive cycle was experimentally studied in *Typhlonectes compressicauda* injected with ^{3}H thymidine. The cycle had three phases: (1) from June until August complete spermatogenesis with spermiogenesis occurs and the stock of germ cells (secondary spermatogonia, spermatocytes, spermatids and spermatozoa) is formed; (2) from August until December-January, during resting period, the stock remains stable; (3) during breeding season lasting from January until May-June, both depletion of spermatozoa and spermatogenesis take place. Variations of the interstitial tissue throughout the sexual cycle have been previously described. In a recent study Measey *et al.* (2008) showed that the African *Boulangerula taitanus* performed a continuous spermatogenic cycle with variations of testis size.

Pituitary

The pituitary was particularly well studied in *Typhlonectes compressicauda* (Doerr-Schott and Zuber-Vogeli, 1984; Exbrayat, 1989, 2006; Raquet and Exbrayat, 2007). Cells of its *pars distalis* display some variations during male annual cycle. Gonadotropic cells produce LH, as was revealed by immunohistochemical studies with anti-LH serum (Doerr-Schott and Zuber-Vogeli, 1984). In January, their cytoplasm contains several scattered granules, among which some larger globules are occasionally seen. At that time the cells and their nuclei reach maximal size and are particularly numerous in the median region of the gland, whereas in May-June they decrease in size, granules are very scanty, and globules become numerous in the cytoplasm. In June, the cells reach their minimal size and no granules can be detected. From July until October, during the resting period, the sizes of cells increase again, accompanied by appearance of new granules. From October until February, the cycle starts again (Fig. 9A).

Lactotropic cells of the pituitary produce prolactin (PRL), as was revealed by immunohistochemical studies with anti PRL-serum (Doerr-Schott and Zuber-Vogeli, 1984). Despite of their size, the cells always contain the same type of big orange-stained granules. The cells are particularly well developed during breeding season (from February until April). Between April and June, at the end of breeding season, the cells decrease in size but retain granules. From June until August, the cells

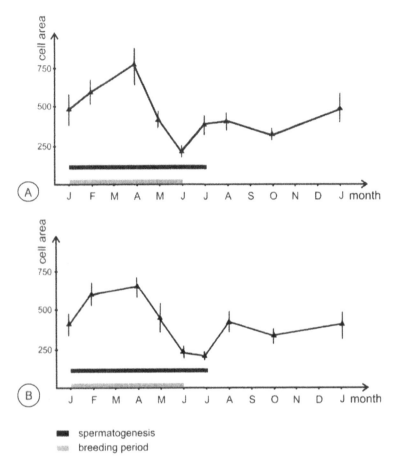

Fig. 9. Annual cycle of the pituitary in male *Typhlonectes compressicauda*. The curves show variation of cell surface mean area on sections (in μm²) according to the period of sexual cycle (after Exbrayat, 1986a). A—Gonadotropic cells; B—Lactotropic cells; vertical bars indicate the standard errors at a threshold of 95%.

increase and reach the maximal size, which remain constant until January, with a slight decrease in October (Fig. 9B). The number of cells is relatively reduced during breeding period (Exbrayat and Morel, 1995). Just after breeding, the number of cells reaches its maximum and remains constant until the next breeding.

The quantification of mRNAs coding for PRL, after their visualization by *in situ* hybridization, allowed showing constant labeling during the whole cycle. A general pattern of lactotropic cells activity seems to be as follows: just after breeding and during resting period the number of lactotropic cells remains constant, synthesizing a constant quantity of PRL which is distributed in the body by circulating blood. At breeding, synthesis level is the same, but it concerns lower number of cells, which produce more mRNAs and—in consequence—increase synthesis of PRL. This hormone can be stored in cells, causing increase of their volume. PRL is released during breeding season. A study using *in situ* hybridization allowed visualizing mRNAs coding for PRL receptors (Exbrayat *et al.*, 1996; Exbrayat and Morel, 2003). The receptors were detected in germ cells, Sertoli cells and Leydig cells. PRL mRNAs were localized on sections in connective tissue, and mainly in glandular structures of the Müllerian ducts. The amount of PRL mRNAs decreases during resting period and increases during breeding season (Fig. 10A,B). The volume of PRL cells, testis size, spermatogenesis, and the secretive activity of the Müllerian gland are well correlated. In *Ichthyophis beddomei*, Bhatta (1987) noticed that granules are present in gonadotropic cells throughout the year. Nevertheless, the cytoplasm contains several vacuoles during the breeding period in January and February. The presence of vacuoles only in March can be correlated with the proliferation of spermatogonia. In this species, like in *Typhlonectes compressicauda*, there is also a strong correlation between the activity of certain cell types of the hypophysis and the reproductive activity of the animal.

Male Müllerian Glands

Müllerian ducts are unique features of male Gymnophiona. They do not degenerate during development, as is the case in other amphibians (Spengel, 1976; Wiedersheim, 1879; Sarasin and Sarasin, 1887-1890; Semon, 1892). Tonutti (1931) described the glandular nature of these organs. Marcus (1930) considered the Müllerian ducts as auxiliary testes. According to other authors, these organs remain rudimentary (Lawson, 1959; Oyama, 1952; Romer, 1955). More recent studies have also been devoted to Müllerian ducts in several species (Wake, 1970a, 1977, 1981; Exbrayat, 1985, 1986a; George *et al.*, 2004a,b, 2005; Akbarsha *et al.*, 2006).

The Müllerian ducts are paired organs parallel to kidneys to which they are attached by the connective tissue. Two morphological types of Müllerian ducts were described (Wake, 1970a). In the first type, as in case of *Uraeotyphlus narayani* (George *et al.*, 2004a), the posterior region of each duct is thick, whereas about 75% of the anterior region remains narrow. In the second type, the Müllerian ducts are evenly thick along their all length, with only the anterior tip being narrow. Only the thick parts of Müllerian duct are glandular, consisting of cells producing granular secretions.

The structure and development of Müllerian glands were studied in details in *Typhlonectes compressicauda* (Exbrayat 1985, 1986a) and in *Uraeotyphlus narayani* (George *et al.*, 2004a,b, 2005), and some aspects of

144

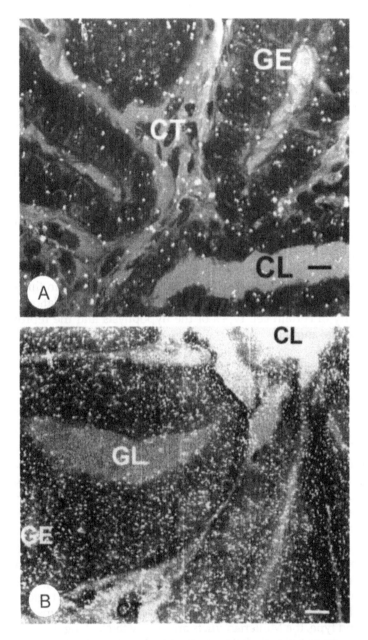

Fig. 10. Localization of mRNAs coding for prolactin receptors (white spots) in Müllerian duct in *Typhlonectes compressicauda* (after Exbrayat and Morel, 2003). A—During sexual quiescence; B—During breeding period. CL—central lumen; CT—connective tissue; GE—glandular epithelium. A—scale bar = 30 μm; B—scale bar = 90 μm.

their development have been described for *Ichthyophis kohtaoensis* (Wrobel and Süß, 2000). In this species, developing Müllerian gland is linked to the corresponding mesonephros. At birth, each Müllerian duct has a central lumen, circular on sections, and lined by columnar epithelium composed of a single layer of cells with big nuclei. Connective tissue underlying epithelium is relatively thick and contains blood vessels. In 12-month old animals, the diameter of the duct increases, the epithelium becomes ciliated and small secondary ducts develop as glandular extensions of the central duct and its lumen. The lumen of the glandular extensions has its own epithelial lining composed of two kinds of cells, secretory with nuclei in their basal parts and granular cytoplasm, and ciliated with nuclei in their apical parts. In 18-month old animals the glands are already well differentiated. Their cells are filled with some spherical granules containing glycoproteinous secretion (3 μm in diameter) and their excretory canals open to the lumen of the central duct. During next months, until males are two years old and reach sexual maturity, glandular activity decreases. Epithelial cells lining both the central and glandular ducts loose their cilia, fragments of cells are detached, and Müllerian ducts start to regress. Few months before breeding Müllerian glands become again active and increase in size. The duration of this period varies between individuals, probably reflecting actual age of males (birth can take place between June and September) (Exbrayat, 1986a).

As it was mentioned above, in adult males Müllerian ducts display seasonal changes. Its diameter reaches its maximum during breeding (January-April), and decreases during resting period (May-November). During breeding period, each duct is lined with pseudostratified epithelium composed of secretory and ciliated cells. The glandular cells contain abundant glycoprotein granules 1 to 3 μm in diameter, in some regions being acidic, but in other regions being neutral. The central duct is filled with granules and fragments of cytoplasm of glandular cells (Fig. 11A).

After breeding season the Müllerian duct decreases. The cells of the central duct loose cilia, glands are reduced to small extensions with only one type of cell without cilia, nor secretions; their cytoplasm is filled with remnants of cilia and fragments of cytoplasm. The connective tissue is relatively abundant (Fig. 11B). Beginning with October the cycle starts again. More recently, George *et al.* (2004a,b) and Akbarsha *et al.* (2006) published a LM and TEM study about male Müllerian gland in *Uraeotyphlus narayani*. They observed a similar morphology with variations according to the sexual cycle, characterized by a testicular quiescence from March until June. These authors described glands with secretions resembling that of *Typhlonectes compressicauda*. In addition, the authors described some amoeboid cells at the base of peripheral glandular tubules.

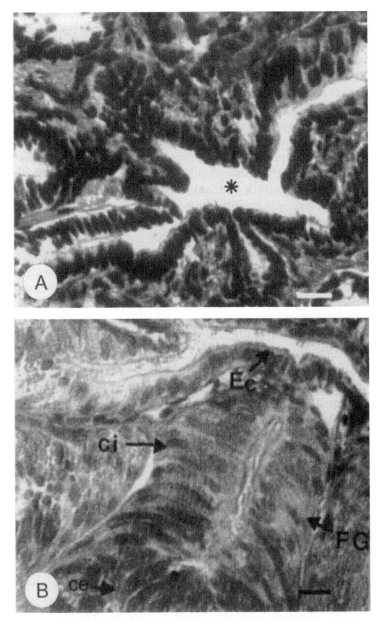

Fig. 11. Müllerian gland of *Typhlonectes compressicauda* (after Exbrayat, 1986a). A—Sexual quiescence (November); B—Breeding period (January). ci—gland cell on the internal part of Müllerian duct; ce—gland cell on the external part of Müllerian duct; Ec—ciliated epithelium; FG—gland tubule, asterisk: central lumen of Müllerian gland. Scale bar = 30 µm.

The glandular secretions of Müllerian glands in *Dermophis mexicanus* and *Typhlonectes compressicauda* contain mucopolysacharides, fructose and acidic phosphatases. Their pH is acidic and similar to that of the sperm. Müllerian ducts of Gymnophiona are supposed to be equivalent to mammalian prostate (Wake, 1981). The Müllerian ducts of Gymnophiona are active glands producing substances being sperm component. The secretion of Müllerian ducts and the spermatozoa probably mix in cloaca, but there are no direct studies confirming this process even George *et al.* (2005) showed the production of Müllerian glands have an effect on spermatozoon motility in *Uraeotyphlus narayani*.

References

Akbarsha MA, MJ George, M Smita, VO Oommen. 2006. Caecilian male Mullerian gland, with Special Reference to *Uraeotyphlus narayani*. In: Reproductive Biology and phylogeny of Gymnophiona (Caecilians), (Ed.) JM Exbrayat, Vol. 5 of Series Reproductive Biology and Phylogeny, BGM Jamieson (series ed.) Science Publishers, Enfield, New Jersey, pp. 157-182.

Anjubault E, JM Exbrayat. 1998. Yearly cycle of Leydig-like cells in testes of *Typhlonectes compressicaudus* (Amphibia, Gymnophiona) In: Current studies in Herpetology, (Eds) C Miaud and R Guyetant, Proceedings of the 9th General Meeting of the Societas Europaea Herpetologica, Le Bourget du Lac, France, pp. 53-58.

Anjubault E, JM Exbrayat. 2000a. Development of gonads in *Typhlonectes compressicauda* Amphibia, Gymnophiona). XVIIIth International Congress of Zoology, Athens, August-September 2000: 51.

Anjubault E, JM Exbrayat. 2000b. Cycle annuel des cellules du tissu interstitiel des testicules chez *Typhlonectes compressicaudus*, Amphibien Gymnophione. Bull. Soc. Zool. Fr. 125: 133.

Anjubault E, JM Exbrayat. 2004a. Contribution à la connaissance de l'appareil génital de *Typhlonectes compressicauda* (Duméril et Bibron, 1841), Amphibien Gymnophione. I. Gonadogenèse. Bull. Mens. Soc. Linn. Lyon. 73: 379-392.

Anjubault E, JM Exbrayat. 2004b. Contribution à la connaissance de l'appareil génital de *Typhlonectes compressicauda* (Duméril et Bibron, 1841), Amphibien Gymnophione. II. Croissance des gonades et maturité sexuelle des mâles. Bull. Mens. Soc. Linn. Lyon. 73: 393-405.

Anjubault E, JM Exbrayat. 2006. Development of gonads. In: Reproductive Biology and phylogeny of Gymnophiona (Caecilians), (Ed.) JM Exbrayat, Vol. 5 of Series Reproductive Biology and Phylogeny, BGM Jamieson (Series ed.) Science Publishers, Enfield, New Jersey, pp. 291-302.

Bhatta GK. 1987. Some Aspects of Reproduction in the Apodan Amphibian *Ichthyophis*. Ph.D. Thesis, Karnataka University, Dharwad, India.

Bhatta GK, E Anjubault, JM Exbrayat. 2001. Structure et ultrastructure des testicules d'*Ichthyophis beddomei* (Peters, 1879), Amphibien Gymnophione. Ann. Mus. du Havre 67: 11-12.

Bons J. 1986. Données histologiques sur le tube digestif de *Typhlonectes compressicaudus* (Duméril et Bibron, 1841) (Amphibien Apode*)*. Mém. Soc. Zool. Fr. 43: 87-90.

Brauer A. 1902. Beitrage zur kenntniss der Entwicklung und Anatomie der Gymnophionen. III. Die Entwicklung der Excretionsorgane. Zool. Jahrb. Anat. 16: 1-176.

Chatterjee BK. 1936. The anatomy of *Uraeotyphlus menoni* Annandale. Part I: digestive, circulatory, respiratory and urogenital systems. Anat. Anz. 81: 393-414.

Cavalho ETC, LCU Junqueira. 1999. Histology of the kidney and urinary bladder of *Siphonops annulatus* (Amphibia, Gymnophiona). Arch. Histol. Cytol. 62: 39-45.

Doerr-Schott J, M Zuber-Vogeli. 1984. Immunohistochemical study of the adenohypophysis of *Typhlonectes compressicaudus* (Amphibia, Gymnophiona). Cell Tissue Res. 235: 211-214.

De Sa R, N Berois. 1986. Spermatogenesis and histology of the testes of the Caecilian *Chthonerpeton indistinctum.* J. Herp. 20: 510-514.

Exbrayat JM. 1985. Cycle des canaux de Müller chez le mâle adulte de *Typhlonectes compressicaudus* (Duméril et Bibron, 1841), Amphibien Apode. C. R. Séanc. Acad. Sci., Paris. 301: 507-512.

Exbrayat JM. 1986a. Quelques aspects de la biologie de la reproduction chez *Typhlonectes compressicaudus* (Duméril et Bibron, 1841), Amphibien Apode. Thèse, Doctorat ès Sciences Naturelles, Université Paris VI, France.

Exbrayat JM. 1986b. Le testicule de *Typhlonectes compressicaudus*; structure, ultrastructure, croissance et cycle de reproduction. Mém. Soc. Zool. Fr. 43: 121-132.

Exbrayat JM. 1988. Variations pondérales des organes de réserve (corps adipeux et foie) chez *Typhlonectes compressicaudus*, Amphibien Apode vivipare au cours des alternances saisonnières et des cycles de reproduction. Ann. Sci. Nat., Zool., 13éme série. 9: 45-53.

Exbrayat JM. 1989. The cytological modifications of the distal lobe of the hypophysis in *Typhlonectes compressicaudus* (Duméril and Bibron, 1841), Amphibia Gymnophiona, during the cycles of seasonal activity. I - In adult males. Biol. Struct. Morph. 2: 117-123.

Exbrayat JM. 1991 Anatomie du cloaque chez quelques Gymnophiones. Bull. Soc. Herp. Fr. 58: 30-42.

Exbrayat JM. 1993. Quelques aspects de la reproduction chez *Typhlonectes compressicaudus* (Duméril et Bibron, 1841), Amphibien Gymnophione. Cah. Inst. Cath. Lyon, Série Sciences 7: 1-263.

Exbrayat JM. 1996. Croissance et cycle du cloaque chez *Typhlonectes compressicaudus* (Duméril et Bibron, 1841), Amphibien Gymnophione. Bull. Soc. Zool. Fr. 121: 99-104.

Exbrayat JM. 2000. Les Gymnophiones, ces curieux Amphibiens. Boubée, Paris, France, pp. 443.

Exbrayat JM, C Dansard. 1992. Ultrastructure des cellules de Sertoli chez *Typhlonectes compressicaudus*, Amphibien Gymnophione. Bull. Soc. Zool. Fr. 117: 166-167.

Exbrayat JM. 2006. Endocrinology of reproduction in Gymnophiona. In:

Reproductive Biology and phylogeny of Gymnophiona (Caecilians) (Ed.) JM Exbrayat, Vol. 5 of Series Reproductive Biology and Phylogeny, BGM Jamieson (Series ed.) Science Publishers, Enfield, New Jersey, pp. 183-229.

Exbrayat JM, C Dansard. 1994. Apports de techniques complémentaires à la connaissance de l'histologie du testicule d'un Amphibien Gymnophione. Rev. Fr. Histotechnol. 7: 19-26.

Exbrayat JM, M Delsol. 1985. Reproduction and growth of *Typhlonectes compressicaudus*, a viviparous Gymnophione. Copeia, 1985: 950-955.

Exbrayat JM, J Estabel. 2006. Anatomy with particular reference to the reproductive system. In: Reproductive Biology and phylogeny of Gymnophiona (Caecilians), (Ed.) JM Exbrayat. Vol. 5 of Series Reproductive Biology and Phylogeny, BGM Jamieson (Series ed.), Science Publishers, Enfield, Jersey, pp. 79-155.

Exbrayat JM, G Morel. 1995. Prolactin (PRL)-coding mRNA in *Typhlonectes compressicaudus*, a viviparous gymnophionan Amphibian. An *in situ* hybridization study. Cell Tissue Res. 280: 133-138.

Exbrayat JM, G Morel. 2003. Visualization of gene expression of prolactin-receptor (PRL-R) by *in situ* hybridization in reproductive organs of *Typhlonectes compressicauda*, a Gymnophionan Amphibian. Cell Tissue Res. 312: 361-367.

Exbrayat JM, P Sentis. 1982. Homogénéité du testicule et cycle annuel chez *Typhlonectes compressicaudus* (Duméril et Bibron, 1841), Amphibien Apode vivipare. C.R. Séanc. Acad. Sci., Paris 294: 757-762.

Exbrayat JM, M Delsol, J Flatin. 1986. *Typhlonectes compressicaudus*, Amphibien Apode vivipare de Guyane. In: Sepanguy-Sepanrit Le littoral guyanais, Actes du colloque Le Littoral Guyanais, Fragilité de l'Environnement, Cayenne, Guyane Française, pp. 119-124.

Exbrayat JM, A Ouhtit, G Morel. 1996. Prolactin (PRL) and prolactin receptor (PRL-R) mRNA expression in *Typhlonectes compressicaudus* (Amphibia, Gymnophiona) male genital organs. An *in situ* hybridization study. 18th Conf. Eur. Comp. Endocr., Rouen, Sept. 1996. Ann. Endocr. 57: Supplement, 55.

George JM, M Smita, VO Oommen, MA Akbarsha. 2004a. Histology and ultrastructure of male Mullerian gland of *Uraeotyphlus narayani* (Amphibia: Gymnophiona). J. Morph. 260: 33-56.

George JM, M Smita, B Kadalmani, R Girija, OV Oommen, MA Akbarsha. 2004b. Secretory and basal cells of the epithelium of the tubular glands in the male Mullerian gland of the Caecilian *Uraeotyophlus narayani* (Amphibia: Gymnophiona). J. Morph. 262: 760-769.

George JM, M Smita, B Kadalmani, R Girija, OV Oommen, MA Akbarsha. 2005. Contribution of the secretory material of Caecilian (Amphibia: Gymnophiona) male Mullerian gland to motility of sperm: a study in *Uraeotyphlus narayani*. J. Morph. 263: 227-237.

Jamieson BGM. 1999. Spermatozoal Phylogeny of the Vertebrata. In: The Male Gamete: From Basic Science to Clinical Applications, (Ed.) C Gagnon, Cache River Press. Vienna, USA, pp. 303-331.

Lawson R. 1959. The anatomy of *Hypogeophis rostratus* Cuvier. Amphibia: Apoda

150

or Gymnophiona. Ph.D. Dissertation Universitry of Durham, King's College, USA.

Lofts B. 1974. Reproduction. In: Physiology of the Amphibia, (Ed.) B Lofts, Vol. 2. Academic Press, New York, USA, pp. 107-218.

Marcus H. 1930. Beitrag zur Kenntniss der Gymnophionen. XIII. Uber die Bildung von Geruchsorgan, Tentakel und Choanen bei *Hypogeophis*, nebst Vergleisch mit Dipnoen und Polypterus. Z. Anat. Ent. 91: 657-691.

Marcus H. 1939. Beitrag zur kenntnis der Gymnophionen. Ueber keimbahn, keimdruusen, Fettkörper und Urogenitalverbindung bei *Hypogeophis* Biomorphosis 1: 360-384.

Measey GJ, M Smita, RS Beyo, CV Oommen. 2008. Year-round spermatogenic activity in an oviparous subterraneus caecilian, *Boulangerula taitanus* Loveridge 1935, (Amphibia Gymnophiona Caecillidae). Trop. Zool. 21: 109-122.

Noble GK. 1931. The biology of the Amphibia. McGraw-Hill, NewYork, USA.

Nussbaum RA. 1985. Systematics of Caecilians Amphibia: Gymnophiona) of the family Scolecomorphidae. Occasional Pap. Mus. Zool., Univ. Michigan 713: 1-49.

Oyama J. 1952. Microscopical study of the visceral organs of Gymnophiona, *Hypogeophis rostratus*. Kumamoto J. Sci. 1B: 117-125.

Pujol P, JM Exbrayat. 2000. Mise en évidence de l'homogénéité des testicules multilobés de deux Amphibiens par des méthodes morphométriques. Bull. Soc. Herp. Fr. 95: 53-66.

Raquet M, JM Exbrayat. 2009. Embryonic development of the hypophysis and thyroid gland in *Typhonectes compressicauda* (Dummeril and Bibion, 1841), Amphibia, Gymnophiona. J. Herpetol. 41: 703-712.

Romer AS. 1955. The Vertebrate Body. Second edition. W.B. Sanders Co, Philadelphia-London.

Roosen-Runge EC. 1977. The process of spermatogenesis in animals. Cambridge University Press, Cambridge, London, New York, Melbourne. pp. 115-121.

Sakai T, R Billo, W Kriz. 1986. The structural organization of the kidney of *Typhlonectes compressicaudus* (Amphibia, Gymnophiona). Anat. Embr. 174: 243-252.

Sakai T, R Billo, W Kriz. 1988a. Ultrastructure of the kidney of a south american Caecilian, *Typhlonectes compressicaudus* (Amphibia, Gymnophiona). II: Distal tubule, connecting tubule, collecting duct and Wolffian duct. Cell Tissue Res. 252: 601-610.

Sakai T, R Billo, R Nobiling, K Gorgas, W Kriz. 1988b. Ultrastructure of the kidney of a South American Caecilian, *Typhlonectes compressicaudus* (Amphibia, Gymnophiona) I: Renal corpuscle, neck segment, proximal tubule and intermediate segment. Cell Tissue Res. 252: 589-600.

Sarasin P, F Sarasin. 1887-1890. Ergebnisse Naturwissenschaftlicher Forschungen auf Ceylon. Zur Entwicklungsgeschichte und Anatomie der Ceylonischen Blindwuhle *Ichthyophis glutinosus*. C.W. Kreidel's Verlag, Wiesbaden.

Scheltinga DM, Wilkinson M, Jamieson BGM, Oomen OV. 2003. Ultrastructure of the mature spermatozoa of caecilians (Amphibia: Gymnophiona). J. Morph. 258: 179-192.

Scheltinga DM, BGM Jamieson. 2006. Ultrastructure and Phylogeny of Caecilian

Spermatozoa. In: Reproductive Biology and phylogeny of Gymnophiona (Caecilians), (Ed.) JM Exbrayat. Vol. 5 of Series Reproductive Biology and Phylogeny, BGM Jamieson (Series ed.) Science Publishers, Enfield, New Jersey, pp. 247-274.

Semon R. 1892. Studien über den Bauplan des Urogenitalsystem der Wirbeltiere. Dargelegt an der Entwicklung dieses organysystems bei *Ichthyophis glutinosus*. Jena Z. Naturwiss. 26: 89-203.

Seshachar BR. 1936. The spermatogenesis of *Ichthyophis glutinosus* (Linn.) I. The spermatogonia and their division. Z. Zellforsch. Mikr. Anat. 24: 662-706.

Seshachar BR. 1937. The spermatogenesis of *Ichthyophis glutinosus* (Linn.). II. The meiotic divisions. Z. Zellforsch. Mikr. Anat. 27: 133-158.

Seshachar BR. 1939. Testicular ova in *Uraeotyphlus narayani* Seshachar. Proc. Indian Acad. Sci. 10: 213-217.

Seshachar BR. 1942a. Stages in the spermatogenesis of *Siphonops annulatus* Mikan. and *Dermophis gregorii* (Blgr) (Amphibia: Apoda). Proc. Indian Acad. Sci. 15: 263-277.

Seshachar BR. 1942b. The Sertoli cells in Apoda. J. Mysore Univ. 3: 65-71.

Seshachar BR. 1942c. Origin of intralocular oocytes in male Apode. Proc. Indian Acad. Sci. 15: 278-279.

Seshachar BR. 1943. The spermatogenesis of *Ichthyophis glutinosus* (Linn.), III. Spermateleosis. Proc. Nat. Inst. Sci. India 9: 271-285.

Seshachar BR. 1945. Spermateleosis in *Uraeotyphlus narayani* Seshachar and *Gegenophis carnosus* Beddome (Apoda), Proc. Nat. Inst. Sci. India 11: 336-340.

Smita M, OV Oommen, MG Jancy, MA Akbarsha. 2003. Sertoli cells in the testis of Caecilians, *Ichthyophis tricolor* and *Uraeotyphlus cf. narayani* (Amphibia: Gymnophiona): Light and electron microscopic perspectives. J. Morph. 258: 317-321.

Smita M, JM George, MA Akbarsha, OV Oommen. 2006. Spermatogenesis. In: Reproductive Biology and phylogeny of Gymnophiona (Caecilians). (Ed.) JM Exbrayat, Vol. 5 of Series Reproductive Biology and Phylogeny, BGM Jamieson (Series ed.) Science Publishers, Enfield, New Jersey, pp. 231-246.

Smita M, OV Oommen, MG Jancy, MA Akbarsha. 2004a. Stages in spermatogenesis of two species of caecilians *Ichthyophis tricolor* and *Uraeotyphlus cf. narayani* (Amphibia: Gymnophiona): Light and Electron microscopic Study. J. Morph. 261: 92-104.

Smita M, MG Jancy, R Girija, MA Akbarsha, OV Oommen. 2004b. Spermiogenesis in Caecilians *Ichthyophis tricolor* and *Uraeotyphlus cf. narayani* (Amphibia: Gymnophiona) Analysis by Light and Transmission Electron microscopy. J. Morph. 262: 484-499.

Spengel JW. 1876. Das Urogenitalsystem der Amphibien. I. Theil. Der Anatomische Bau des Urogenitalsystem. Arb. Zool. Inst. Wurzburg. 3: 51-114.

Taylor, EH. 1968. The Caecilians of the world. A taxonomic review. University of. Kansas Press. Lawrence, Kansas, U.S.A. pp. 848.

Taylor EH. 1969. A new family of African Gymnophiona. Univ. Kansas Sci. Bull. 48: 297-305.

Tonutti E. 1931. Beitrag zur Kenntnis der Gymnophionen. XV. Das Genital-system. Morph. Jahrb. 68: 151-292.

Tonutti E. 1932. Vergleichende morphologische Studen uber Eddarm und Kopulations-organe. Morph. Jahrb. 70: 101-130.

Tonutti E. 1933. Beitrag zur Kenntnis der Gymnophionen. XIX. Kopulations-organe bei Weiteren Gymnophionenarten. Morph. Jahrb. 72: 155-211.

Wake MH. 1968a. The comparative morphology and evolutionary relation-ships of the urogenital system of Caecilians. Ph.D. Dissertation, University of California, USA.

Wake MH. 1968b. Evolutionary morphology of the Caecilian urogenital system. Part I: The gonads and fat bodies. J. Morph. 126: 291-332.

Wake MH. 1970a. Evolutionary morphology of the caecilian urogenital system. Part II: The kidneys and urogenital ducts. Acta Anat. 75: 321-358.

Wake MH. 1970b. Evolutionary morphology of the caecilian urogenital system. Part III: The bladder. Herpetologica 26: 120-128.

Wake MH. 1972. Evolutionary morphology of the caecilian urogenital system. Part IV: The cloaca. J. Morph. 136: 353-366.

Wake MH. 1977. The reproductive biology of Caecilians. An evolutionary perspective. In: The Reproductive Biology of Amphibians, (Eds.) DH Taylor and S I Guttman, Miami Univ., Oxford, Ohio, pp. 73-100.

Wake MH. 1980. Reproduction, growth and population structure of the central american Caecilian *Dermophis mexicanus*. Herpetologica 36: 244-256.

Wake MH. 1981. Structure and function of the male Mullerian gland in Caecilians (Amphibia: Gymnophiona), with comments on its evolutionary significance. J. Herp. 15: 17-22.

Wake MH. 1995. The spermatogenic cycle of *Dermophis mexicanus* (Amphibia: Gymnophiona. J. Herp. 29: 119-122.

Wake MH. 1998. Cartilage in the cloaca: Phallodeal spicules in caecilians (Amphibia: Gymnophiona). J. Morph. 237: 177-186.

Wiedersheim R. 1879. Die Anatomie der Gymnophionen. Jena, Gustav. Fisher.

Wrobel KH, F Süß. 2000. The significance of rudimentary nephrostomial tubules for the origin of the vertebrate gonad. Anat. Emb. 201: 273-290.

5

Oogenesis and Female Reproductive System in Amphibia—Anura

Maria Ogielska and Jolanta Bartmańska

Structure of Ovaries in Anura

Structure of Ovaries in Adults

Anatomy: Ovaries in anuran amphibians are paired organs lying on the ventral side of the kidneys. In adult females ovaries are much bigger than kidneys, because they are filled with diplotene oocytes at various stages of development. Ovaries are composed of several lobes called also ovarian sacs (Figs. 1A,C,E and 6A). The left ovary is usually bigger than the right one, and contains 1-3 sacs more. Ovarian sacs are well seen in juvenile females and become less distinct in the adults. The number of sacs is species specific and in most cases ranges from one to more than twenty. Anatomical studies on the number of ovarian sacs were carried out in several species by Bhaduri and Basu (1957). They recorded 1 lobe in *Arthroleptis sylvaticus*; 3 in *Chrysobatrachus cupreonitens*; 4-5 in *Phrynobatrachus natalensis*, *Afrixalus fluvovittatus leptosomus*, *Hemisus marmoratum guinnense*; 5-6 in *Rana (Conraua) cassipes*, *Phrynobatrachus versicolor*, *Cacosternum boettgeri*, *Chiromantis rufescens*, *Hyperolius viridiflavus coerulescence*, and 6-7 in *Rana subsigillata*. More recently Ogielska and Kotusz (2004) described the number of ovarian sacs as 6-9 (and sometimes 10) in *Rana lessonae*, *R. ridibunda*, *R. temporaria*, *R. arvalis*; 8 in *Bombina bombina*; 10-12 in *Hyla arborea*, *Bufo bufo*, *B. viridis*; 13 in *Pelobates fuscus*, and the highest number of 23 lobes in *Xenopus laevis*. The shape of ovarian lobes is variable (Fig. 1A,C,E), and can be pyramidal with tops oriented towards the mesovarium (*P. fuscus*, *R. lessonae*), irregular (*R. temporaria*) or strongly folded (*B. viridis*).

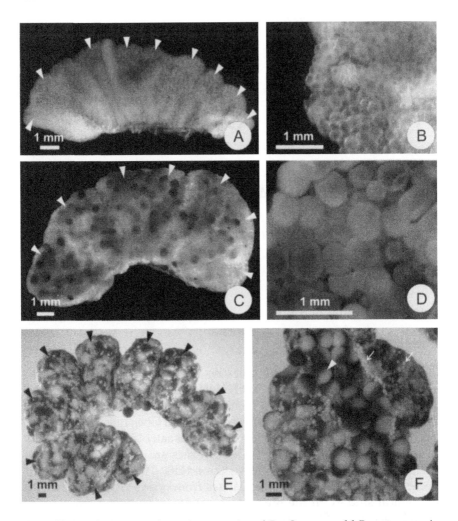

Fig. 1. Shape of ovaries and ovarian sacs. A and B—One year old *Rana temporaria*; the ovary is composed of 9 ovarian sacs (arrowheads) filled with class 1 and class 2 diplotene oocytes seen in higher magnification in B. C and D—Two year old *Rana temporaria*; the ovary is composed of 7 ovarian sacs filled with class 1, 2, and 3 oocytes, which are opaque due to the presence of forming yolk. E and F—Two year old *Rana lessonae* before the first ovulation; the ovary is composed of 10 ovarian sacs filled with class 1 and 2 oocytes distributed in the external layer of ovarian cortex (arrows), and full developed class 5 oocytes (arrowhead in F) with heavy pigmented animal hemisphere, and unpigmented vegetal hemisphere.

The ovaries are enveloped by peritoneum, which fuses at the dorsal part, thus forming a double-layered sheet called *mesovarium*. It attaches each ovary to the dorsal body cavity wall. The peritoneal envelope is composed of the outer ovarian epithelium and internal connective tissue layer. These two layers form the *theca externa*. The connective tissue is also distributed throughout the cortex and forms a stroma composed of fibrocytes, collagen bundles, numerous blood vessels, smooth muscle fibers, and nerves. The inner lining of the ovarian cortex is composed of one-layer inner ovarian epithelium, which faces the ovarian lumen (Rugh, 1951; Dumont and Brummett, 1978). Each cell of the outer ovarian epithelium posses a short cilium, which project into coelom from the approximate center of the cell; the cells of the inner epithelium have no cilia. In both kinds of ovarian epithelia the adjacent cells overlap and are connected by desmosomes (Dumont and Brummett, 1978).

Between the outer and the inner ovarian epithelium several layers of diplotene oocytes form the ovarian cortex. The cortex thickness depends on both the number and size of the oocytes (depending on age of a female and stage of ovarian cycle) and also on the total number of ovarian sacs. As a rule, species with numerous ovarian sacs, as in *X. laevis*, have a thinner cortex.

Ovarian Follicles: The cortex of an adult ovary contains hundreds or thousands (depending on species) of ovarian follicles, each of which envelops a diplotene oocyte. Each follicle is composed of one layer of epithelial follicle cells and a thin layer of connective tissue (*theca interna*), which is continuous with the stroma (Figs. 7C and 9A,B,F). The connective tissue of *theca interna* in vitellogenic oocytes contains tiny blood vessels, but not muscle fibers and nerves (Dumont and Brummett, 1978).

The morphology of follicle layers changes during oocyte growth. The follicle cells of class 1 oocytes (Dumont, 1972; for oocyte classes see the section *Oogenesis in Anura*) are flat and closely apposed to the oocyte surface (Fig. 9A) with basal membrane lying externally toward the *theca* (Dumont, 1978). The *theca interna* is thin, composed of flat fibrocytes, collagen bundles and small blood vessels; it becomes enhanced with more collagen bundles when oocytes enter class 2 (Fig. 9B,D). Follicle cells of class 2 oocyte become higher, with more abundant mitochondria and reticulum. Follicle cells and *theca* increase in height and thickness until the oocyte ends class 4. Afterwards they become flatter and thinner, and the space between them enlarges (Fig. 17C). At that stage the follicle cells are stellate and their processes contact the neighboring cells by desmosomes, but large spaces intervene between the regions of contact. These spaces form channels between the oocyte surface and *theca interna*. Functionally, the channels provide a pathway between the circulatory system and the surface of the oocyte. This is especially important for the uptake of yolk precursors (vitellogenin), which is synthesized in the liver and transported with blood to the oocyte surface, where is sequestered by endocytosis.

The number of follicle cells in class 6 oocyte of *Xenopus laevis* reaches about 6,000 (+/- 1,000) (Horrel, 1987). The apical part of follicle cells, facing oocyte surface, form extensions (macrovilli). They penetrate the acellular matrix secreted by the oocyte (oocyte or vitelline envelope) and are in structural and functional contact with the surface of underlying oocyte (Fig. 9B,C). Contacts of the both cell membranes are reinforced by formation of gap junctions situated tip-to-tip, and side-to-side between microvilli of the oocyte and macrovilli of the follicle cells (Fig. 6C) (van den Hoef, 1984). Gap junctions in *Bufo arenarum* and *Ceratophrys carnwelli* are composed of connexin C×43 and C×32. The number of gap juntcions is high during oocyte growth and decreases markedly (to as little as none) during the quiescent state of the oocyte and before oocyte maturation (Sánchez and Vilecco, 2003). Fully-grown oocytes are ovulated at metaphase of the second meiotic division by small rupture of *theca externa*. Follicle cells together with *theca interna* form post-ovulatory follicles, which soon become filled with proliferating follicle cells and finally degenerate.

Development of Ovaries and Differentiation of Ovarian Cortex

Witschi (1929) was among the first authors, who described ovary differentiation and development at the light microscope level. He named several structures in developing ovaries and this nomenclature is still in use. Ovary differentiation has been studied in a number of amphibian species. The studies were performed in the following genera and families: *Rana* (Ranidae) (Witschi, 1929; Cheng, 1932; Grant, 1953; Kemp, 1953; Mizell, 1964; Jørgensen, 1981; Wang and Hsu, 1974; Ogielska, 1990; Ogielska and Wagner, 1990, 1993; Wagner and Ogielska, 1990, 1993; Gramapurohit *et al.*, 2000; Saidapur, 2001; Falconi, 2001); *Bufo* (Bufonidae) (Tanimura and Iwasawa, 1987; Jørgensen, 1973c, 1974, 1984, 1976), *Xenopus* (Pipidae) (Al-Mukhtar and Webb, 1971; Dumont, 1972; Kress and Spornitz, 1972, Coggins, 1973; Iwasawa and Yamaguchi, 1984); *Bombina* (Discoglossidae) (Lopez, 1989); *Rhacophorus* (Rhacophoridae) (Tanimura and Iwasawa, 1989). Recently a comparative study on ovary differentiation and development of 12 species belonging to 6 families (Pipidae, Bombinatoridae, Pelobatidae, Bufonidae, Ranidae and Hylidae) was published by Ogielska and Kotusz (2004).

Based on observations from the above studies it seems clear that the pattern of ovary development is the same in all species studied so far. Ogielska and Kotusz (2004) divided the whole process into ten stages according to presence and abundance of successive generations of germ cells, beginning with primordial germ cells (PGCs) until fully-grown diplotene oocytes. Schematic drawings of the ten stages is shown in Figure 2: I-III—undifferentiated gonad, IV—sexual differentiation, V—first nests of meiocytes, VI—first diplotene oocytes, VII-IX—increasing number of diplotene oocytes and decreasing number of oogonia and nests, X—fully developed ovary composed of diplotene oocytes and rudimental patches of oogonia. Stages I–III are described in more details in the part 'Undifferentiated Amphibian Gonad' in this volume.

Stage I: Primordial germ cells (PGCs) have already immigrated from the forming gut into gonadal ridges. The gonadal ridges are formed as two lateral folds of the dorsal gut mesothelium, below the mesonephros. In all species studied so far, tadpoles achieve this stage at somatic Gosner stages 24-25. The PGCs are localized inside the ridges in one or few rows, later forming groups. In comparison to somatic cells of the forming gonad, the PGCs are large and loaded with yolk platelets. They start to multiply mitotically and gradually loose yolk platelets (vitellolysis).

Stage II: PGCs continue to multiply. After vitellolysis is completed, the germ cells are called primary gonia (in the case of ovary development the gonia will be called primary oogonia despite the fact that the gonads are not yet sexually differentiated). The primary oogonia are individual cells, like single PGCs, which are still present.

Stage III: Somatic cells of the external gonad epithelium proliferate intensively and migrate toward the central part of the gonadal ridges. The gonad is now composed of a central core (medulla) of somatic cells filling a "primary gonadal cavity" (according to Witschi's nomenclature, 1929), and a cortex still containing a few PGCs, but mainly composed of the growing number of primary oogonia. The medulla is solid, but the cells are not uniformly distributed, thus forming metameric "knots" reflecting the arrangement of future ovarian sacs (Fig. 3A).

Stage IV: The earliest features of sexual differentiation can be detected beginning with this stage. The somatic cells of the solid medullar knots gradually disappear (Fig. 3B) and the forming lumen of the secondary ovarian cavity becomes lined by the epithelium. The mitotic activity of the primary oogonia is high; some of them enter the last mitotic cycle with incomplete cytokinesis, thus forming cytoplasmic bridges. Such oogonia form a group (nest) of secondary oogonia (Fig. 3B; 15A). The nest is separated from neighboring nests and primary oogonia by a layer of somatic prefollicular cells, which do not penetrate between the secondary oogonia within the nest. Beginning with this stage, blood vessels and mesenchyme cells invade the ovarian cortex through a narrow space (remnants of the primary cavity) between the cortex and the medulla.

Stage V: The most proximal part ("progonad", Witschi, 1929) of an

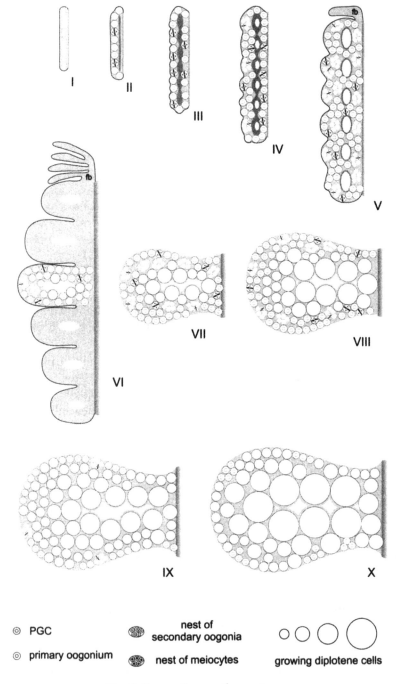

| ⊚ PGC | ⬤ nest of secondary oogonia | ○ ○ ◯ ◯ |
| ⊚ primary oogonium | ⬤ nest of meiocytes | growing diplotene cells |

Fig. 2. See caption on the next page.

ovary differentiates into a fat body (Fig. 4A). The lumen and epithelium of the secondary ovarian cavity are well differentiated at this stage (Fig. 4B). The size of the ovary increases due to the enlargement of the lumen and the growing thickness of the cortex. The secondary oogonia enter meiosis and the cortex is now composed of primary oogonia and nests of secondary oogonia or meiocytes. Nuclei of meiocytes are round and smooth in outline.

Stage VI: The ovary still grows in size, the fat body is distinct and becomes more and more separate from the gonad proper (Fig. 5A). First diplotene oocytes appear, no longer arranged in nests. The prefollicular cells of the nest wall proliferate and separate the individual diplotene oocytes, thus giving rise to oocyte follicles. The cortex is now composed of diplotene cells, nests of meiocytes, nests of secondary oogonia, and the outermost layer of single primary oogonia (Fig. 5B).

Stage VII: Beginning with this stage, the ovarian growth depends mainly on the increasing number and size of diplotene oocytes generated from the meiocyte nests. Once pachytene oocytes of a nest transform into diplotene cells, each oocyte starts to be enveloped by its own follicle cells. Thus the nest does not exist any longer, giving rise to a group of individual diplotene oocytes. The diplotene cells protrude into the ovarian lumen (secondary cavity), diminishing its size and changing its shape from spherical into an increasingly thin and flat space.

Fig. 2. Schematic illustration of ovary development and differentiation in anuran amphibians. The model represents an individual with 6 ovarian sacs. Stage I—undifferentiated gonad containing primordial germ cells (PGCs) loaded with yolk. Stage II—undifferentiated gonad; PGCs transform into gonial cells with high mitotic activity. Medulla (deep gray) is formed inside the central part of gonad. Stage III—undifferentiated gonad; medulla increases in size and forms metameric "knots". The cortex is filled with increasing number of dividing gonia. Stage IV—Sexual differentiation of an ovary; medulla starts to diminish by forming a lumen inside each knot. Some gonial cells (primary oogonia) enter the cell cycle, which ends in formation of nests of secondary oogonia. Stage V—First nests of secondary oogonia enter meiosis, thus forming nests with early primary meiocyes (leptotene-pachytene). The cortex becomes thicker, whereas the medulla is replaced by a lumen of ovarian cavity. Stage VI—First pachytene meiocytes start to transform into diplotene cells, which protrude into the ovarian cavity. VII—Diplotene oocytes grow in number and size, and are situated in the inner part of the cortex, whereas the outermost layer of the cortex is composed of primary oogonia and nests. VIII—Further growth of diplotene oocytes accompanied by thinning of the outermost layer of the cortex. IX—Primary oogonia and nests are present in thin areas. Stage X—Full-grown ovary. The cortex is filled with diplotene oocytes of various classes, and primary oogonia are restricted to small germ patches.

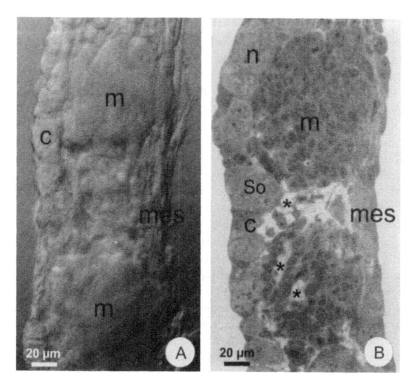

Fig. 3. *Rana temporaria.* Part of early ovary at stage IV with two future ovarian sacs now seen as two metameric thickenings with medullar "knots" inside, and peripheral cortex. A—Total preparation in the Nomarski optics. B—Longitudinal semithin section stained with toluidine blue; the lumen is formed inside medullar tissue (asterisks). c—cortex; m—medulla; ne—nests of secondary oogonia; mes—mesovarium.

Stages VIII and IX: The growth of the ovary continues as a result of the increasing number of diplotene cells, which also grow in size (20-50 μm in diameter). The fat body grows as well and becomes divided into several finger-like processes (Fig. 6A). Primary oogonia and nests of secondary oogonia and early primary oocytes are restricted to decreasing areas in the external parts of the cortex (Fig. 7A,B) or in septae between the ovarian sacs. They still multiply, although the rate of mitotic divisions decreases. The thickness of the cortex increases gradually and becomes about tenfold as thick as in ovarian stage II. The metameric ovarian sacs are more or less distinct, depending on the species.

Stage X: This stage represents the fully developed ovary, i.e., the cortex is composed almost entirely of diplotene cells (Figs. 1B,D,F; 7C). In juvenile ovaries, diplotene oocytes are previtellogenic; before first ovulation in a

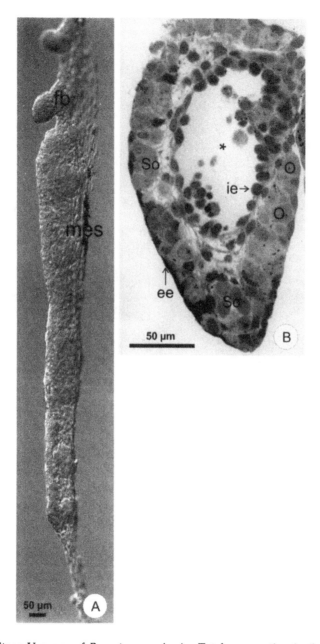

Fig. 4. Stage V ovary of *Rana temporaria*. A—Total preparation in the Nomarski optics. B—Cross semithin section stained with toluidine blue of one of the ovarian sacs. The lumen of the ovarian cavity is marked by an asterisk; O—primary oogonia; So—secondary oogonia; ie—internal epithelium lining of the lumen; e—external epithelium; fb—fat body; mes—mesovarium.

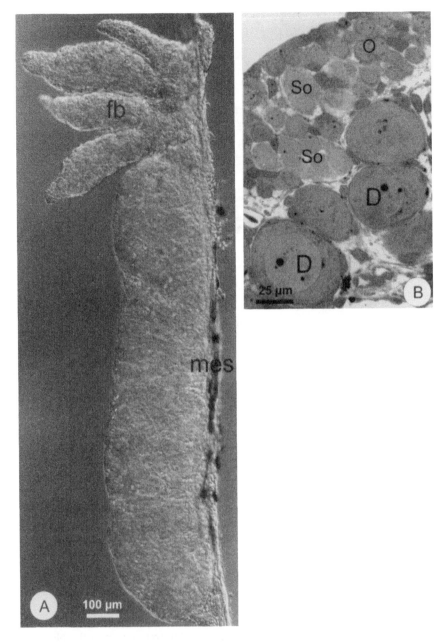

Fig. 5. Stage VI ovary of *Rana temporaria*. A—Total preparation in the Nomarski optics. B—Cross semithin section stained with toluidine blue of a part of an ovarian sac. First diplotene oocytes appear. O—primary oogonia; So—secondary oogonia; D—diplotene oocytes; fb—fat body; mes—mesovarium.

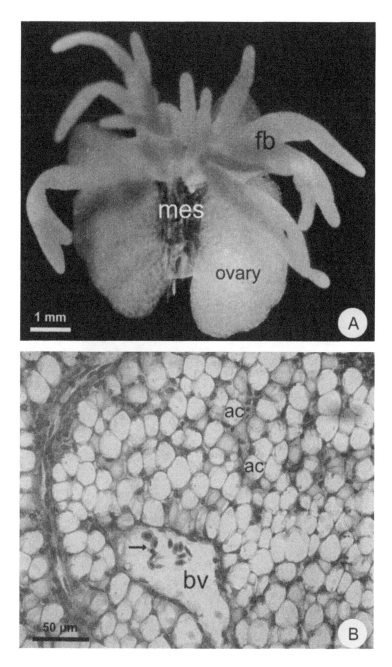

Fig. 6. Stage X ovary of *Rana temporaria*. A—Total view. B—Section of the fat body stained with HE. fb—fat body; mes—mesovarium; bv—blood vessel with erythrocytes (arrow); ac—adipose tissue cells.

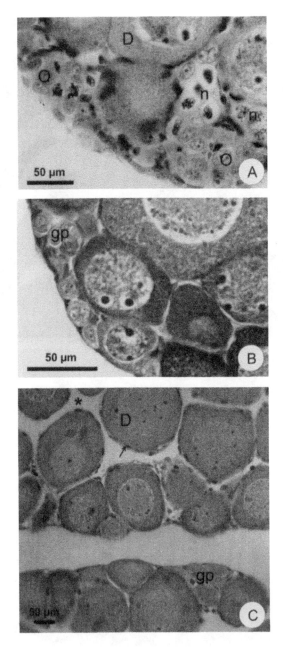

Fig. 7. Stage VIII (A) and IX (B) ovary of *Rana temporaria,* and stage X (C) ovary of *Rana lessonae.* D—diplotene oocyte inside a follicle (arrow); gp—germ patch; O—primary oogonia; ne—nests of meiocytes; the lumen of the ovarian cavity is marked by asterisk. A and B stained with HE, C stained with iron hematoxyline.

young female frog, the cortex is composed of diplotene previtellogenic and vitellogenic oocytes (classes 1-5 according to Dumont, 1972). The oocytes are arranged in a gradient from the smallest in the most external part of the cortex to the largest adjoining the ovarian cavity wall (lumen). Each diplotene oocyte is enveloped by a follicle composed of a layer of epithelial (follicle) cells and theca, composed of connective tissue, fibers of smooth muscle cells, small blood vessels and nerves. The lumen gradually decreases in size and becomes a thin and flat space. The oogonia (both primary and secondary) are restricted to the outermost parts of the cortex. They usually form groups ("germ patches", Witschi, 1929), scattered among the loose connective tissue (Fig. 8A,B). They differ in size and number, from one to several dozens of cells. The germ patches are enveloped by somatic cells, which also penetrate inside the patch, thus dividing it into smaller groups. Some of these groups are nests of secondary oogonia, between which cytoplasmic bridges are observed. The nests of leptotene zygotene and pachytene cells are rarely observed. The secondary oogonia of the mature ovary are ovoid cells with irregular nuclei. One of the most striking features of oogonia in the germ patches of a mature ovary is their degeneration (Fig. 25D). In the light microscope they appear as dense masses (Fig. 8B). Their cytoplasm is filled with vacuoles, folded membranes, annulate lamellae, and large space between the nucleus and the cytoplasm. Mitochondria degenerate and gradually transform into lamellar bodies.

The rate of somatic development of anuran amphibians is only roughly correlated with the rate of gonad differentiation and varies among species. The relative difference between the differentiation of the somatic and germ line cells is one of the major problems of heterochrony (for review see Raff, 1996). The duration of somatic development of tadpoles, as well as the time of metamorphosis, differs among species and may be additionally affected by many environmental conditions. For ovary differentiation, tadpole age is more crucial than stage of somatic development. Generally, three rates of ovary differentiation, as related to somatic growth, were distinguished: basic, accelerated, and retarded (Ogielska and Kotusz, 2004). Most species studied so far represent the basic rate; accelerated rate was described in green frogs (subgenus *Pelophylax*), and retarded rate was described for genus *Bufo*. According to results published by del Pino *et al.* (1986), the egg-brooding hylid frog *Gastrotheca riobambae* represents the accelerated pattern of ovary differentiation. The same authors also mentioned that no clear relationship was observed between the external characteristic of larvae and the gonad development. The comparison between somatic and gonadal stages for a basic rate of gonadal development in *Rana temporaria* is shown in Table 1.

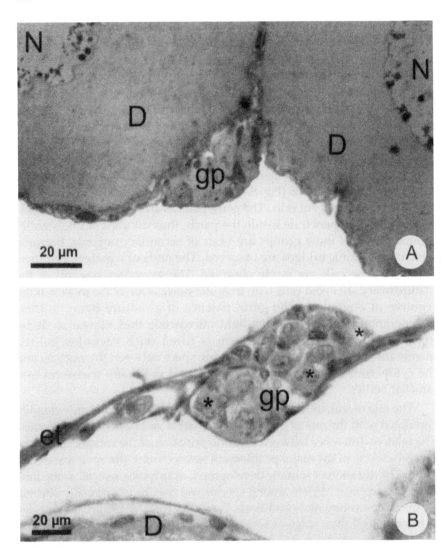

Fig. 8. Germ patches in stage X ovaries of *Rana temporaria*. A—Germ patch in the connective tissue close to internal theca of the ovarian follicle containing diplotene oocytes. B—Germ patch in the outer theca. Degenerating primary oogonia are marked by asterisks. et—external theca; gp—germ patch; D—diplotene oocyte; N—nucleus. A—Semi-thin section stained with toluidine blue; B—HE staining.

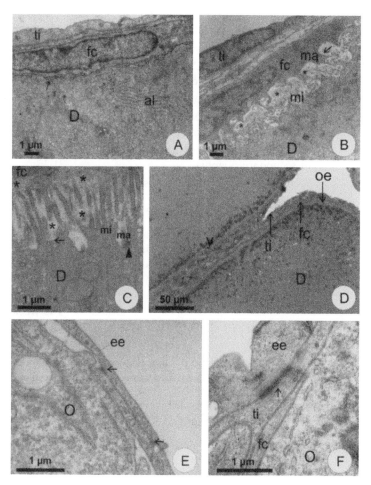

Fig. 9. Structure of follicle and ovarian walls. A—Class 1 diplotene oocyte; cells of theca interna (ti) and follicle cells (fc) are flat, and microvilli on the oocyte surface (arrows) are not well developed. B—Class 2 oocyte. Follicle cells produce large macrovilli (ma), and the oocyte surface has microvilli. Amorphous material for oocyte envelope accumulates between in the space between follicualr cell and oocyte surface (asterisks). C and D—Early vitellogenesis in class 3 oocytes. The distance between follicle cells and oocyte surface enlarges and is penetrated by highly elaborated microvilli and macrovilli; the oocyte envelope becomes thicker (aterisks). The material for yolk (Y) is accumulated by oocyte by endocytosis (arrow in C). E—External epithelium of a young gonad. Epithelial cells are reinforced by desmosomes (arrows). F—External wall of a more advanced ovary. External epithelial cells and theca cell are connected by desmosomes (arrow). al—annulate lamellae; ee—external epithelium; fc—follicle cells; mi—microvilli; ma—macrovilli; O—oocyte; ti—theca interna. A-C and F— TEM; D—Semi-thin section stained with toluidine blue.

Table 1. Comparison of somatic and gonadal stages in *Rana temporaria*, the temperate zone species with the basic rate of ovary differentiation

Gonadal stage[1]	Description of gonads	Somatic stage
I	PGCs in gonadal ridges	25-27[G]
II	Gonia in undifferentiated gonad	28[G]
III	Somatic cells form medulla and cortex; gonia multiply in cortex	29-30[G]
IV	Sexual differentiation of gonad; first nests of secondary oogonia	31-35[G]
V	Secondary oogonia enter meiosis	36-40[G]
VI	First diplotene oocytes	41-46[G] metamorphosis
VII	Nests and oogonia in diminishing external part of cortex; Increasing number of diplotene oocytes; Gradual loss of nests decreasing mitotic activity of primary oogonia;	1-5 weeks AM
VIII		5-13 weeks AM
IX		1-year-old juveniles
X	Cortex composed of diplotene oocytes; Primary oogonia restricted to germ patches	2-years-old juveniles and adults

1—According to Ogielska and Kotusz (2004).
G—According to Gosner (1960).
AM—after metamorphosis.

Bidder's Organ

The most anterior part of the gonad in species belonging to the family Bufonidae differentiates in a different way and is known as the Bidder's organ (Fig. 10A). It is much more prominent in males than in females, because in females its structure is similar to the rest of ovarian lobes. For this reason the Bidder's organ has been primarily described as characteristic of males. However, the morphology and results of some experiments make the Bidder's organ more suitable to describe in the section devoted to females. For this reason the following paragraphs will involve the information gathered for both males and females.

Although the presence of the Bidder's organ in male bufonids has been known since the 19th century (Spengel, 1879, in: Farias *et al.*, 2002), its role, differentiation and structure are still not fully understood. In *Bufo marinus* the Bidder's organ has a yellow-red color, is slightly translucent, and measures 3.2 mm in length. Recently Farias *et al.* (2002) reported the histological and ultrastructural study of adult males of the Brazilian species *Bufo ictericus*. The Bidder's organ in this species is encapsulated by the sheet of a mesothelium composed of the loose connective tissue. It is composed of the cortex containing diplotene oocytes at various stages of development and the medulla with large blood vessels (coiled arteries and convoluted veins). The connective tissue of the medulla is rich in acid carboxylated polymers and collagen fibers. The collagen fibers are localized also around the follicles, as was reported for *Nectophrynoides malcolmi* (Wake, 1980), *Bufo japonicus formosus* (Moriguchi *et al.*, 1991; Tanimura and Iwasawa, 1992), and *B. woodhousii* (Pancak-Roessler and Norris, 1991). The ultrastructure of Bidderian oocytes resembles that of the ovarian diplotene oocytes of the female, with multiple nucleoli and lampbrush chromosomes. The smooth endoplasmic reticulum is well developed, whereas the rough reticulum is rarely observed. The annulate lamellae are occasionally observed, and the Golgi structure is poorly developed. The mitochondria are elongated with flattened cristae. Peroxisomes and lipid droplets were also observed. The yolk platelets were not described. Each diplotene oocyte is surrounded by the follicle cells. The space between the oocyte and the follicle is penetrated by microvilli. The oocyte surface is covered by the PAS-positive amorphous oocyte envelope.

Pancak-Roessler and Norris (1991) studied the relationship between the testes and the Bidder's organs in adult males of *Bufo woodhousii* from Colorado, with emphasis on the mechanism of inhibition of oogenesis in Bidderian diplotene oocytes. One group of animals had bilaterally removed testes (orchidectomy), whereas the other was intact. They treated both groups with gonadotropins (pregnant mare serum gonadotropin, PMSG, and human chorionic gonadotropin, hCG) and found that diplotene oocytes in the Bidder's organ of the castrated males begun to develop. Normally the oocytes are in pre-vitellogenic stages and never grow further; in orchidectomized males—even without administration of gonado-tropins—the oocytes started the uptake of yolk and the weight of the Bidder's organ increased about twofold as a result of both the growth of the preexisting oocytes and recruitment of new small oocytes into growth phase. The comparison between the ovarian and bidderian oocytes from the adult male *Bufo marinus* has been recently studied by Brown *et al.* (2002). These authors undertook the molecular analysis of the nuclear-lamina-associated polypeptide 2 (LAP2), which is differentially expressed in somatic cells (LAP2β) and oocytes (LAP2ω). The somatic LAP2β was expressed also in bidderian oocytes, but after orchdectomy LAP2ω was detected and the bidderian oocytes start to grow. These results suggest that the testes play a major role in the inhibition of oogenesis, and the high

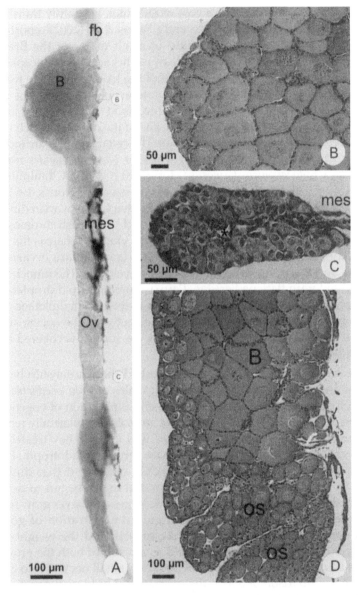

Fig. 10. Bidder organ of juvenile female *Bufo viridis*. A—Total view of the Bidder organ (B) and ovary (Ov) in Nomarski optics. The dotted lines indicate the levels of cross secions shown in B and C, respectively. B—Cross section of the Bidder organ containing diplotene oocytes. C—Cross section of the ovary, below the Bidder organ with nest of meiocytes (stage V ovary). D—Longitudinal section of a part of ovary of a specimen older than in A-C, with Bidder organ (B) and two ovarian sacs (os). Mes—mesovarium; fb—fat body. B-D—HE staining.

level of circulating gonadotropins did not overcome the inhibitory effect of androgens. As was suggested by Pancak-Roessler and Norris (1991), the Bidder's organ probably possesses receptors for gonadotropin and is able to respond in a manner similar to ovary. Hypophysectomy causes a decrease in the Bidder's organ weight suggesting that it is influenced by pituitary hormones (Penhos and Cardeza, 1952 *fide* Pancak-Roessler and Norris, 1991).

The Bidder's organ develops from the anterior part of the gonad proper, next to the fat body (Viertel and Richter, 1992). Its differentiation with comparison to the gonad proper was studied in *Bufo japonicus formosus* by Tanimura and Iwasawa (1986, 1987). In this species the Bidder's organ starts to differentiate in larvae 6 days after hatching, at stage corresponding to Gosner's 24/25. At this time the germ cells were scattered among the somatic cells of a gonad. During premetamorphic stages the Bidder's organ gradually increases, and rapidly enlarges at the time of metamorphosis. Neither definite cortico-medullary structure, nor primary gonadal cavity was observed throughout the further development, although a small space appears in the center of the Bidder's organ at the same time when the secondary ovarian cavity is formed in ovaries. At the premetamorphic stages, the germ cells grow rapidly in volume, resembling young diplotene oocytes (Fig. 10B), whereas the rest of the gonad follows normal pattern of ovarian development (Fig. 10C). Their number grows until metamorphosis and does not change thereafter. Normal meiocytes appear at the periphery of the Bidder's organ later, i.e., about the same time as in developing ovary (Fig. 10D) (in females, and in corresponding time in males). At the time when ovarian diplotene oocytes appear, the Bidderian oocytes start to degenerate. The role and nature of these cells are unknown.

Fat Body

The fat bodies are composed of adipose tissue (Fig. 6A,B) and serve as nutritional reserve both in males and females. They are formed from the most anterior part of the undifferentiated gonads, the progonads (for details see the part 'The Undifferentiated Amphibian Gonad' in this volume). This morphological observation was recently supported by molecular analysis of germ-cell related mRNAs. Chan *et al.* (1999, 2001) isolated and characterized the *fatvg* gene from the *Xenopus laevis* germinal plasm. *fatvg* is a homologue of a mammalian gene expressed in adipose tissue and in a tadpole gonad is expressed in the fat body. *fatvg* is originally associated with the germ plasm located in germ cell precursors, but before the colonization of germinal ridges (during larval tail bud stages) the two cell lineages segregate. One of them will form PGCs, whereas the other will form precursors of the fat body. The fat body precursor cells, which form two masses of cells at the dorso-anterior part of the endodermal margin, colonize the anterior part of the genital ridge and later transform into fat bodies.

Prasadmurthy and Saidapur (1987) studied the role of the fat body on oogenesis and vitellogenesis in the non-hibernating skipper frog *Rana cyanophlyctis*. In this species the fat bodies represent from about 1% (in pre-breeding phase) to about 2% (in post-breeding phase) of the body weight. In a series of 30 or 60 day-long experiments with the bi- or unilateral fat body excision they revealed that in frogs without fat bodies the weight of ovaries and the number of large-sized second growth phase diplotene oocytes decreased, whereas the number of atretic follicles increased. The same treatment followed by hCG or fat body extract administration reduced the effect of the fat body ablation. On the other hand, the long-term observation of frogs with removed fat body revealed that finally the ovary mass increased and the number of atretic follicles decreased. The bilateral ovariectomy caused the increase of the fat body mass. On the basis of these findings the authors concluded that fat bodies in *R. cyanophlyctis* have a supporting role in vitellogenesis as an immediate source of lipids needed for follicle growth. However, when frogs were fed properly, they continued vitellogenesis after an initial decrease of the ovary weight, apparently using another source of lipids. Any role of fat bodies in oogenesis and vitello- genesis was also reported by Jørgensen (1986) in *Bufo bufo*. After unilateral fat body excision there was no effect on the growth of oocytes in the ipsilateral ovary. Our own observations (Ogielska, unpublished) of deve- lopment of gonads in the European hybridogenetic water frog *Rana esculenta* are in agreement with the observation of Prasadmurthy and Saidapur (1987), who stated that the lack of gonads causes the increase of fat body weight. In *R. esculenta* the gonads are often poorly developed or sterile; in such cases the fat bodies are abnormally big.

Female Reproductive Tract

Oviduct and Cloaca

After ovulation the oocytes are released to the body cavity and must find their way to oviducts. The body cavity, oviduct and cloaca were described in details in the frog *Rana pipiens* by Rugh (1951). Contrary to the male, the body cavity of a female is almost entirely lined by the ciliated epithelium. Cilia are absent from the peritoneum of the gut and kidneys. The action of cilia enables the hundreds or thousands of eggs ovulated by large ovaries to reach any of the two oviductal openings. As was described by Barth (1953), the cilia literally sweep the eggs all the way from the ovary to the opening of the oviduct. Experiment carried out by Rugh (1951) with artificial object of similar size as eggs revealed that the cilia are active all the year round and can transfer the object into the oviducts. The eggs reach the oviducts within 2 hours after ovulation. Each oviduct is suspended by a fold of peritoneum and attached to the dorsal side of body cavity.

At the anterior end, the oviduct is open by an ostium, slit-like when empty, and round at the moment of egg entrance. The ostium leads to the infundibulum, which walls are ciliated and highly elastic. Next to the infundibulum there is a relatively long oviduct. The function and size of oviducts depends on steroids produced in the ovaries, especially E2 (see the section *Regulation of oogenesis in Anura*). Each oviduct is composed of two parts: the short and relatively straight *pars recta*, and long, highly coiled *pars convoluta* (Fig. 11). These two parts are anatomically and functionally distinct. The posterior end of the oviduct is enlarged and forms the ovisac (*uterus*). The walls of the entire oviduct are composed of 3 layers: internal epithelium, fibrous connective tissue, and external peritoneal coat.

Pars Recta: This portion is 1.5 cm long in *Rana japonica* and constitute 1/20 (5%) of the total length of the oviduct (Yoshizaki and Katagiri, 1981), and about 3 cm long in *Bufo japonicus* (Hiyoshi *et al.*, 2002) and *Bufo arenarum* (Fernandez *et al.*, 1997). In *Xenopus laevis* the *pars recta* occupies the anterior 5.5% of the total length (about 40 cm) of the oviduct and is divided into two portions: anterior one fifth and posterior four fifths, responsible for the formation of the pre-fertilization layer (see below for details) (Yoshizaki and Katagiri, 1984). In *Rana japonica*, Yoshizaki and Katagiri (1981) showed in SEM and TEM that the inner surface of the *pars recta* has longitudinal folds, which run spirally along the axis of the tube. The surface of epithelial cells is covered with dens cilia and microvilli. The cilia are higher in the uppermost part, near the ostium. Secretory cells (type I) are scattered among the ciliated cells. The secretory cells contain tiny granules of carbohydrate material 0.5 μm in diameter. This material is added to the vitelline envelope of the egg as it moves through the *pars recta*. After oviposition the secretory cells are empty. The secretory function of the *pars recta* is hormonally controlled by 17 β-estradiol and progesterone (Fernandez *et al.*, 1997).

Pars Convoluta: This portion is the longest of the entire oviduct. In *Bufo arenarum*, Winik *et al.* (1999) divided it into 4 parts: the intermediate proximal zone (IPZ), the pars preconvoluta (PPC), the pars convoluta (PC), and the intermediate distal zone (IDZ). The inner surface of the entire oviduct is longitudinally folded and the epithelium consists of secretory and ciliated cells. Columnar ciliated cells are localized on the folds and also posses few short microvilli between the cilia. The nuclei are located medially and numerous mitochondria are located apically. Ciliary movement is responsible for propelling and rotation of the oocytes when

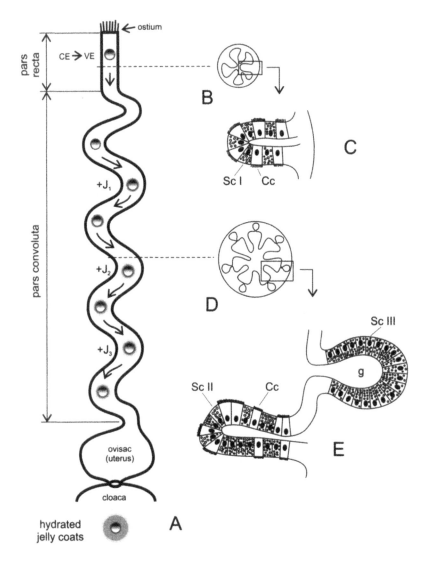

Fig. 11. Schematic representation of anuran oviduct (A) with an oocyte passing through the *pars recta* and *pars ovulate*. In the *pars recta*, the coelomic envelope (CE) of the oocyte is converted into the vitelline envelope (VE). In the *pars convoluta* the oocyte is invested with 3 jelly coats (J_1-J_3). B—Cross section of *pars recta* showing the folds of the inner oviductal wall. C—represents one of the folds composed of ciliated (Cc) and secretory cells type I (Sc I). D—Cross section of *pars convoluta* with folds of a wall, and glandular crypts (g) between the folds. E—represents the histological structure of the fragment inside the frame with ciliated cells (Cc) and secretory cells type II (Sc II) covering the folds, and glandular crypts composed of secretory type III cells (Sc III).

passing through the oviduct. The numerous secretory cells are scattered among the ciliated cells. These cells are large and columnar with cytoplasm filled with various kinds of secretory granules and produce material for jelly coats. The granules are released into the oviductal lumen by exocytosis. However, some of the cells release secretory products in the aporcine maner, in which cells bud off membrane-enclosed portions of the cell, or in holocrine manner, in which the entire contents of the cells are released following rupture of the plasma membrane. The nuclei of secretory cells are situated medially or basally and apical surface is covered with microvilli. Mitochondria, rough and smooth endoplasmic reticulum, and Golgi complexes are numerous. The secretory cells are fully developed during the pre-ovulatory period and degenerate by autolysis in the post-ovulatory phase.

In *Rana japonica* (Yoshizaki and Katagiri, 1981) and in *Bufo bufo japonicus* (Katagiri *et al.*, 1982) two types (II and III) of secretory cells have been described. Winik *et al.* (1999) described also the two types in *Bufo arenarum*: the secretory cells of the apical and lateral ridges on the folds and the pyramidal ciliated secretory cells at the bottom of the folds. The latter authors described also the basal cells located in the basal region of the epithelium, close to, or in contact with the basal lamina. These cells resemble the lymphocytes, most probably engaged in phagocytosis of cell remnants and further restoration of epithelial cells.

Between the folds of the surface epithelium the single and tubular glands are situated. Their cells are covered by microvilli and the secretion is released directly to the oviduct lumen. During the pre-ovulatory period the thickness of the glandular layer increases in the cranio-caudal direction (Yoshizaki and Katagiri, 1981; Winik *et al.*, 1999). Superficial epithelium and secretory glands in all parts of the oviducts undergo marked changes associated with ovulation. The height of cells decreased and the secretory granules are reduced in number. These changes are most probably under the hormonal control of the ovarian cycle.

The number of convolutions of the *pars convoluta* is various and species-specific, and ranges from few dozens to nearly two thousand (reviewed by Tyler, 2003), and its shape also varies from species to species (Bhaduri and Basu, 1957). It is usually highly coiled, but in some species (*Hyperolius viridiflavus coerulescens*) has only few coilings. In some species (*Rana (Ptychadena) mascarenicus, Phrynobatrachus versicolor, P. natalensis*) the *pars convoluta* can gradually enlarges until it reaches the uterus. In others (*R. (Aubria) subsigillata, Chiromantis rufescens*) the diameter of the anterior part is much narrower than in the posterior part. As was shown for *Ascaphus truei* (Sever *et al.*, 2001), the diameter of the oviduct is reduced in females after spawning season. The eggs are flattened and highly distorted when they pass through the oviduct, which takes from 2 to 4 hours in *Rana pipiens* (Rugh, 1951). Then they reach the highly elastic uterus (ovisac), where they are stored for about 1 day until oviposition during amplexus.

Uterus: Each oviduct enlarges in its most posterior part to form the uterus (ovisac). The two uteri of a female can fuse, thus forming one common uterus. The separate uteri are most common in Anura and were described in *Rana pipiens* (Rugh, 1951). The common uterus was described in *Rana subsigillata, Cacosternum boettgeri, Arthroleptis sylvaticus, Hyperolius viridiflavus coerulescens, Chrysobatrachus cupreonitens,* and *Hemisus marmoratum guineense* (Bhaduri and Basu, 1957). The separate uteri have two independent openings to the cloaca on the both sides of the uro-genital papilla, whereas the common uterus has only one opening. In young females the openings do not function and appear in the fully mature female frogs (Bhaduri and Basu, 1957).

During development of a female tadpole the oviducts appear around the time of metamorphosis as the Müllerian ducts. In the frog *Rana esculenta* their differentiation begins at Gosner stage 43 (Viertel and Richter, 1992). They extend posterior and lateral to the Wolffian ducts and are attached to the dorsal body wall. The Müllerian ducts originate near the pronephros funnel and reach the cloaca by their posterior ends. In tadpoles, the Mülerian ducts have a thin wall and are not coiled

Egg Envelopes and Their Role in Fertilization

The role of the oviduct is not only to transfer the ovulated oocyte from the body cavity to the cloaca, and then to the outside, but also to prepare the female gamete for further fertilization and to equip it with additional jelly envelopes. Recently the egg envelopes have attracted much attention, mainly because they play roles in sperm-egg interactions. They have specific sites for sperm binding and are necessary for the acrosome reaction and formation of a block to polyspermy.

Egg envelopes are generally classified according to the site of origin. The original classification of Ludwig (1874), which is also accepted today (Browder *et al.*, 1991), classifies the envelopes according to the cells engaged in their production. Hence, the primary envelope is produced by the oocyte, the secondary envelope—by the follicular cells, and the tertiary envelopes—by the oviduct. However, another classification is also commonly accepted, according to which the envelopes produced inside the ovary are primary envelopes and these produced by the oviduct—the secondary envelopes (Balinsky, 1965; Berrill and Karp, 1976). Unfortunately, these classifications together with various names given by authors for the same structures cause some confusion. Additionally, in amphibians the primary (i.e., vitelline) envelope is modified during migration through the oviduct. In order to clarify the nomenclature and to focus on the structural and functional modifications of the primary envelope, the following names will be used, irrespective of the original names given by the authors: oocyte envelope (OE); coelomic envelope (CE); vitelline envelope (VE), and fertilization envelope (FE).

Primary Oocyte Envelope: Eggs are ovulated from the ovaries directly to the body cavity. They are released from the ovarian follicles together with a thin (several µm) oocyte envelope. The oocyte envelope (OE), known also as vitelline envelope or vitelline membrane, is apposed on the oocyte surface (Fig. 9B,C; 17C). It is composed of glycoproteins and is secreted by the diplotene oocyte itself, as revealed by ultrastructure and immuno-cytochemical studies (Pinto *et al.*, 1985; Yamaguchi *et al.*, 1989; Wallace and Selman, 1990; Tian *et al.*, 1997). In *Xenopus laevis* it starts forming in late class 1 oocytes before cortical granules and yolk formation. At that time the follicle cells are pushed away from the oocyte surface, thereby forming the perivitelline space where the OE is formed. OE precursors are extruded from large vesicles, which are abundant in the cortical region and are observed only during early diplotene. This first secretory wave is followed by the next one in class 1 and 2 oocytes, when smaller vesicles appear, extrude new, more fibrillar material, which finally (class 5) forms bundles filling the gap (perivitelline space) between oocyte surface and follicle cells. Four major glycoproteins (gp120, gp69/64, gp43/41, and gp37) were detected in OE of *Xenopus laevis* (Yamaguchi, 1989; Tian, 1997). On the basis of sequence similarity to mammalian *zona pellucida* (ZP) components, they are also named ZPAX, ZPA, ZPB, and ZPC, respectively (Barisone *et al.*, 2003). They will form binding sites for cortical granule contents release by exoocytosis during fertilization.

The perivitelline space is penetrated by macrovilli of follicle cells and microvilli of the oocyte (Fig. 9B). Beginning with class 2 oocytes, the OE becomes continuous. Microvilli of class 1 oocytes are scattered and short, in classes 2 and 3 they become longer, more numerous and branched, being the longest and most elaborated in class 4 oocytes (Fig. 9A-C) (Dumont and Brummett, 1978; Pinto *et al.*, 1985). In fully grown oocytes, OE is bi-layered: the inner layer between microvilli is composed of the loosely packed fibrils, and the outer layer, in which fibrils are densely packed and oriented parallel to the oocyte surface. The fibrils are about 15 nm in diameter and form bundles 75-100 nm in diameter (Larabell and Chandler, 1989). The space occupied by microvilli looks like a light zone beneath OE when observed under the light microscope and for this reason it is sometimes called *zona pellucida*. Some of the glycoproteins of OE are proteolytically processed after ovulation and fertilization (Tian *et al.*, 1997).

The envelope of an ovulated oocyte is called the coelomic envelope (CE) and is composed of two kinds of filaments. In *Rana japonica* one kind forms a network parallel to the cell surface, whereas the other runs in perpendicular bundles (Yoshizaki and Katagiri, 1981). Similar composition of CE, but with predominant bundles (40-70 Å in diameter) parallel to the oocyte surface was also described in *Xenopus laevis* (Grey *et al.*, 1974), where the total thickness of the CE was about 1 μm. In *Bufo bufo japonicus* the CE is 6 μm thick and consists of filament bundles running both parallel and perpendicular to the egg surface (Katagiri *et al.*, 1982).

Dramatic changes in structure and molecular composition of CE are observed when an egg passes through the *pars recta* of the oviduct. The substances secreted by this portion of the oviduct modify CE and converts it into the vitelline envelope (VE). Experiments with pars recta extracts in *Bufo arenarum* showed that after 8 min of treatment the CE is affected, and after about 1 hour its partial dissolution is observed (Miceli and Fernandez, 1982; Mariano *et al.*, 1984). Katagiri *et al.* (1999) observed similar changes during CE to VE transition in *Xenopus laevis*. During that time the perpendicular bundles become less conspicuous and the network of parallel fibrils is destroyed. Similar changes were also described in *Bufo bufo japonicus* (Katagiri *et al.*, 1982). In this species the VE is about a half thinner than CE after passing the *pars recta*. The filament bundles are less conspicuous and fine electron-dense particles are seen in the interstices between them. Concomitantly, the carbohydrates produced by the type I secretory cells, unique to the pars recta, cover the egg surface and eventually penetrate into the space around oocyte surface (perivitelline space) (Yoshizaki and Katagiri, 1981; Katagiri *et al.*, 1982). At the same time the proteolytic conversion of glycoprotein components of CE was observed: from 40-52 kDa characteristic of CE to 39 kDa characteristic of VE in *Bufo* (Takamune *et al.*, 1986; Katagiri, 1987), and from 43 kDa to 41 kDa in *Xenopus* (Katagiri *et al.*, 1999). This biochemical change is controlled by the oviductin (66 kDa trypsin-like protease) (Hardy and Hendrick, 1992; Lindsay and Hendrick, 1998; Lindsay *et al.*, 1999; Hiyoshi *et al.*, 2002). Oviductin is transcribed in a hormonally-dependent way as a large mRNA, which is later extensively processed post-translatoinally. Oviductin mRNA is expressed in type I secretory cells localized in the bottom of the epithelial folds of the *pars recta* (but not in the *pars convoluta*). The *Bufo* and *Xenopus* oviductins have the same general molecular structure with two protease domains (α and β). It seems that only the α protease domain is biologically active. These changes, both structural and biochemical, are crucial for further fertilization, as was experimentally shown for *Rana pipiens* (Elinson, 1971b, 1973), *Rana japonica* (Yoshizaki and Katagiri, 1981), *Bufo japonicus*, *B. arenarum* (reviewed by Katagiri, 1987), and *Xenopus laevis* (Katagiri *et al.*, 1999). These authors demonstrated that the coelomic eggs are not penetrable by sperm and thus cannot be fertilized. On the other hand, the ceolomic eggs treated with the extract of the *pars recta* gain the fertilization ability. The results of these experiments lead to the conclusion that the oviductin is responsible for the conversion of CE to the sperm-penetrable form, VE.

Apparently the molecular composition and structure of the VE of the eggs, which have passed through the *pars recta*, are different from those of the coelomic eggs and represents a mixture of the primary and secondary envelopes. This change is a prerequisite for the sperm acrosome reaction to occur during fertilization. Anuran eggs are fertilized externally in fresh water immediately after spawning. The sperm acquire motile ability in water hypotonic in relation to the body fluids (see the Part 'Spermatogenesis and Male Reproductive System in Anura'). The sperm must penetrate jelly envelopes (see below) and VE to fuse with the oocyte plasma membrane. A sperm is incorporated almost perpendicular to egg surface and the entire process usually takes 3 - 8 min (Katagiri, 1987). After reaching the VE, the acrosome reaction starts in sprematozoa. In *Bufo bufo japonicus* the reaction involves the breakdown of the membrane-bounded acrosomal cap, the release of the acrosome vesicle lysine (spermatolysines) onto the egg surface, and the fusion of the inner acrosomal membrane with the oocyte plasma membrane (Yoshizaki and Katagiri, 1982). After the action of spermatolysins, the thickness of the VE increases about 3 times (Yoshizaki and Katagiri, 1981). The spermatolysins are tryptic or chymotryptic proteases, depending on species (Katagiri, 1987).

In *Xenopus laevis*, in which acrosome reaction takes place in contact with the VE, it was shown that VE posses a substance added in the *pars recta*, which has the abitlity to induce the acromsome reactions, and which is heat-unstable and Ca^{+2} dependent with a molecular weight greater than 10 kDa (Ueda *et al.*, 2002). The sperm bound to the VE undergoes the acrosome reaction within 12 min (Ueda *et al.*, 2002). The site of sperm entry is crucial for early embryogenesis in amphibians. It must take place in the animal hemisphere; otherwise the main axes of an embryo would not be properly induced (for review see Gerhart *et al.*, 1986; Ruiz i Altaba and Melton, 1990). Cabada *et al.* (1989) demonstrated in *Bufo arenarum* that both the VE and the plasma membrane in the animal and vegetal hemisphere differ considerably in ultrastructure and in response to spermatolysins. They distinguished 3 regions of the VE in uterine eggs: the outermost composed of fibrils with an irregular arrangement; the middle zone formed by a dense net of bundles of fibrils containing electron-dense granules; and the region limiting the perivitelline space. The middle zone is much thicker in the animal hemisphere. Also microvilli of the ooplasm were longer and more numerous in the animal hemisphere. After spermatolysins action, the microvilli of the animal hemisphere decreases in number and length, while in vegetal hemisphere no significant modifications were observed. The VE of the animal hemisphere becomes disorganized and only a thin middle zone was observed.

Within seconds after the sperm entry a rapid transient depolarization of oocyte membrane potential takes place (fast block to polyspermy), and then a wave of exocytosis of oocyte cortical granules spreads from the point of sperm entry over the entire egg surface (slow block to polyspermy) (Yamasaki and Katagiri, 1991). The latter phenomenon was originally described in *Rana pipiens* (Kemp and Istock, 1967; Goldenberg and Elinson, 1980) and *Xenopus laevis* (Grey *et al.*, 1974, 1976; Wolf, 1974). The average rate of the "activation wave", i.e., the movement of the front of cortical granules breakdown, was calculated as 10-16 μm/sec (Hara and Tydeman, 1979). Cortical granules are described in the section devoted to oogenesis. Five minutes after fertilization in *Bufo bufo*, the VE becomes smooth at the site of sperm entrance and this change spreads gradually over the whole egg surface within next 3 min (Semik and Kilarski, 1998). Upon fertilization VE is transformed into the fertilization envelope (FE). This transformation is possible in result of the release of cortical granule content (lectins), which penetrates through VE and is bound to the receptors localized in the innermost jelly layer (Yoshizaki, 1989). FE is preserved during early development of an embryo. Structurally, FE is composed of thin electron dense layer (fertilization layer) and more dispersed internal layer. Morphological observations at the electron microscope level of Grey *et al.* (1974) gave evidence that FE is formed of the material extruded from the cortical granules (lectins). These lectins are composed of two kinds of polypeptides, 46 kDa and 42 kDa (Yoshizaki, 1986). The extrusion of the cortical granules content during the cortical reaction begins 1.5-2 min after the sperm entry. After next 3-5 min. the cortical material is seen on the outer surface of VE, and the entire reaction is completed within 5-9 minutes (Wolf, 1974). However, more detailed studies on the fertilization layer in *Xenopus laevis* carried out by Yoshizaki and Katagiri (1984) revealed that in this species the precursor of the fertilization layer (pre-fertilization layer) is formed in the distal one fifths of the *pars recta* length, which has special secretory cells. The formation of the pre-fertilization layer is unique for *Xenopus laevis* and was not reported for any other species. During early phases of VE to FE conversion, the VE hardens, become resistant to deformation, proteolysis, thermal dissolution, and impenetrable to sperm (Tian *et al.*, 1997). FE hardening caused by cortical granule exocytosis is immediately followed by a limited hydrolysis of glycoprotein, gp69, playing the key role in sperm reception, to gp61, which is inactive in sperm binding. These structural and molecular alternations, which result in biological properties of the egg envelopes, were also described for *Bufo arenarum* (Barisone *et al.*, 2003), although the molecular weights of glycoproteins differ and are obviously species-specific. After sperm entry, FE becomes again insensitive to sperm in a number of species: *Rana japonica, R. nigromaculata, Hyla arborea* and *Xenopus laevis* (Iwao and Katagiri, 1982). In conclusion we can summarize that sperm do not bind to OE, CE, and FE, but only to VE.

The unique organization of the oolemma and egg envelopes in respect to the sperm entry site is known in *Discoglossus pictus* (Klag and Ubels, 1975; Denis-Donini and Campanella, 1977; Andreucetti and Campanella, 1980, 1982). It is probably the only amphibian species, in which the site of sperm entry during fertilization is restricted to the central portion of the animal hemisphere in the "dimple" region, and not to entire animal half. The animal dimple is formed from the germinative area, i.e., a slightly depressed disc that surrounds and includes the animal pole of the oocyte at the end of vitellogenesis. After ovulation, the coelomic eggs are still flattened at the animal pole. When the oocyte enters the *pars convoluta* of the oviduct, the maturation occurs and the first polar body is expelled. In more distal parts of the oviduct, the germinative area invaginates and gives rise to the concave dimple. A glycoprotein content of the dimple (jelly plug) has four fucosylated glycoproteins (200, 270, 260, and 270 kDa), not present in the rest of egg surface (Maturi *et al.*, 1998). During passage through the oviduct, the animal dimple is filled by heavy gelatinous animal plug and afterwards covered by three jelly envelopes (Denis-Donini and Campanella, 1977). A few minutes after fertilization, cortical granules (restricted exclusively to the germinative area and the resulting dimple region) undergo exocytosis. Thirty minutes after fertilization the concavity and the dimple regress completely and the egg become spherical.

Secondary Envelopes (Jelly Coats): The secondary envelopes are known in amphibians as jelly coats and are composed mainly of mucin-type glycoproteins (60-70% carbohydrates and 30-40% proteins). Acquisition of jelly coats is necessary for normal fertilization and further development of the embryos. Several concentric capsules (jelly layers) are added to eggs as they pass through in the posterior part of the oviducts. The number of the jelly layers is variable and species specific, and ranges from one in *Rana catesbeiana* to several in *Alytes* and *Pipa*; also their thickness varies from 1 mm in *Gastrotheca cornuta* to 5.8 mm in *Rana spinosa*, and is not correlated with the egg size (reviewed by Duellman and Trueb, 1986). The jelly is secreted by the secretory cells of type II and III (Yoshizaki and Katagiri, 1981; Katagiri *et al.*, 1982) and epithelial glands of the *pars convoluta*. Each of the layers is secreted by different parts of the *pars convoluta*. In *Rana pipiens*, Elinson (1971a) designated them V1, V2 and V3, from inner to outer. The *pars convoluta* in this species is divided into two portions: the proximal thin-walled tube and the distal thick-walled tube. The proximal part produces V1 and V2, whereas V3 is added in the thick-walled tube. Katagiri (1965) described four jelly layers (J1-J4) in the toad *Bufo bufo formosus*, Carroll *et al.* (1991) found two layers (J1 and J2) in *Lepidobatrachus laevis*, Denis-Denini and Campanella (1977) described three layers (J1-J3) for *Discoglossus pictus*, and del Pino (1973) found also three jelly layers (J1-J3) in *Xenopus laevis*. In the latter species del Pino (1973) revealed that the proximal part of the oviduct secrete sulfate, which is incorporated exclusively into J1. Sulfated glycoproteins were also detected in J1 of *Lepidobatrachus laevis* (Carroll *et al.*, 1991). After a number of experiments with various media, del Pino (1973) postulated that small diffusible molecules or ions present in jelly coats are necessary for normal fertilization. The presence of divalent ions (Mg^{+2} and Ca^{+2}) in jelly coats of *Bufo bufo japonicus* was demonstrated by Ishihara *et al.* (1984), who suggested the role of the balanced salinity for gamete fusion. Bonnell *et al.* (1998) provided the detailed study of the composition of jelly coats in *Xenopus laevis*, together with a short review of the earlier findings. Each layer has a unique fibrillogranular structure and consists of a fiber network, to which globular components are bounded. The fiber network is composed of high molecular weight fibrous glycoconjugates: three acidic of 450, 630 and 900 kDa, and two neutral of 450 and 700 kDa. These high-molecular-weight molecules are stable and seem to be structural components of the jelly. The globular component contains proteins from 75 to 250 kDa in size, and is diffusible. J1 and J2 consist of acid glyco-proteins, whereas J3 consists of neutral carbohydrates. The glycoconjugates are uniquely associated with different jelly layers. The major components of J1 are unbranched linear molecules with less amount of globular component. J2 contains long parallel fibers devoid of cross-linking, but with large amount of globular material. J3 contains branched fibrous molecules of various lengths, decorated by single globular particles. The diffusible low-molecular-weight globular proteins can be easily washed out, especially from J3. This released proteins of molecular weight less than 50 kDa were shown to act as chemoattractants for the sperm and responsible for fertilization (Olson and Chandler, 1999). Recently Olson *et al.* (2001) purified and characterized allurin, acidic 21-kDa protein, 184 amino acids in length, from the *Xenopus* egg jelly. Its effect on sperm is direction-dependent and gradient-dependent: the steeper the gradient the stronger the attraction. Al-Anzi and Chandler (1998) suggest that this small chemo-attractnat protein is released within minutes during spawning and plays an important biological role in guiding short-living sperm toward egg mass.

The growing number of studies on the chemical composition of amphibian jelly coat suggests the role of carbohydrates in fertilization. The species specific variety of O-linked oligosaccharides were reported in a number of species, e.g. *Xenopus laevis* (Guerardel *et al.*, 2000); *Rana dalmatina*

(Morelle *et al.*, 1998; Florea *et al.*, 2002); *R. temporaria* (Florea *et al.*, 1997; Coppin *et al.*, 1999); *R. urticularia* (Morelle and Strecker, 1998); *R. palutris* (Maes *et al.*, 1999); *R. arvalis* (Coppin *et al.*, 1999); *R. ridibunda* (Mourad *et al.*, 2001); *Bufo arenarum* (Arranz *et al.*, 1997; Morelle *et al.*, 1998); *B. bufo* (Morelle *et al.*, 1997), and are supposed to play an important role in molecular recognition system. Even in the very closely related species *Bombina bombina* and *B. variegata*, which easily hybridize in the nature (Szymura, 1993; Szymura and Barton, 1986), Coppin *et al.* (2003) revealed different glycan structures. In *Rana clamitans* the egg jelly is a major factor in preventing cross-fertilization with *R. pipiens* sperm, as was experimentally shown by Elinson (1974). The oligosaccharides containing a terminal a-linked galactose residue seem to be bounded by the cortical granule lectin (Quill and Hedrick, 1996). The acrosome reaction seems to occur in the jelly layers in *Discoglossus pictus* (Campanella *et al.*, 1997). In other species studies so far (*Xenopus, Bufo, Rana*) the jelly has weak or none activity for inducing the acrosome reaction (Ueda *et al.*, 2002).

Larval Development Inside Egg Envelopes and Hatching

The early embryo develops in a chamber inside a capsule composed of the fertilization envelope (FE) surrounded by jelly coats. After gastrulation the embryos start to elongate and grow in size. At that time the FE disappears, the chamber enlarges, and is limited by the inner jelly layer, which soon liquifies. The thick jelly coats are believed to absorb some of the ultraviolet-B light, especially in species, which lays eggs in shallow waters exposed to high radiation. However, recent experiments with de-jelled eggs of *Rana temporaria* revealed no evidence that jelly acts as a UV-B filter (Räsänen *et al.*, 2003).

The egg jelly protects an embryo until hatching (Gosner stage 19). At that time the embryo produces some specific enzymes (proteases) (Katagiri, 1974, 1975), which digest the egg envelopes (both fertilization envelope and jelly coats) before active hatching of early larvae. The hatching glands are distributed on the head and their secretions allow the embryo to escape from the jelly capsule.

Females of several species produce some additional empty jelly capsules devoid of eggs. It was reported for tropical species such as *Phyllomedusa tarsius* (Crump, 1974) and temperate zone *Rana temporaria*. In the latter species, empty capsules were situated on the surface of the clutch and constituted about 1-2% of the total number of capsules (eggs) of a clutch; the presence of empty egg capsules was not correlated with age of a female or with size of a clutch (Ogielska *et al.*, unpublished data). It has been suggested (reviewed by Duellman and Trueb, 1986) that jelly capsules act as lenses, raising the temperature inside the egg clutch by focusing the sunrays. The measurements of the temperature inside and outside the clutch in R. sylvatica, which is the species of very similar biology to R. temporaria, revealed no significant differences (Cornman and Grier, 1941 in Duellman and Trueb, 1986); therefore that idea is probably not correct. The other possible role of additional capsules is the supply of additional water to developing eggs in arboreal egg masses of Phyllomedusa (Agar, 1910; Pyburn, 1980 in Duellman and Trueb, 1986). Egg jelly capsules may also serve as a mechanical (and in some species also chemical) protection of the eggs. The jelly capsules of species, which deposit eggs on leaves and fold the leaf over the eggs (Agalychnis and Phyllomedusa), serve also as glue that keeps the leaf edges together. Egg capsules in species with direct development, i.e., in which the entire development until metamorphosis takes place inside egg capsules, are tougher and the inner jelly layer (J1) does not liquefy. Also hatching is modified; there is no hatching glands producing proteolytic enzymes, but metamorphs open the egg capsule mechanically with the aid of special epithelial egg tooth, which disappear soon after hatching (Crump, 1974; Duellman and Trueb, 1986). The remnants of the egg capsules after hatching may serve as the substratum, to which freshly escaped larvae adhere until they will be able to swim.

Foam Nests: Many tropical species supply the clutches of eggs with foam. The foam nests may be constructed in water, or on ground or vegetation. This mode of reproduction protects the eggs and embryos against desiccation until hatching. The peripheral portion of the nest forms a thin viscous layer, whereas the interior keeps moisture. The foam is beaten by fore- and hindlimbs of both or one parent, depending on species. The foam is made of various substances, such as water, air, oviductal secretions, and seminal fluid (reviewed by Duellman and Trueb, 1986; Brizzi et al., 2003). Although the external morphology of the foam nests is rather well known, the role of oviductal excretions in their construction is poorly explored.

Internal Fertilization and Oviductal Gestation

The unique adaptation of the oviduct and uterus to internal fertilization was described in the tailed frogs Ascaphus truei and A. montanus (Sever et al., 2001, 2003). These are the only known anuran species with the internal fertilization, which posses the male copulatory organs actively inserted into the female cloaca during amplexus. This adaptation is connected with the fertilization of eggs in fast moving streams (Stebbins and Cohen, 1995). The female reproductive tract is also adapted to the internal fertilization. Distally to the *pars convoluta* (*ampulla* according to Sever et al., 2001) there is the ovisac. The two ovisacs join and form the oviductal sinus (described in other species as the common uterus—see above). The walls of ovisacs have simple tubular exocrine glands in the anterior part. These glands serve as the sperm storage tubules (SSTs). SSTs are composed of ciliated cells and secretion cells with microvilli, and are filled with the PAS-positive glycoprotein granules. Sperm deposited during amplexus is stored in bundles in SSTs and in lumen of oviduct, is present in females several weeks after the eggs have been laid, and can probably be stored for one or two years. Although the SSTs have the sperm storage functions, they do not resemble the urodelan spermatheca (see *'Oogenesis and Female Reproductive System in Urodela'* in this volume).

Oviductal retention of developing embryos inside oviducts (ovovivipartity and viviparity) is a rare phenomenon in Anura and occurs in *Eleutherodactylus jasperi* (Leptodactylidae) and 5 species belonging to the former genus *Nectophrynoides* (Bufonidae) (*Altiphrynoides malcolmi, Nectophrynoides tornieri, N. vivpara, N. liberiensis,* and *Nimbaphrynoides micranotis*) (for reviews see Wake, 1993; Sever *et al.*, 2003). In case of ovoviviparity, embryogenesis relays exclusively on the yolk material deposited in the developing egg, and only gaseous exchange between the mother and embryo takes place. In the case of viviparity, the developing fetuses additionally ingest orally the secretions of the oviductal epithelium. Although the fertilization in ovoviviparous and viviparous anurans is internal, no intromittent organs are developed and sperm is transmitted by cloacal apposition.

Eleutherodactylus jasperi is an ovoviviparous species. Its posterior parts of the 12 mm long oviducts fuse and form a 5 mm long common uterus, inside which 3-5 embryos are retained for 33 days until the metamorphosis is completed (Wake, 1978). The pregnant uterus is thin-walled and its epithelial lining differs from the more anterior part of the oviduct. The epithelial cells have no cilia and microvilli, and the secretory cells are inactive. Capillaries lie close to the external side of the epithelium, apparently enabling the gaseous exchange.

Nectophrynoides tornieri and *N. vivpara* are also ovoviviparous. The former species produce 9-37, and the latter 114-135 large yolky ova. *N. occidentalis* and *N. liberiensis* are viviparous. The former species produce 4-35, and the latter 6-24 small ova. Wake (1993) reviewed the literature about the reproductive biology of *Nectophrynoides* species, and Xavier (1977) summarized the numerous studies on the best-known *N. occidentalis*. The gestation lasts 9 months, partly during dry season, when the toads aestivate. During active life the embryos start to feed actively on the mucopolysaccharide secretion of the oviductal epithelium. The young are born as fully metamorphosed froglets. After parturition the hypertrophied epithelium exfoliates and degenerates, then regenerates again for the next season. The state of epithelium is under hormonal control of progesterone produced by the corpora lutea. The high level of progesterone ceases the growth phase of oocytes and allows the epithelium to increase its secretory activity (see the section *Regulation of oogenesis*).

Oogenesis in Anura

Oogenesis in amphibians is traditionally described as a continuous process, in which resting primary oogonia are gathered in germ patches of an ovary of a mature female and act as stem cells, thereby renewing the pool of germ-line cells before each breeding season. During next 2-3 years the resulting diplotene oocytes accumulate RNAs and yolk, grow in size, mature, and are ready for ovulation. After ovulation the mid-size oocytes grow and fill the space left by already ovulated portion of eggs. Then the next generation of oogonia is recruited from germ patches and the process is repeated every year during the entire reproductive life span of a female. This general model of amphibian oogenesis was proposed by Witschi (1929, 1956) and is cited in the textbooks on developmental biology (Browder *et al.*, 1991; Gilbert, 2000). However, studies of Jørgensen (1973b), Billeter and Jørgensen (1976), Callen *et al.* (1986), and our own (Ogielska *et al.*, 2007; in preparation) show that the definite pool of early diplotene oocytes is established during juvenile period and is sufficient for the whole life span of a female. After each ovulation a new portion of diplotene oocytes (not oogonia) is recruited from the pool, and undergoes vitellogenesis and growth. According to the proposed model, primary oogonia multiply effectively ony during the juvenile period, and become rudimetary and resctricted to germ patches during mature life. They retain the ability of mitotic divisions, but the resulting cells soon degenerate and even not enter meiosis.

Fig. 12. Oogenesis in anurans. Preoogenesis is a period of high mitotic activity of primary oogonia, which occurs only in an early ovary of tadpoles and young juveniles. Oogenesis starts when primary oogonia enter the unique cell cycle resulting in a nest of secondary oogonia (A); secondary oogonia tranform into primary oocytes and enter meiosis (B). Until the late pachtytene, the oocytes are interconnected by cytoplasmic bridges, which disappear when oocytes achieve diplotene stage, and start to emerge from the nest (C); the prefollicular cells of the nest wall proliferate and become follicular cells covering each oocyte individually. The diplotene oocyte grows until maturation and resumption of meiosis (D and E). The first meiotic division (F) results in formation of one secondary oocyte and one polar body (G). The oocyte is ovulated at metaphase II (H), and the second meiotic division is completed after activation of a spermatozoon during fertilization (I).

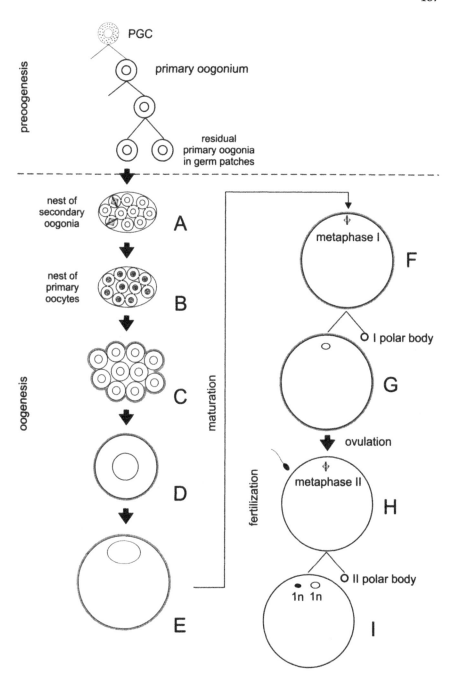

Fig. 12. See caption on page 186.

In the light of the above results, I suggest using the term "pre-oogenesis" in analogy to pre-spermatogenesis (see *'Spermatogenesis and Male Reproductive System in Anura'* in this volume). The pre-oogenesis would be resctricted to the period, when mitotically dividing juvenile primary oogonia create a great number of precursor cells for secondary oogonia, which enter the unique synchronous mitotic cycle and give rise to nests of secondary oogonia. Contrary to sermatogenesis, during oogenesis meiosis starts before metamorphosis.

Oogenesis is usually staged according to phases of meiotic division into primary and secondary oogonia, primary and secondary oocytes, and ova. Each of these stages is characterized by various morphological features, which are described below. Additional sub-stages are usually applied for primary oocytes undergoing a sequence of phases of the first meiotic division (leptotene, zygotene, and pachytene). Until early pachytene, the morphology of male and female germ cells is similar. Diplotene in female meiosis is exceptional, because during this phase meiosis is halted and is not restored until the progesterone signal triggers maturation, which ends the meiotc arrest. Oocytes are ovulated at metaphase II and meiosis is completed when sperm head activates the resumption of meiosis. The schematic illustration of oogenesis is shown in Fig. 12.

Morphology of Germ Cells

Female germ cells have various names depending on the stage of their differentiation (Raven, 1961). All of them are descendants of primordial germ cells (PGCs) of a larval gonad. Primary and secondary oogonia are diploid cells, which multiply mitotically. After the last mitotic division secondary oogonia transform into primary oocytes, which enter meiosis. The prophase of the first meiotic division in amphibian oogenesis has two phases: short leptotene, zygotene and pachytene, and long diplotene. The nucleus of the diplotene oocyte is also named the germinal vesicle. Metaphase, anaphase, and cytokinesis of the first meiotic division, as well as the second meiotic division, are also short, and result in one big cell (ovum) and two or three small abortive cells (polar bodies).

Primary Oogonia: After sexual differentiation of the gonad (ovary stage IV), "juvenile" primary oogonia are localized in the ovarian cortex. They are relatively big cells (15-20 μm in diameter), easy to distinguish from somatic cells of an ovary (Fig. 3B). They are single cells surrounded by one or few flat prefollicular somatic cells. Primary oogonia of early ovaries (up to ovarian Stage VI) are easily recognizable, numerous, and have a high mitotic activity (Fig. 13D). Beyond Stage VI, when diplotene oocytes grow in number and size, primary oogonia are situated in the most external parts of the cortex. In adult females (ovarian Stage X), primary oogonia are restricted only to small "germ patches" scattered on the surface of ovary in the loose connective tissue of the external theca (Fig. 8A,B) (Witschi, 1929; Ogielska and Kotusz, 2004).

The cytoplasm of the primary oogonia contains a rather sparse collection of organelles (smooth endoplasmic reticulum, ribosomes, few

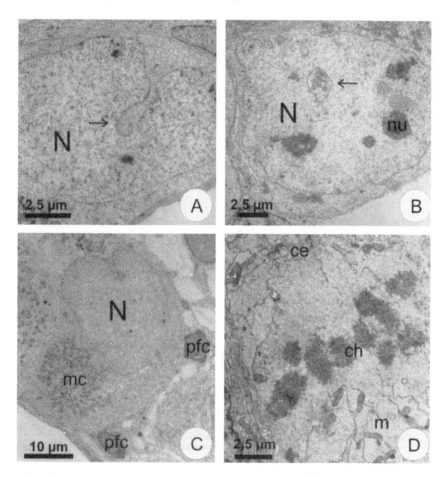

Fig. 13. Primary oogonia of the developing ovary of *Rana temporaria* (A, B), *R. ridibunda* (C), and *R. lessonae* (D). The shape of nuclei in primary oogonia of Ranidae, unlike in other families, is smooth in outline, but usually have several narrow "finger-like" invaginations (arrows; longitudianal section in A, and cross section in B) often containing mitochondria. Mitochondria are usually gathered as a cloud at one side of the nucleus (C). D—Metaphase in the primary oogonium. pfc—prefollicular cells; N—nucleus; nu—nucleolus; ch—chromosomes; mc—mitochondrial cloud.

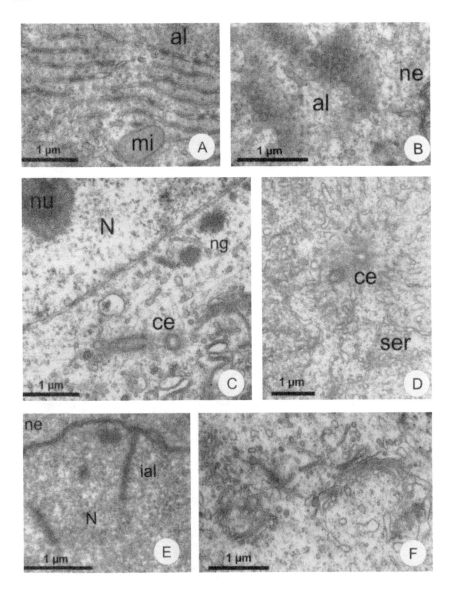

Fig. 14. Ultrastructure of amphibian germ cells (*Rana temporaria, R. ridibunda,* and *R. lessonae*). A—Stocks of annulate lamellae in longitudinal section (A) and cross section (B). C and D—Centrioles in primary oogonia. D—Smooth endoplasmic reticulum around centrioles. E—Intranuclear annulate lamellae. F—Dictyosomes. TEM. ce—centriole; ial—intranuclear annulate lamellae; mi—mitochondrium, nu—nucleolus; N—nucleus; ng—nuage materal; ne—nuclear envelope; ser—smooth endoplasmic reticulum.

dictyosomes, some microtubules, small vesicles, annulate lamellae and centrioles) (Fig. 14A-D,F) (Al-Mukhtar and Webb, 1971; Wang and Hsü, 1974; Coggins, 1973; Lopez, 1989; Ogielska and Wagner, 1990, 1993). Nuclei of primary oogonia contain few nucleoli and are highly lobulated in most of species, except of those belonging to the family Ranidae, in which nuclei are smooth in outline, with the exception of a few deep finger-like invaginations of the nuclear envelope containing cytoplasm and mitochondria (Fig. 13A) (Ogielska and Wagner, 1990, 1993; Ogielska and Kotusz, 2004). Mitochondria are the most prominent organelles and are usually grouped asymmetrically at one side of the nucleus (Fig. 13C), thereby forming the mitochondrial cloud with a pair of centrioles inside (Al-Mukhtar and Webb, 1971; Coggins, 1973; Heasman et al., 1984; Tourte et al., 1981; Kloc et al., 1998, 2001). In tadpoles, at early stages of gonadal differentiation (up to Stage VI), mitochondria are long with transverse cristae. In the genera *Rana* and *Rhacophorus* (Kress and Spornitz, 1974; Ogielska, 1990) some of them contain crystalline inclusions in the external or rarely internal mitochondrial space. These mitochondria have been transferred to germ-line cells of a larval ovary together with the germ plasm. Oogonia retain the mitotic activity in sexually mature females, but most of the primary oogonia within the germ patches degenerate (Figs. 8B, 25D). Their cytoplasm in filled with vacuoles, folded membranes, annulate lamellae, and large spaces separated nuclei from cytoplasm; mitochondria degenerate and gradually transform into lamellar bodies (Ogielska and Kotusz, 2004).

Secondary Oogonia: After several cell cycles some primary oogonia become the precursor cells of secondary oogonia and enter a unique cycle of synchronous mitotic divisions. During subsequent mitoses of the precursor cell cytokineses are not complete and the resulting cells are connected by cytoplasmic bridges (Fig. 15D). These cells form a cluster surrounded by somatic follicular cells, which do not penetrate between secondary oogonia (Fig. 15B,C). This morphological and most probably also functional unit is called "a nest" in amphibians and resembles a cyst of secondary spermatogonia in testes. There is also an intriguing similarity of the nests of secondary oogonia in amphibians to the cysts in mammals and insects (reviewed by Pepling et al., 1999). Thereby, the hypothesis is considered that cyst formation is a step, which precedes the onset of meiosis. The cycle of mitotic divisions of the precursor cell involves an alteration in the cell cycle, which can be regulated by signaling from the somatic follicle cells and intrinsic genetic information of the germ-line cells (reviewed by Pepling et al., 1999).

The ultrastructure of the secondary oogonia is similar to that of primary oogonia, with exception of the shape of nuclei and presence of intercellular bridges. Nuclei become spherical and smooth in outline, and

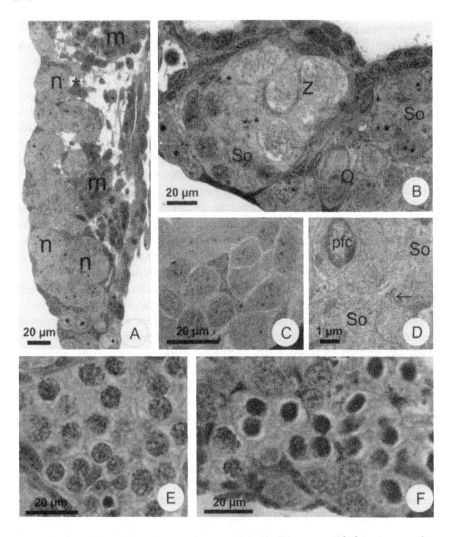

Fig. 15. A—Longitudinal section of a part of Stage IV ovary with forming ovarian cavity (asterisk) between medullar "knots"(m) and groups of primary oogonia and nests of secondary oocytes. B—A nest of secondary oocytes transforming into early meiocytes. C—Nests of secondary oocytes. D—Cytoplasmic bridge between two secondary oocytes (arrow). E—Nests of secondary oogonia and leptotene oocytes. F—Nests of leptotene/zygotene oocytes. n—nests of secondary oogonia; So—secondary oogonia; Z—zygotene oocytes; pfc—prefollicular cells; m—remnants of the medulla; O—primary oogonium; m—medullar cells. A, C, E, F—*Rana lessonae*; B, D—*R. temporaria*.

neither lobes nor finger-like invaginations are observed any longer. Intercellular bridges are short canals with a thin bundle of microtubules lying longitudinally in the lumen. The molecular composition and spatial arrangement of intercellular bridges were recently studied in *Xenopus laevis* (Kloc *et al.*, 2004b). Their diameters range from 0.5 to 1 µm, and a ring of actin filaments supports their walls. Additionally, the actin cross-linking protein kelch and hts protein (both characteristic of *Drosophila* intercellular bridges) were also detected inside the bridges in *Xenopus*. In secondary oogonia and early meiocytes, Kloc *et al.* (2004b) also described the presence of spectrin, which is a protein characteristic of a specialized portion of cytoplasm (fusom) shared via cytoplasmic bridges by secondary oogonia in insects.

The spatial arrangement inside the nest, the position of mitochondrial clouds and centrioles, and the number and position of cytoplasmic bridges interconnecting secondary oogonia (and early meiocytes) have been discussed in relation to polarity of egg, which is associated with the proper formation of embryonic axial structures (Gerhart, 1986; Klymkowsky and Karnowsky, 1994). Recent detailed studies by Kloc *et al.* (2004b) strongly suggest that polarity of *Xenopus* egg is established as early as in primary oogonia in form of asymmetrical localization of mitochondrial cloud containing centrioles and fusom-like material. This asymmetry is conserved in secondary oogonia and early meiocytes.

Early Primary Oocytes: Early primary oocytes (known also as early meiocytes) at leptotene, zygotene, and pachytene stages of the first meiotic prophase constitute the nest (Fig. 16A, 15B,E,F) and stay connected by cytoplasmic bridges (Coggins, 1973). Their cytoplasm is organized similarly to the cytoplasm of secondary oogonia and the main difference concerns the nuclei. At the beginning of leptotene, the chromosomes start to condense and form axial elements, which will form synaptonemal complexes during zygotene. Chromosomes are attached by their ends to the inner lamina of the nuclear envelope at the side adjacent to the mitochondrial cloud. At the end of pachytene the cytoplasmic bridges disappear and each individual germ cell become enveloped by proliferating follicle cells, thereby forming its own individual follicle. In that way the nest is replaced by a group of early diplotene cells.

Diplotene Oocytes: Diplotene oocytes are the best studied female germ cells. Although they are still primary oocytes, and thereby dividing cells in phase M of the cell cycle (meiotic prophase I), their metabolism resembles that of the interphase cell (G2 according to the DNA content) (Masui and Clarke, 1979). For this reason, as they are relatively easy to manipulate, they are used as models for a variety of cytological studies. Diplotene is the longest stage of oogenesis and its duration is species-specific. It can lasts from few month to several years, depending on the age of sexual maturity and reproductive time span of a female.

Diplotene oocytes are fast growing cells, which enlarge their volume tremendously. They are usually classified into six stages (or classes), according to size, degree of yolk accumulation, pigmentation, position of

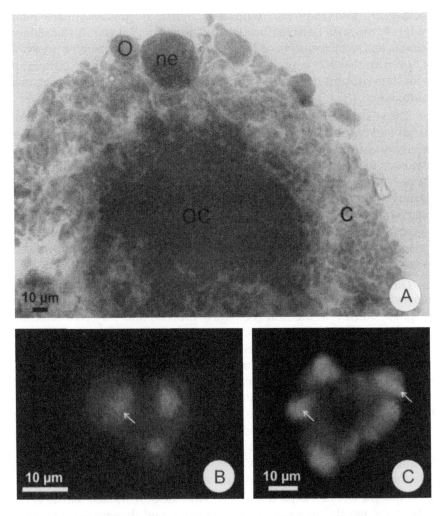

Fig. 16. A—Stage V ovary of *Xenopus laevis* digested with dispase and squashed under the coverslip. The cortex containing oogonia and nests of early meiocytes surrounds the wall of ovarian cavity. B and C—Isolated nests of pachytene oocytes with the nuclei containing caps of amplified rDNA stained with DAPI (arrows). O—primary oogonium; ne—nest; oc—ovarian cavity; c—ovarian cortex.

the nucleus (germinal vesicle), and overall structure of the follicle (Kemp, 1953; Grant, 1953). The most commonly used are classes defined and described by Dumont (1972) for *Xenopus laevis*. Stages 1 and 2 are previtellogenic, stages 3-5 are vitellogenic, and stage 6 is postvitellogenic. The diameter of stage 1 diplotene oocytes ranges from 50 to 300 μm, and they are transparent with rather relatively large nuclei of smooth appearance. Golgi complexes and reticulum are scanty. The ooplasm is rich in ribosomes. The nuclear envelope has a smooth contour and nucleoli become numerous. Stage 2 oocytes range from 300 to 450 μm. They become white in color and opaque. Microvilli appear on the surface of oolemma and first small cortical granules form in the peripheral layer of the cell. Also mitochondria and premelanosomes appear in this region. The nuclear envelope becomes irregular in outline and more nucleoli are located at the periphery of the nucleus. This arrangement of nucleoli is present also in stage 3 oocytes, which range from 450 to 600 μm. Pigment granules appear in melanosomes and the coloration of oocytes is light brown. The pigmentation is evenly distributed, without difference between animal and vegetal hemisphere. Accumulation of yolk by micropinocytosis and formation of primordial yolk platelets at the periphery of the oocyte are clearly visible. The nucleus is still highly irregular in outline, with many nuclei at the periphery, and lampbrush chromosomes in the center. In stage 4 oocytes (600-1000 μm in diameter), the pigmentation of the animal hemisphere is heavier than in the vegetal one. The cytoplasm is filled with yolk platelets, which form a gradient, being bigger in the vegetal hemisphere, and the nucleus moves toward the animal pole. The nuclear envelope is still irregular in outline, but the lampbrush chromosomes condense and move along with nucleoli, to the center of the nucleus. Vitellogenesis is at its most rapid phase during stage 4 of oogenesis. In stage 5 oocytes, the difference in vegetal and animal hemispheres pigmentation is distinct, and the diameter ranges from 1000 to 1200 μm. The nucleus becomes further displaced toward the animal pole and the polarization of the nuclear envelope begins. It is more convoluted at the vegetal, and rather smooth at the animal side. The nucleoli and condensed chromosomes form a mass in the center of the nucleus (karyosphere). The postvitellogenic stage 6 oocytes are fully-grown and their diameter ranges from 1200 to 1300 μm. The nucleus is eccentrically situated in the animal pole. During oocyte maturation the nuclear envelope begins to break down at the convoluted vegetal side.

The enlargement of oocytes is not uniform over time. In *Xenopus laevis*, Callen *et al.* (1980) described four phases of diplotene oocyte growth, each of different duration and kinetics. The first phase oocytes (pre-stage 1) are present in young females and lasts for a few weeks. Then oocytes enter the second phase (stage 1 and early stage 2), in which they can be arrested for several years. The third phase is characterized by rapid growth due to active vitellogenesis (stages 2-4), which lasts several months. During the fourth phase (oocyte stage 5 and 6), the growth rate slows again and oocytes mature within about 12 hours after progesterone stimulation.

Ultrastructure of the Diplotene Oocyte: The ultrastructure of amphibian oocytes was studied in many anuran species: *Rana erythrea, R. graeca, R. occipitalis, Bombina bombina, Bufo bufo, Ptychadena mascariensis, Hyla arborea* (Kress, 1982), *Rana pipiens* (Kessel and Ganion, 1980), *R. esculenta* and *R. temporaria* (Kress and Spornitz, 1972), *R. ridibunda* and *R. lessonae* (Ogielska and Wagner, 1990, 1993;) *Bufo marinus* (Richter, 1987), *Discoglossus pictus* (Denis-Donini and Campanella, 1977; Andreucetti and Campanella, 1980, 1982). However, most of our knowledge about the organization and function of diplotene oocyte, especially with respect of underlying molecular mechanisms, is based on studies on one species, *Xenopus laevis.* The diplotene oocyte is equipped with organelles common in animal cells, such as nucleus, diktyosomes of the Golgi complexes, endoplasmic reticulum, ribosomes, and mitochondria. But, on the other hand, the diplotene oocyte is a highly specialized cell and as such has a variety of modified or unique structures. It is also the polarized cell, and thereby various organelles differ in spatial distribution. The nucleus is big when compared to most of the somatic cells, and for this reason it was given a special name and is commonly known as the germinal vesicle. The cytoplasm, named also the ooplasm, is filled with yolk, which is a special protein and lipid reserve for the future embryo. Some kind of genetic information in already transcribed form (mRNA) is localized to various ooplasmic domains, which will be unevenly distributed to blastomeres of the developing embryo, thereby initiating different cell lineages. The cortical region of the cytoplasm is highly specialized for the reception of sperm during fertilization and is involved in establishing the future anterio-posterior, and dorso-ventral axes of the embryo. The more detailed description of the specialized structures of the diplotene oocyte is provided below.

The Nucleus (Germinal Vesicle): Along with growth of the oocyte the nucleus increases in volume, and the ratio between the volumes of oocyte and nucleus remains constant (Callen *et al.*, 1980). The chromosomes of the diplotene oocyte, although dividing meiotically, are highly transcriptionally active. As early as at stage 1, when oocytes have 40 μm in diameter, the chromosomes transform into the so-called lampbrush state (Fig. 17A,B) (Mitchell and Hill, 1986). The lampbrush chromosomes were described in a number of animals, amphibians among others (for reviews see Callan, 1963; MacGregor, 1980; Browder *et al.*, 1991; Gilbert, 2000). They consist of two homologue chromosomes, still joined together by chiasmata, which were formed during crossing-over in pachytene. Each chromosome consists of two parallely arranged sister chromatids. In transcriptionally active parts of chromosomes the chromatin fibers are loosely arranged and form loops. Inactive chromatin is condensed and forms chromomeres. Each loop is decorated by primary transcripts (hnRNAs) at various stages of transcription. The high activity of chromosomes during diplotene is necessary for providing various RNAs for the oocyte maintenance and for long-term storage. The number of nuclear pores in diplotene oocyte is high and ranges from 50 to 60 per μm² in *Xenopus* (Sánchez and Vilecco, 2003).

RNA Synthesis and Storage: A large amount of various types of RNA is stored in the ooplasm during amphibian oogenesis. This stock of RNA is

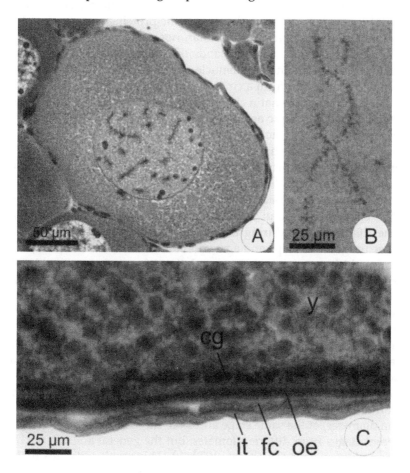

Fig. 17. Dilpotene ooctes in of *Rana ridibunda*. A, B—Lampbrush chromosomes. C—Structure of ovarian follicle wall and cortical granules in full-grown class 5 oocyte. cg—cortical granules; it—internal theca; fc—follicular cells; oe—oocyte envelope; y—yolk platelets.

a necessary "dowry" for the early development of a resulting embryo. During cleavage until the mid-blastula transition, the translation of the zygote genes does not take place and the embryos do not synthesize their own RNA (for reviews see Browder et al., 1991; Gilbert, 2000). In full-grown Xenopus oocyte only 2% of ribosomes are found in polysomes; the rest will be activated during early development of the embryo (for review see Smith, 1986). When compared to a somatic cell which synthesizes 10-100 ribosomes per second, amphibian oocyte produces approximately 300 000 ribosomes during the same time (Scheer, 1973). This process is strongly connected with the formation and activity of nucleoli, which are the sites of pre-ribosome synthesis. Nucleoli are formed around the rDNA coding for rRNA, which is (after binding to specific proteins) a precursor of two ribosome subunits. During amphibian oogenesis, rDNA is selectively amplified in hundreds and thousands of copies, which—detached from chromosomes—are found free in nucleoplasm. From few to about a dozen rDNA copies fuse and form a circle, around which a functional nucleolus is generated. Such chromatin circles were also observed in Odontophrynus americanus previtellogenic oocytes (Beçak and Pizzocaro, 1980). The minor wave of rDNA amplification is observed in primary oogonia (and spermatogonia), but the major wave starts in early meiotic oocytes (but not spermatocytes). The process is best known in Xenopus laevis. Prior to amplification, in leptotene/zygotene nuclei, only two large nucleoli situated on the opposite side to site of chromosome attachment to the inner nuclear envelope are found. rDNA amplification starts in late zygotene and reaches its maximum during pachytene. During this time, the nucleoli become irregular in shape and a mass of fibrillar material is formed, called the cap, which represents the accumulation of extrachromosomal rDNA (Coggins, 1973). The cap is particularly big in X. laevis, (Fig. 16B,C) whereas in other species may be less distinct or absent, as is the case of Gastrotheca riobambae (del Pino et al., 1986), and Rana ridibunda, and R. temporaria (Ogielska, unpublished). In Bufo the amplified rDNA is distributed in a number of round bodies of variable size (reviewed by MacGregor, 1982). In late pachytene and very early diplotene the multiple nucleoli are found inside the nucleus, and later they move toward the periphery close to the nuclear envelope. The number of amplified nucleoli is variable. In X. laevis their number per nucleus ranges from 500 to 2.500 (ranging from 2 to 12 μm in diameter) at the same stage and from the same female, as well as of different stages from different females, but the general amount of rDNA seems to be constant (Thiébaud, 1979a,b). This result indicates that the varying number of nucleoli is a result of their fusion and fission.

The formation of multiple nucleoli is still poorly understood. The problem was recently studied by Mais et al. (2002). In previtellogenic oocytes, the amplified rDNA genes are intensively transcribed. This allows producing of unusually high numbers of ribosomes. Multiple nucleoli are formed of three major prenucleolar components characteristic of nucleoli in general: spherical fibrillar bodies (composed of fibrillarin and nucloelolin characteristic of dense nucleolar component), granular bodies (composed of nucleolin characteristic of granular component), and rod-like elements (composed of nucleolar organizing region containing rDNA). These three prenucleolar components fuse, thus giving rise to a functional nucleolus. At the beginning of vitellogenesis, in stage IV oocyte, the rate of ribosome production start to decrease and the multiple nucleoli gradually cease their activity. They change in morphology and move toward the center of oocyte nucleus, surrounding the forming karyosphere (see below).

The high amplification of rDNA in oocyte nuclei is not the only strategy of rRNA accumulation in amphibians (MacGregor, 1980). In 14 species of marsupial hylids (genera: *Hemiphractus, Stefania, Amphignathodon, Flecto-notus,* and *Gastrotheca,* belonging to the subfamily Hemiphractinae) oocytes have not one, but multiple germinal vesicles ranging from 4 to 3,000, depending of species (del Pino, 1989a). This multinucleate oogenesis was studied in detail in *Flectonotus pygmaeus* (MacGregor and del Pino, 1982). Each of the growing oocytes of *F. pygmaeus* has from 1000 to nearly 3000 meiotic nuclei with 4C DNA content. It is believed that the multinucleate diplotene oocyte is formed during several cycles of oogonial multiplication within one cyst followed by fusion of oogonia. When the oocytes grow, the nuclei that are located peripherally are markedly enlarged and contain lampbrush chromosomes. The multiple nuclei gradually disappear before vitellogenesis and only one will form the definitive germinal vesicle. Ribosomal genes are amplified in each of the multiple nuclei and the overall amplification is about 280 times higher than in *Xenopus* (MacGregor and del Pino, 1982). A similar strategy is observed also in another family Ascaphidae (MacGregor and Kezer, 1970). Oocytes of *Ascaphus truei* have eight nuclei, probably resulting from 3 oogonial mitotic cycles within one cyst. Each of the nuclei contains lampbrush chromosomes and amplifies rDNA. However, the overall amount of rDNA constitutes about 1/8 of that in *Xenopus*. Seven nuclei degenerate before vitellogenesis and only one become the definitive germinal vesicle. The lower amount of rDNA amplification was also reported in *Gastrotheca riobambae*, another species of egg-brooding hylids. Although in this species diplotene oocytes have only one nucleus, the rDNA level is lower than in *Xenopus laevis* (del Pino *et al.,* 1986). The level of rDNA amplification is reflected in general amount of ribosomes, and—in consequence—influences the rate of early embryonic development (del Pino, 1989b).

Ribosomes are composed of rRNA synthesized in nucleoli (28S, 18S, and 5.8S), but also of 5S RNA, which is coded by genes situated outside the rDNA region. The synthesis and accumulation of special oocyte-type

5S RNA (other than somatic-type 5S RNA) takes place prior to rRNA synthesis (Denis and Wegnez, 1977). The large stock of ribosomes is necessary for the translation of various mRNA during early embryogenesis. New (zygotic) nucleoli formation starts after mid-blasutula transition and until that stage the embryo uses exclusively the maternally derived stocks of RNAs. In *Xenopus*, maternal-derived ribosomes are functional until hatching and support newly formed ribosomes. Rapid cleavage and embryogenesis, as well as time of hatching, is essential for surviving of amphibian larvae, and is correlated with the amount of rDNA and number of ribosomes. In an egg-brooding hylid, *Gastrotheca riobambae*, the number of nucleoli (300 per oocyte nucleus) and the amplification of rDNA are lower than in *Xenopus*, and this results in prolonged time of early development (gastrulation is completed within 14 hours in *Xenopus* and many other amphibians, compared to 2 weeks in *Gastrotheca*) (del Pino, 1989a; del Pino *et al.*, 1986). Such a long period of development is possible because eggs and embryos are incubated in special dorsal pouches of a *Gastrotheca riobambae* female (del Pino, 1989b).

Special attention should be paid to various kinds of mRNAs synthesized on the lampbrush chromosomes. They are generally known as maternal mRNAs, produced and accumulated in growing diplotene oocytes. They are stored as Poly(A)RNAs throughout the entire cytoplasm. Aggregations of germ-line cells specific RNAs form "nuage" (in French "cloud"), which is a common name used for the electron-dense material often seen as exported from the nucleus by nuclear pores (Figs. 14C, 18B). It is composed of proteins and RNAs, and forms granules of the germ plasm. Modified nuage material is also associated with mitochondria as intermitochondrial cement (Fig. 18A,B). Some RNAs are also localized to specific regions of the oocyte, and bounded to cytoskeletal elements. The proper localization of various RNAs is essential for further embryonic development (for review see King, 1995; King *et al.*, 1999). The main function of mRNA localization is to restrict the synthesis of the encoded proteins to specific subcellular domains, which will be later transmitted to selected blastomeres, thereby specifying selected parts of the embryo to establish various cell lineages. The RNAs localized to the animal pole are different than those localized to the vegetal pole. Almost all our knowledge about the oocyte RNAs comes from various studies on *Xenopus laevis* (for review, see Kloc *et al.*, 2001). Altogether, 14 RNAs characteristic of the vegetal and 12 RNAs characteristic of the animal hemisphere are described. All of them are either non-coding RNAs (*e.g. Xlsirts*) or coding mRNAs. The latter represents genetic information for future nuclear proteins (transcription factors), signal transduction proteins (both signals and receptors), and enzymes. The roles of several localized RNAs still remain unknown.

In the vegetal hemisphere, two classes of localized RNAs are disting-

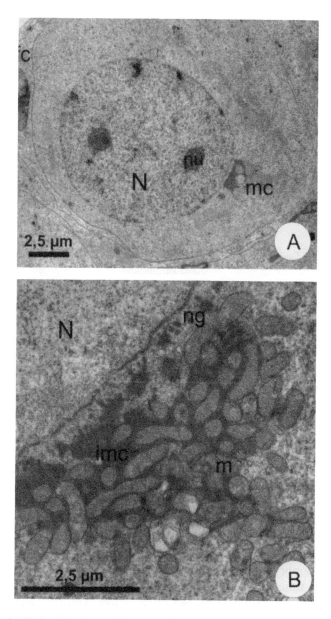

Fig. 18. Early diplotene oocytes of *Rana ridibunda*. A—Mitochondria are localized at one side of the nucleus. B—Germ plasm is emanating from the nucleus, and is gathered in form of intermitochondrial cement. N—nucleus; nu—nucleolus; mc—itochondrial cloud; fc—follicle cell; ng—nuage material; imc—inter-mitochondrial cement.

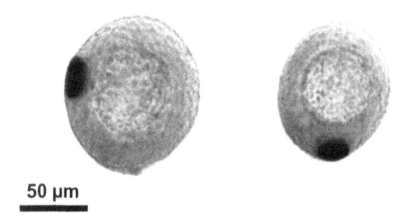

50 µm

Fig. 19. Localization of *Xcat2* mRNA in isolated diplotene oocytes of *Rana catesbeiana*.

uished: one class will be involved in the formation of primary embryonic layers (*VegT*) and proper early embryonic induction and mesoderm formation (*Vg1*) (for reviews see Ruiz i Altaba and Melton, 1990; Yisraeli *et al.*, 1995; Smith and White, 2003). The other class segregates with the germ plasm and will specify further germ-line cells of an embryo (Fig. 19). The germ cells related RNAs (*Xcat2, Xwnt11, Xpat, Xdazl, Xlsirts, fatvg, DEAD South, XFACS, Fingers, Hermes, Xtox1*) are described in Part 1 of this volume 'The Undifferentiated Amphibian Gonad'.

mRNAs localized to the animal hemisphere (*An1a, b; An2; An3; An4a, b; βTrCP; G protein; oct 60; PABP; PKCα; Xl-21; Xlan-4; Xlcaax-1*) are more difficult to study and not much is known about their role and way of localization. For example *An1, oct60,* and *xl-21* are transcription factors, *An2* is a mitochondrial ATPase, *An3* is RNA helicase, and *xlcaaX* involved in G-protein transduction pathway (Weeks *et al.*, 1995; King *et al.*, 1999).

All types of RNAs (rRNAs, mRNAs, tRNAs) described above, either those necessary for the functioning of the oocyte itself, or those stocked for the future embryo, need complicated transcription machinery. Each type of RNAs is transcribed by different polymerases (pol I, pol II, and pol III, respectively) and respective transcription factors (TFI, TFII, and TFIII), and many other gene activation and repression molecules. Small nucleolar RNAs (snoRNAs) and specific protein, fibrillarin, are molecules connected with rRNA metabolism and ribosome production. Primary transcripts and pre-mRNAs are processed and spliced by various small nuclear RNAs (snRNAs). The sites of assembly of the transcription machinery of the oocyte nucleus are localized in multi-function bodies known as "spheres" or "sphere organelles" (Gall and Callan, 1989; Phillips *et al.*, 1992). The spheres are homologues of the Cajal bodies of somatic nuclei (for review see Gall, 2000). They contain factors involved in pre-mRNA splicing, pre-rRNA processing, and histone pre-mRNA cleavage. They are bodies of various size (1-10 μm in diameter), scattered among multiple nucleoli, and their number in stage 5-6 oocytes ranges from 50 to 100. The spheres consist of matrix and a variable number of small bodies (B-snurposomes), some embedded in the matrix, some extruding from the surface, and some found free in the nucleoplasm. B-snurposomes are aggregates of a variety of molecules connected with mRNA processing. They are especially abundant in amphibian oocytes, owing to a massive synthesis and storage of various kinds of mRNAs, as described above.

Oocyte Cortex and Cytoskeleton: The cytoskeleton of the diplotene oocyte is composed of three major filamentous components (microtubules, intermediate filaments and microfilaments) and a variety of related proteins (talin, spectrin, vinculin, myosin, γ-tubulin (for reviews see Larabell, 1995; Gard *et al.*, 1995; Klymkovsky, 1995; Ryabova and Vassettzky, 1997). The cortex of a cell is usually defined as a thin peripheral layer composed of the plasma membrane and underlying microfilaments. The cortex of oocytes is usually complex and well elaborated. Moreover, the oocyte cortex changes its structure and function during maturation, and later after fertilization and early development. The superficial layer of the cortex is composed of the plasma membrane (oolemma) with abundant microvilli strengthened by actin filament network, and the deeper subcortical layer containing pigment and cortical granules (Merriam *et al.*, 1983). More recently, Ryabova and Vassetzky (1997) proposed the classification of the cortex into the external (or membrane-containing) and the internal (or pigment-containing) layers. The external layer is responsible for the sperm entry, whereas the functions of the internal layer include exocytosis of cortical granules and isomeric contraction of the egg surface after fertilization . The two layers are well elaborated in the animal half of the oocyte, where the internal layer of the cortex is thicker. The external layer is composed of the plasma membrane with structurally connected microfilaments. The parallel bundles of actin filaments constitute the skeleton of microvilli. The cortex of the diplotene oocyte before maturation is not able of contraction. The arrangement of microfilaments is rigid, as a result of cross-linkage mediated by spectrin (Ryabova and Vassetzky, 1997). The matrix of the internal layer of the cortex is composed of a rigid cytoskeletal network of intermediate filaments, microtubules, and actin filaments. Pigment granules are embedded in polymerized actin (Ryabova and Vassetzky, 1997) and the internal layer contains cortical granules and mitochondria, membranous vesicles and endoplasmic reticulum. The cortex is thinner in the vegetal hemisphere and yolk is in contact with the plasma membrane. The cortex is thicker in the animal-hemisphere which does not contain yolk platelets. The thickness of the cortex in *Xenopus* ranges from 3 (vegetal hemisphere) to 7 (animal hemisphere) μm.

Beneath the cortex, actin filaments extend into regions of the yolk-free cytoplasmic "corridors" in the animal hemisphere, between the plasma membrane and the nucleus. The corridors contain membranous cisternae, mitochondria and annulate lamellae (reviewed by Larabell, 1995). The corridors are reinforced by other cytoskeletal elements, *i.e.*, microtubules and intermediate filaments (Gard *et al.*, 1995). These corridors disappear during maturation when the cytoskeletal network is deeply rearranged. The intermediate elements in the cytoplasm are composed of type I and type II keratins (Klymkovsky, 1995). They are present throughout all stages of oogenesis. In fully-grown stage VI oocyte the keratin network is organized in an asymmetric fashion: in the animal hemisphere it forms an irregular network, whereas in the vegetal hemisphere it forms a regular net.

Microtubules are present in all stages of oogenesis (Gard *et al.*, 1995). They are present in the mitochondrial cloud of the oogonia and early meiocytes, then rearrange and are found in the cortex around the nucleus and Balbiani body. In larger diplotene oocytes they also radiate from a yolk-free cytoplasm around the nucleus, to the cortex. Beginning with stage 4–6 oocytes, the microtubule array becomes polarized along the animal-vegetal axis.

Microtubules and actin microfilaments are required during the endocytosis and intracellular transport of yolk precursors. Microtubules participate in transport and translocation of yolk platelets inward from the oocyte surface to the center of the oocyte, and in RNAs localization. A cage of cytoskeletal filaments surrounds the Balbiani body (Tourte *et al.*, 1991; Klymkovsky, 1995; Gard *et al.*, 1995). The microtubule organizing centers (MTOC) are also the sites of cell-division spindle organizers. Stage 6 oocytes lack functional centrioles and centrosomes (Gard *et al.*, 1995). Maternal centrioles are present in oogonia and early meiotic oocytes, but disappear at stage 1 diplotene oocytes. The MTOC disappears at the early stage 1 oocytes.

Mitochondria and Balbiani Body: Mitochondria are the most prominent organelles observed in the ooplasm throughout the entire oogenesis. They originate from the mitochondria associated with germ plasm, which is inherited by primordial germ cells (PGCs), and resulting oogonia and early primary oocytes (see 'The Undifferentiated Amphibian Gonad' in this volume). The mitochondria in PGCs, oogonia, and early meiocytes are usually gathered as a mitochondrial cloud at that side on the nucleus that contains centrioles (Fig. 20). The structure, fate, and role of the mitochondrial cloud is best known in *Xenopus laevis*, and poorly studied in other amphibians (reviewed by Kloc *et al.*, 2004b). At early diplotene stages in juvenile females, when oocytes have the diameter about 200 µm, the mitochondria form a compact "cap" closely apposed to one side of the nucleus. The mitochondria inside the cap are long, and—as was rightly described by Billet and Adam (1976)—"spaghetti-like" in shape. As was described for *Xenopus laevis*, the cap splits into two groups during early

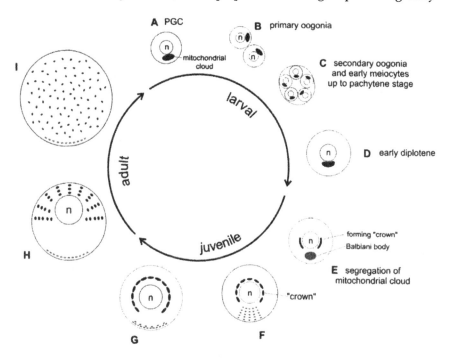

Fig. 20. Schematic illustration of mitochondrigenesis in a *Xenopus* germ line cells. In primordial germ cells (PGC), primary oogonia (B), and secondary oogonia and early meiocytes (C) mitochondria form an aggregation at one side of the nucleus. In early diplotene oocytes (D) mitochondria form a cap juxtaposed to the future vegetal side; the cap is a site of active mitochondriogenesis. Soon two classes of mitochondria are formed: one transforms into the Balbiani body, where mitochondria are inactive and do not multiply, whereas another one starts to form a "crown" around the nucleus (E). The Balbiani body together with the germ plasm is translocated to the cortical layer of the vegetal pole, and the "crown" is composed of actively dividing mitochondria (F, G). In growing diplotene oocyte the mitochondria derived from the crown are now localized in the animal hemisphere inside the corridors of cytoplasm strengthened with cytoskeletal elements; the mitochondria derived from the Balbiani body are still localized to the vegetal pole. After maturation and germinal vesicle break down, mitochodria are distributed throughout the entire ooplasm (I).

diplotene stage: one forms the Balbiani body and the other surrounds the nucleus, forming a "crown" (Tourte *et al.*, 1981, 1984, 1991). Both the cap and the crown are sites of intense mitochondriogenesis, as was shown in experiments with H³-thymidine incorporation and measurements of mtDNA amount (Tourte *et al.*, 1984; Mignotte *et al.*, 1987). About 12 rounds of mtDNA replication occur before vitellogenesis and only 4-5 at later stages of oogenesis. It is worth noting that the shape of the mitochondrial cap and intense mtDNA replication occur only in young females and were not observed in diplotene oocytes of the same size in adult females. This supports the hypothesis that germ cells are stocked in an ovary in form of class 1 and class 2 oocytes and are not generated after each ovulation (for details see the section *Pool of diplotene oocytes*). The part, which migrates toward the vegetal pole, forms a transient structure known as mitochondrial cloud or Balbiani body (for reviews see Guraya, 1979; Kloc *et al.*, 2004). The structure, fate, and role of the mitochondrial cloud is best known in *Xenopus laevis* (reviewed by Kloc *et al.*, 2004), and poorly studied in other amphibians. The mitochondria of the Balbiani body gradually become inactive in mtDNA replication (Tourte *et al.*, 1984; Mignotte *et al.*, 1987). These mitochondria form aggregates around electron-dense material characteristic of germ line cells, known as "intermitochondrial cement" or "nuage" containing germline-specific localized RNAs. Then the mitochondria form a big mass known as Balbiani body (Fig. 21A,C), in which the mitochondria and the intermitochondrial cement segregate in such a way that the cement (now transformed into germinal granules) is directed toward the oocyte periphery. The shape, size, and number of Balbiani bodies differ among species, as does the size of oocytes containing Balbiani body. Finally the Balbiani body reaches the cortex of the vegetal pole and forms the germinal plasm.

In *Xenopus* there is no morphological difference between mitochondria active in mitochondriogenesis and those, which are inactive and located in the Balbiani body (Tourte *et al.*, 1984). However, these two classes of mitochondria differ in staining affinity to toulidine blue (Mignotte, 1987) and haematoxylin-eosine in that way that one class is more acidophilic (blue), whereas the othae is basophilic (red) (Fig. 21D) (Rozenblut and Ogielska, unpublished). The morphological difference between the two classes of mitochondria is particularly distinct in the ranid frogs. In all *Rana* species studied so far, and also in *Rhacophorus maculatus*, mitochondria located within the Balbiani body (more or less distinct, depended on the species) and its derivatives, posses characteristic crystalloid inclusions (Intramitochondrial Crystalline Inclusion Bodies, ICIB) (Fig. 21B), primarily believed to be involved in yolk formation (Ward, 1962a,b; Lanzavecchia, 1965). However, later cytological studies revealed the differences in the crystalline lattice (Lanzavecchia, 1965; Spornitz, 1972; Kress and Spornitz,

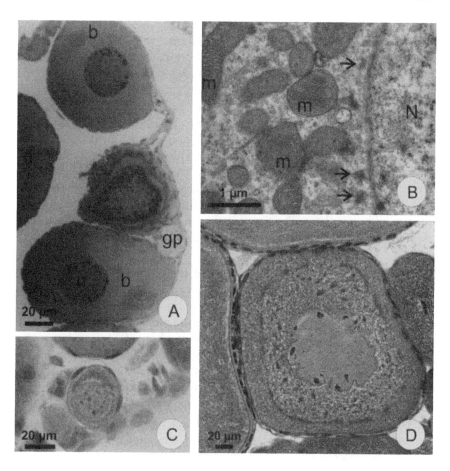

Fig. 21. Balbiani body in (A) *Rana temporaria*, and (C) *R. lessonae*. B—Crystals in mitochondria of primary oogonium (arrowhead), arrows indicate the nuage material emanating from the nucleus. D—Mitochondriogenesis in *R. ridibunda*.
b—balbiani body; gp—germ patch; m—mitochondria; n—nucleus.

1972; Ward, 1978a,b), as well as different biochemical composition (Ward *et al.*, 1985) of the two structures. The only suggestion is that ICIBs are the morphological feature of inactive mitochondria, which segregate later to the Balbiani body. The intramitochondrial crystals are formed inside cristae or in the intermembrane space (Kress and Spornitz, 1974). The number of mitochondria with ICIBs decreases significantly when meiosis starts, and are observed sporadically until zygotene oocytes, absent in pachytene, and become visible again in early diplotene (Ogielska, 1990). The origin and function of ICIBs are unknown, but most probably they play a role in inactivation of mitochondria.

Cortical Granules: Cortical granules are commonly found in many invertebrates and vertebrates and play essential role during fertilization. Among amphibians they are absent in Urodela and Gymnophiona, and are well defined in Anura. They are small (1.5-2.5 μm) membrane-bounded vesicles formed by the Golgi complexes and situated in the cortical region just beneath the oolemma (Fig. 17C) (Ward and Ward, 1968; Charbonneau *et al.*, 1986). Grey *et al.* (1974) described two types of cortical granules in *Xenopus laevis*. The predominant type, averages 1.5 μm in diameter, is homogenous of moderate electron density. This type of cortical granules is predominant in the animal hemisphere. They form a single layer at a density about 18 granules per 100 μm² of egg surface area. The second type of cortical granules is larger (up to 2.5 μm) and distributed mainly in the vegetal hemisphere. They are filled with more flocculent material and are distributed in irregular row in various distances from the oocyte surface.

Each cortical granule is surrounded by cisternae of endoplasmic reticulum. The cisternae lying next to the oocyte surface are in close contact with plasmalemma among the microvilli (Grey *et al.*, 1974; Gardiner and Grey, 1983). These plasma membrane-reticulum junctions are 3.5 times more abundant in the animal then the vegetal hemisphere of *Xenopus laevis* egg. The number of cortical granules increases dramatically during oocyte maturation. Cortical granules contain glycoproteins (lectins), which are extruded onto the oocyte surface after sperm entry, and transform the oocyte envelope into the fertilization envelope, thereby preventing polyspermy.

The egg of *Discoglossus pictus* is exceptional, because it does not content typical cortical granules (Andreucetti and Campanella, 1980). Instead, it contains dense granules formed by the Golgi complex, which contain a core composed mainly of proteins and only small amount of polysaccharides. The granules are driven by cytoskeletal elements toward a specialized region of yolk-free cytoplasm, i.e., the germinative area. The germinative area is a unique structure of the *Discoglossus pictus* egg and after ovulation will form a concave dimple at the animal pole. The animal dimple is the only sperm-binding site, where the granules undergo exocytosis after sperm activation in the way similar to cortical exocytosis in other anuran species (Klag and Ubbels, 1975; Denis-Donini and Campanella, 1977; Andreucetti and Campanella, 1980).

Annulate Lamellae: Annulate lamellae (AL) are commonly present at all stages of amphibian oogenesis, from primary oogonia to late diplotene oocytes (Fig. 14A,B) (Kress and Spornitz, 1972; Wang and Hsü, 1974; Kress, 1982; Ogielska and Wagner, 1990, 1993). AL are parallel membranes, often stacked, perforated by regular and precisely aligned pores ("annuli"), which structurally resemble pore complexes of the nuclear envelope. The number of AL comprising a stack range from several to more than one hundred, and their length also varies. AL are embedded in electron dense material containing fibrils and granules, most probably composed of ribonucleoproteins. In amphibian oocytes the AL are present in all regions of the ooplasm, but their distribution changes in time (Imoh *et al.*, 1983). In fully-grown amphibian oocyte the AL are distributed mainly in the animal pole between the nucleus and oolemma, but are also present in the vegetal subcortical region. The differentiation, origin, and role of AL are still a matter of discussion (for review see Kessel, 1992). AL have a close relationship with numerous cell organelles. The structural resemblance to the nuclear envelope, as well as the juxtanuclear position suggest that AL are generated by the nuclear envelope and constitutes the storage form of the envelope, or of the nuclear pore complexes. The membranes of AL often exhibit continuity with cisternae of smooth or rough endoplasmic reticulum. The AL are sometimes associated with nucleolus-like bodies, nuage and intermitochondrial cement in early oocytes of a tadpole ovary.

Yolk Platelets: Almost a half of the volume of a fully-grown diplotene oocyte is occupied by yolk. Originally this name was used for all storage material, especially lipid droplets ("fatty yolk"), glycogen, and lipoprotein granules ("protein yolk") (Ward, 1962a,b). The best known and the most complex is the protein yolk, which is commonly known as the yolk platelets, although they are not flat but ovoid in shape (Fig. 22C). Each yolk platelet is composed of the crystalline core surrounded by an amorphous superficial layer and encapsulated by a membrane (Fig. 22A,B).

The core is a main part of a yolk platelet and is composed of lipovitellines (Lv I and Lv II), phosvitin and fhosvettes (Wallace, 1963a,b; Ohlendorf *et al.*, 1978; Redshaw and Follet, 1971; Wiley and Wallace, 1981; Schneider, 1996). Lipovitellines are lipoproteins and are the major components (91%) of the core. Molecular weight of Lv I range from 111 to 121 kDa, and that of Lv II ranges from 30 to 33 kDa. The other main component of the crystalline lattice is a highly phosphorylated protein of a molecular weight 32-34 kDa (phosvitin), composed of about 50% of serine. Phosvitin may be substituted by two other phosphoproteins, phosvette-1 (19 kDa) and phosvette-2 (13-14 kDa). Both lipovitellins and phosvitin form dimmers (Birrelli *et al.*, 1982), which are arranged in a highly regular orthorhombic crystalline lattice, forming about 70% of the platelet volume; water together with a variety of ions (K, Ca, Na, Mg, Cl) constitutes the remaining 30%. Similar crystalline lattice was also reported for *Bufo arenarum* and *Ceratophrys carnwelli* (Sánchez and Vilecco, 2003). The other protein synthesized in the liver and transported to the yolk platelets by the bloodstream is the vitronectin-like protein, which was detected in *Bufo arenarum* (Sánchez and Vilecco, 2003); however, its role is not known.

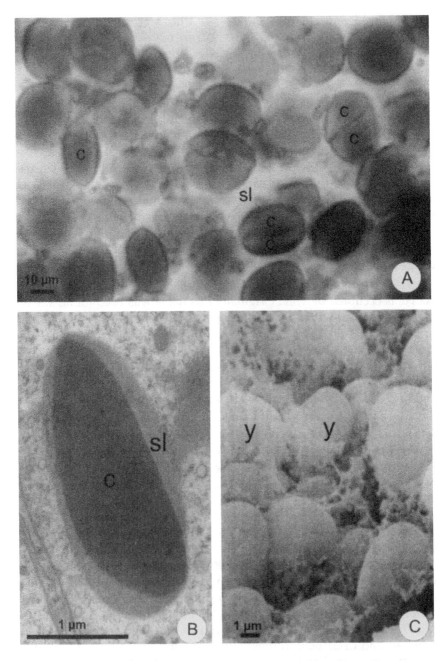

Fig. 22. Yolk of amphibian oocyte. A and B—*Rana esculenta*. C—*Xenopus laevis*. c—core; sl—superficial layer; y—yolk platelet. A—paraffin section stained with fast green FCF and safranine 0. B—TEM. C—SEM.

The superficial layer is composed of the amorphous material, with some tiny fibrils and granules. In *Xenopus laevis* it contains acid polysaccharides, histones and nucleic acids, mainly high molecular weight DNA (Horn, 1962; Wallace, 1963a; Ohno *et al.*, 1964; Hanocq *et al.*, 1972; Tandler and la Torre, 1976). The superficial layer of *Rana temporaria* and *Bufo bufo* is bi-layered: a layer adjacent to the crystalline core is more homogenous, whereas the external part is less electron-dense and more variant (Romek and Krzysztofowicz, 2001). In early stages of vitellogenesis in *Bufo marinus, B. arenarum*, and *B. paracnemis* (Richter, 1987; Sánchez and Vilecco, 2003) the superficial layer is densely packed with lipid droplets, which eventually fuse with the core.

Yolk is accumulated during vitellogenesis, which is a long-lasting process and can be divided into the auto- and heterosynthesis. Authosynthesis occurs in previtellogenic oocytes and results in formation of the primary yolk. This process was described in species other than *Xenopus leavis* (Kessel and Ganion, 1980; Kress, 1982). The primary yolk is formed of vesicles produced by various oocyte organelles, such as annulate lamellae, nuclear envelope, Golgi complexes, and endoplasmic reticulum. The vesicles fuse and form multi-vesicular bodies, which later transform into primary vesicles with a dense crystalline core inside. However, the main bulk of yolk is formed by heterosynthesis from a precursor complex protein, vitellogenin. Vitellogenin is synthesized in liver cells under the hormonal control of estradiol 17 β produced by the follicle cells (Follet *et al.*, 1968; Wallace and Dumont, 1968; Wallace and Jared, 1969; Redshaw and Follet, 1971; Tata, 1976; Wiley and Wallace, 1981). Vitellogenin is a lipophosphoglycoprotein, a member of low-density lipoproteins (LDL). In *Xenopus*, vitellogenins are coded by several, most probably 4-6 genes (Tata *et al.*, 1980). They are synthesized as large (about 210-220 kDa) primary translation product of a single mRNA, which is extensively modified post-translationally. Vitellogenins of all egg-laying animals, both invertebrates and vertebrates, are believed to arise from a common ancestral lineage. Each vitellogenin molecule is composed of a receptor-binding region (Lv I), and a phosvitin-phosvette region (reviewed by Schneider, 1996).

Vitellogenin dimmers are transported by bloodstream to the ovary, where the oocytes sequester them by receptor-mediated endocytosis (Dumont, 1978; Opresko and Wiley, 1987a; Schneider, 1996). Vitellogenin must exit the blood vessels and traverse the theca, follicle cell layer, and the oocyte envelope. Capillary blood vessels form a dense network in the theca. Adjacent endothelial cells of capillary form gaps, through which the vitellogenin disperses in the connective tissue, then traverse the basement membrane of the follicle cells, and pass through channels between arms of the stellate follicle cells. After passing across the oocyte envelope, it reaches oocyte surface. Experiments with tracers injected into the circulatory system in *Xenopus* indicate that they reach oocyte surface about two hours following injection (Dumont, 1978). Similar results were obtained for *Bufo arenarum* and *Ceratophrys carnwelli* (Sánchez and Vilecco, 2003). The surface of vitellogenic oocytes (Stage 4) posses many deep crypts located in the base of microvilli. Along the walls of the crypts and between the microvilli, a great number of endocytotic pits and vesicles are localized. The pits are equipped with clathrin coat (cytoplasmic side) and vitellogenin receptors on the membrane surface. The receptors are highly conserved integral membrane glycoproteins composed of five domains; the most external is a ligand (vitellogenin)-binding domain (Schneider, 1996). After binding vitellogenin, the endocytotic pits with receptors inside transform into coated vesicle with internalized vitellogenin. The coated vesicles fuse and form endosomes, in which vitellogenin is proteolytically cleaved into lipovitellin and phosvitin, or phospvettes. Afterwards, endosomes transform into vesicles of secondary yolk precursors with small yolk core inside. According to Massover (1971), about 1000 secondary vesicles form one small (about 3 µm in long axis) yolk platelet.

The regional difference in the pattern of vitellogenesis was described in *Discoglossus pictus* (Andreucetti and Campanella, 1982). In this species the cytoplasm of the animal pole in full-grown oocyte is equipped with small yolk platelets formed exclusively by autosynthesis, whereas big yolk platelets formed by heterosynthesis are present only in the vegetal half.

The yolk platelets grow in size during the entire vitellogenesis; thereby the biggest platelets are the oldest. During active vitellogenesis, the lipovitelline is sequestered by the entire oocyte surface, both animal and vegetal. However, at the end of vitellogenesis the distribution of big and small yolk platelets is uneven, and the vegetal-animal gradient is established, with big platelets (12-14 µm) occupying the vegetal half (Callen, 1986; Danilchik and Gerhart, 1987). In stage V oocytes, platelets originating in the animal hemisphere during stage III and IV move almost entirely to the vegetal hemisphere, whereas platelets originated during late vitellogenesis occupy the animal hemisphere.

The endocytosis of vitellogenin is mediated by gonadotropins (Wallace and Jared, 1968, 1969; Bergink and Wallace, 1974; Wallace *et al.*, 1972; Opresko and Wiley, 1987b). The vitellogenin is sequestered and transported into the oocyte cortex within about one hour, and proteolytically transformed into lipovitellin and phosphitin within the next four hours. The crystallization of the platelet core starts after 18 hours.

The big stock of yolk proteins is utilized by a developing embryo in a process of vitellolysis, which starts after the mid-blastula transition. A platelet starts degradation from the periphery, i.e., first the superficial layer and then the core is dissolved (Karasaki, 1963b; Jurand and Selman, 1964).

Vitellolysis is mediated by phosphatases released from lysosomes stocked during oogenesis, which remove phosvitin. The released lipovitellin is a precursor of cytoplasmic membranes of an early embryo (Robertson, 1979).

Pigment Granules: The accumulation of pigment starts in class 3 oocytes (about 500 µm in diameter) and increases during class 4 oocytes, until they reach about 800 µm in diameter (Dumont and Eppig, 1971). Pigment granules containing melanin are situated beneath the layer of cortical granules (Charbonneau *et al.*, 1986). They are formed as membrane-bounded round premelanosomes, which grow in size by fusion of small vesicles that add the electron-dense material until it fills the premelanosome. In *Rana temporaria* and *R. esculenta* the pigment granules in the fully-grown oocytes reach a diameter of approximately 0.5 µm (Kress and Spornitz, 1972).

The amount and localization of pigment granules depend on a species. In *Rana lessonae*, it is localized as a thin layer in the animal hemisphere (Fig. 1F); in *R. temporaria* it is more abundant and distributed randomly within the entire ooplasm. In this species the pigment granules of the animal hemisphere are about twice as large as those from the vegetal hemisphere (Wischnitzer, 1966). In species, which lay eggs in places protected from the UV, the eggs are devoid of melanin (Crump, 1974; Duellman and Trueb, 1986). The melanin accumulated in the oocyte during oogenesis is not a precursor of the melanin in melanophores of growing larvae. The tadpoles developed from batches of pigmentless eggs produced by non-albinotic *Rana temporaria* and *R. esculenta* females have normal melanophores, which differentiate at the same time as in larvae derived from normally pigmented eggs (Sladeèek, 1964; Ogielska-Nowak, 1985) (Fig. 23). The egg-derived pigment accumulates in ectodermal layer of the embryo and is eliminated during larval development. In neural tube and developing brain, the pigment granules are excreted from the cells to the fluid inside the lumen and are phagocyted by specialized big cells floating inside the lumen, as was described in *Xenopus laevis* (Kordylewski, 1983, 1984). Round oocyte melanosomes are retained only in pigment layer of the eye; they differ from larval elongated melanosomes (Eppig, 1970).

Diakinesis: Diakinesis is the last phase of the first meiotic prophase. It takes place at stage 6 oocytes, prior to maturation. The lampbrush chromosomes start to condense and retrack the loops at the end of stage 4 and in stage 5 oocytes. Condensing chromosomes move to the center of the nuclei, along with accompanying nucleoli, which detach from the periphery of the nucleus. At the end of this process, the highly contracted chromosomes form a knot (karyosphere) separated from the rest of the nucleus by a karyosphere capsule. The karyosphere was reported in egg-brooding hylids *Gastrotheca riobambae* and *Flectonotus pygmaeus* (del Pino *et al.*, 1986; del Pino, 1989). The formation and structure of karyosphere were described in the ranid frogs *Rana ridibunda* and *R. temporaria* (Gruzova and Parfenov, 1973, 1977; Zbarsky and Filatova, 1979; Parfenov and Gruzova, 1984). The karyosphere and the capsule are firmly bound together. Externally to the capsule, there are numerous nucleoli, which formerly were located in the periphery of the nucleus, close to the nuclear membrane. Accumulation of nucleoli starts at the vegetal side of the oocyte, where the nuclear envelope starts to disintegrate during maturation. The capsule has a diameter of 150-200 µm and is composed of sheets of pore complexes (annuli) forming the pseudomembranes, microfibriles, membrane vesicles, and intranuclear annulate lamellae (see also Kessel, 1968). The capsule contacts with nuclear envelope by narrow invaginations. The role of the karyosphere capsule is not clear, but it may separate nuclear DNA from the extrachromosomal rDNA of the nucleoli. The nucleoli of the karyosphere cease the production of preribosomes and are composed of fibrillar centers and fibril component, but devoid of the granular component; they do not incorporate H^3-uridine.

Maturation of Diplotene Oocytes: About 12 hours after stimulation by progesterone, the oocyte starts to continue meiosis and converts from an unfertilizable oocyte into a fertilizable egg. The gap junctions between the follicle cells and the oocyte become uncoupled and this state interrupts the inflow of the oocyte maturation inhibition factor (Sanchez and Villecco, 2003).

The most dramatic changes during maturation concern the rearrangement of cytoskeletal architecture. The trigger for the disassembly of the cytoskeleton is connected with phosphorylation caused by MPF kinase. The oocyte cortex undergoes structural and functional changes, which are essential for fertilization. The surface of the oocyte becomes relatively smooth, since the microvilli shorten due to reorganization of microfilament cytoskeleton. The actin microfilaments of the external cortex rearrange from oriented (rigid) to randomly arranged (contractile) network (Ryabova and Vassetzky, 1997). This is accompanied by the spectrin dispersal, which is detectable in the cortex of oocytes before, but not present after maturation. The isotropic network of actin filaments is ready to interact with myosin, and thereby able to contract. The contraction is generated by interaction between antiparallel actin microfilaments and myosin. The parallel assembly of microfilament is probably triggered by Ca^{+2}. The cytoskeletal system of the internal cortex also rearranges. The keratin filaments are destroyed; microfilaments cross-link and form three dimensional network, probably enabling the contraction. Along with the cortex reorganization, also RNAs are relocalized, especially those involved in the dorsal patterning of the embryo. Keratin filaments disassemble into soluble oligomeres (Klymkovsky, 1995) and actin microfilaments become reoriented

(Ryabova and Vassetzky, 1997), liberating the cortex from the rigid connection with the ooplasm. Upon activation during fertilization, the keratin system begins to reappear, forming a fishnet structure (Klymkovsky, 1995).

Corridors of yolk-free cytoplasm between the plasma membrane of the animal half and the nucleus disappear. As a result the cortex (in form of a gel) acquires the ability to rotate in reference to the interior, ungelled ooplasm (Ryabova and Vassetzky, 1997). However, the most spectacular change of the oocyte organization is caused by microtubules rearrangement (Hausen et al., 1985; Danilchik and Denegre, 1991; Gard et al., 1995). The yolk platelets are relocated and the nucleus (germinal vesicle) opens and releases its content. The nuclear envelope breaks and nuclear proteins (i.e., nucleoplasmin) are released from the nucleus, some of which are probably involved in gene regulation during early development (Malacinski, 1974; Hausen et al., 1985). Microtubules and their organizing centers gather near the vegetal part of the nucleus and display a transient array connected with germinal vesicle break down (GVBD). This complex rapidly migrates toward the animal pole, where condensed chromosomes are released.

The diplotene oocytes are cells arrested at the prophase of the first meiotic division. During maturation, the oocytes reenter the first meiotic division, enter metaphase I, and complete the first meiotic division by eliminating one set of chromosomes to the first polar body. Soon afterwards, they form mitotic spindles of the second meiotic division and are ready for ovulation and fertilization (Ferrell, 1999). *Discoglossus pictus* is the only amphibian species, in which the first polar body extrusion occurs not inside the ovary before ovulation, but during the passage thought the oviduct (Denis-Donini and Campanella, 1977).

The meiotic spindles are formed despite of the absence of classical centrosomes and centrioles, which have disappeared in early meiocytes. Meiotic chromosomes, when released from the nucleus during GVBD, are surrounded by γ-tubulin, which serves as MTOCs. The meiotic spindles of oocytes after maturation are acentriolar. The centrioles are provided later, together with the spermatozoon (Gard et al., 1995). Parthenogenetically activated eggs are capable of forming a microtubular aster from a large pool of maternal centrosomal components (Gard et al., 1995). However, the lack of paternal centrioles result in arrested development of an early embryo. *Xenopus* oocytes have a substantial pool of centrosomal proteins (γ-tubulin). After GVBD, each chromosome is surrounded by a sheet of γ-tubulin, which organizes microtubules into a spindle. Assembly of acentriolar meiotic spindles follows a complex pathway that is characterized by four stages: aggregation of microtubules and chromosomes; establishment of a bipolar spindle axis; prometaphase elongation and orientation parallel to the oocyte surface; rotation of a spindle perpendicular to the oocyte surface. The spindle of the second meiotic metaphase is oriented parallel to the surface and seem to be anchored to the cortex (Gard et al., 1995).

After 12 h of progesterone induction and GVBD, the stocks of AL start to decompose and the AL membranes give rise to a number of small vesicles, which form the ER cisternae around the cortical granules (Imoh et al., 1983; Larabell and Chandler, 1988; Kessel, 1992). The resumption of meiosis results in alignment of the cortical granules in one layer beneath the ooplasm. Cortical granules become wrapped by elaborated cisternae of endoplasmic reticulum (Charbonneau and Grey, 1984), which releases Ca^{+2} after sperm entry. There is also a change in water content after maturation. The oocytes become more voluminous and the animal hemisphere is about 10% more hydrated than the vegetal one (Lau et al., 1984). After ovulation the oocytes may enlarge up to about 30%, most probably as a result of water uptake (Callen, 1986).

Polarization of an Oocyte: Establishing of the Animal-vegetal Axis: A fully-grown amphibian diplotene oocyte is a polarized cell composed of two hemispheres: animal and vegetal, and two respective poles (Fig. 24). The final polarization is achieved during maturation, when the content of the nucleus is released into the cytoplasm and the re-arrangement of cytoskeleton occurs (for review see Larabell, 1995), followed by the rearrangement of deposited yolk and some organelles (Gerhart et al., 1986; Daniltchik and Gerhart, 1987). The animal pole is indicated by the localization of nucleus (germinal vesicle) and—after its breakdown—by the localization of the second meiotic metaphase spindle. The vegetal pole is indicated by the localization of germinal plasm and other kinds of localized RNAs, as was described above. In many amphibian species a difference between the two moieties is additionally reinforced by the presence of pigment in the animal hemisphere. However, eggs of many species are devoid of pigment and in such cases their external appearance does not reflect the internal polarity.

Pool of Diplotene Oocytes

The number of primordial germ cells (PGCs) is established during early larval development and in amphibians their number in gonadal ridges ranges from several dozens to two hundreds, depending mainly on the age and stage of a tadpole, but usually is about 60-70 of PGCs (reviewed by Hardisty, 1966). PGCs are precursor cells for primary oogonia, which—after several cycles of mitotic divisions—give rise to secondary oogonia, transform into oocytes, which enter meiosis and finally give rise to a large pool of early diplotene oocytes. It is worth noting that just after sexual differentiation of gonads the proliferation of primary oogonia is more extensive in comparison to primary spermatogonia in the testes, and the number of oogonia is much higher than the number of spermatogonia (Ijiri and Egami, 1975).

After metamorphosis and during juvenile period the number of oogonia and early oocytes decreases, and in adult females they are restricted to several "germ patches" scattered in the connective tissue in the most external parts of the ovarian cortex (Witschi, 1929; Ogielska and Kotusz, 2004). Because amphibians produce rather large numbers of eggs during several breeding seasons, there is a common belief, originally formulated by Witschi (1929, 1956), that after each spawning the oogonia from the germ patches become activated and renew the pool of oocytes by a wave of oogonial mitoses (Smith, 1955; Mathews and Marshall, 1956; Franchi *et al.*, 1962; Redshaw, 1972; Lofts, 1974; Tokarz, 1978; Blüm, 1986; Wallace and Selman, 1990; Browder *et al.*, 1991; Gilbert, 2000). However, studies of Jørgensen (1973b), Billeter and Jørgensen (1976) on *Bufo bufo*, and Callen *et al.* (1986) on *Xenopus laevis* suggest that the definitive pool of early diplotene oocytes is established during juvenile period and is sufficient for the whole life of a female. After each ovulation a new portion is recruited from the pool, undergoes vitellogenesis and maturation. This is consistent with data provided by Bounoure *et al.* (1954) for *Rana temporaria*, that the total number of oocytes in ovaries after metamorphosis is about 12,000. Our own studies on the same species (Ogielska *et al.*, 2007; Augustyńska and Ogielska, unpublished data) are in agreement with the latter hypothesis. The ovaries of female frogs before first spawning contained in average 16,000 (from 9,000 to 23,000) diplotene oocytes at various classes. The average number of eggs oviposited by a female from this population is about 2,000, which gives a pool of oocytes for 9-10 breeding seasons. *R. temporaria* females mature when they are 3 years old. The oldest female from this population was 9 years old, and the oldest *R. temporaria* from another population reported by Płytycz *et al.* (1995) was 12 years old, which gives the reproductive life span ranging from 6 to 9 years (in this case the same number of clutches). Oogonial mitoses observed in the germ patches give rise to primary and secondary oogonia, which degenerate before they transform into oocytes. In conclusion we can say that the number of oocytes is established during the second year of a female life and is sufficient for about a dozen of spawnings, *i.e.*, for the whole life span.

Degeneration of Germ Cells and Intrafollicular Atresia

Degeneration affects germ cells at all stages of oogenesis, but is most conspicuous in vitellogenic diplotene oocytes, especially those that have not been ovulated during breeding season. Atresia is a specific kind of physiological intraovarian digestion of a diplotene oocyte by its own follicle cells that proliferate, hypertrophy, and become phagocytic. Earlier stages of germ cells, i.e., oogonia and nests of leptotene-pachytene meiocytes can also degenerate, but prefollicular cells, which surround them, are not hypertrophic and probably not phagocytic. However, degeneration of oogonia and early meiocytes is poorly studied. First of all, it occurs rarely in normally developing ovaries during the period when germ cells at these stages predominate. In adult ovaries, oogonia are restricted to germ patches scattered at the periphery of an ovary, and most of them degenerate (Ogielska and Kotusz, 2004).

Atresia of diplotene oocytes is a common phenomenon and many authors who studied amphibian oogenesis mentioned about the presence of degenerating oocytes. Dumont (1972) described intraovarian degeneration of oocytes as an effect of starvation of a female *Xenopus laevis*. As a result, stage 3-6 oocytes degenerated, whereas stage 1 and stage 2 oocytes were intact. Atretic oocytes were also observed by Callen (1986) in females of the same species 3-4 weeks and 1.5 month after spawning, when atretic oocytes represented about 5% of the total vitellogenic oocytes population in an ovary. The first more detailed study devoted to atresia in the amphibian ovary was undertaken by Guraya (1969) in sexually mature female *Bufo stomaticus*. Since that time there have been no studies directed on this problem.

Recently we studied the problem of germ cell degeneration in frogs *Rana temporaria, R. lessonae, R. ridibunda,* and *R. esculenta* (Augustyńska *et al.,* 2007), where degeneration appears in that portion of oocytes, which start vitellogenesis. Because degeneration was not uniform in vitellogenic diplotene oocytes, we classify them into 3 main types. Type I and II were observed in vitellogenic oocytes, whereas type III was observed in both previtellogenic and early vitellogenic ones (Fig. 25).

Type I (Fig. 25A-C) was divided into four stages (A-D). At stage A nucleoli fuse and form big aggregates and the nucleus shrinks. After disintegration of the oocyte envelope (vitelline membrane), the follicle cells start to proliferate and invade the ooplasm, where they start phagocytosis of the ooplasm. At stage B the proliferating phagocytes invade the follicle until its entire volume is filled. Follicle cells that were juxtaposed to the *theca interna* did not hypertrophy. In regions where ooplasm has been already digested, the phagocytes started to degenerate and accumulate dark pigment. At stage C the follicle cells adjacent to *theca interna* start to hypertrophy and degenerate. Concomitantly with degeneration of oocytes, the follicles shrink and at the end of stage C their diameter attained 150-200 μm. The follicles were still filled with phagocyting cells, all of them containing dark pigment. At stage D the condensed black pigment formed a mass within the follicle. The pigment-containing atrectic vesicles were well seen in total preparations. The *theca* becomes folded, thicker, and penetrated by blood vessels. The volume of the follicle strongly decreases.

In type II the most characteristic feature was a release of free ooplasm that dispearsed between adjacent follicles. Groups of phagocyte cells were sometimes seen inside the ooplasm.

Type III of degeneration occurred in previtellogenic and early vitellogenic oocytes. Follicular cells did not invade the ooplasm; lampbrush chromosomes condense in the germinal vesicle and form irregular aggregates. The nucleoli fuse and the volume of the germinal vesicle shrinks. In some oocytes the condensed ooplasm was fragmented. The ooplasm degenerates and aggregates of organelles were well seen as eosinophilic masses.

Degeneration of PGCs was never observed, and that of primary oogonia was extremely rare. Similarly, degeneration of germ line cells in a nest phase, as well as previtellogenic class 0 and 1 oocytes in tadpoles and juveniles before first hibernation, was rare.

Atresia in previtellogenic diplotene oocytes was also studied by Wang (1980), who reported that in *Rana catesbeiana* ooplasm become shrunken, mitochondria start to degenerate and form a distinct, abnormally big mass.

The degenerating follicles in sexually mature amphibians have been sometimes designated as "corpora atretica" or "corpora lutea", because they were believed to be involved in production and secretion of steroid hormones (Perry and Rowlands, 1962). However, Guraya (1969) and Jørgensen (1992) evidenced that atretic follicles in amphibians did not perform any endocrine function, but the remnants of thecal cells of a vitellogenic follicle constitute a part of interstitial tissue.

Number and Size of Deposited Eggs: The relation between the number and size of eggs focuses the attention of many scientists acting in various fields of biology, such as developmental biology, ecology, evolutionary biology, and many others, and was reviewed thoroughly by Duellman and Trueb (1986). There is a great variety among anuran amphibians concerning the number and size of eggs in a clutch. The number of clutches deposited during one breeding season also varies from one, which contains all oocytes produced in an ovary for a given breeding season, to several, which are ovulated in portions. Shape and number of clutches also vary from species to species. Deposited eggs may form several irregular masses, exemplified by *Rana lessonae* or *R. arvalis*, or one more spherical big mass, as in *R. temporaria*. Females belonging to the genus *Bufo* produce eggs arranged in two long strings, representing the content of the two oviducts. The number of eggs in a clutch ranges from 1 in *Siminthillus limbatus* to about 48,000 in *Rana catesbeiana*. The smallest eggs among anurans are produced a pipid *Hymenochirus boettgeri* (0.75 mm in diameter), whereas

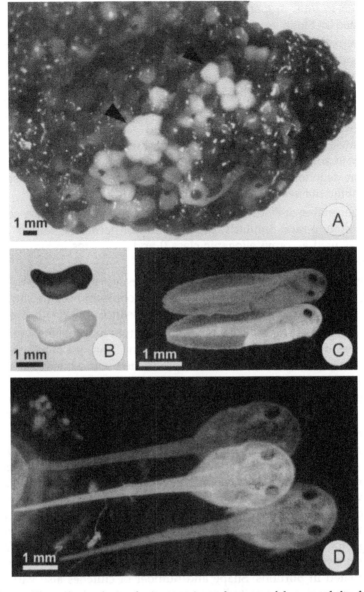

Fig. 23. Fate of egg-derived pigment in embryos and larvae of the frog *Rana esculenta*. A—Cluster of pigmentless oocytes in the ovary (arrowheads). B—Stage 18 embryos derived from normally pigmented (upper) and pigmentless eggs. C—The same individuals at stage 22. *De novo* synthesized melanin is seen in the pigment layer of the eye retina and in skin melanophores. D—Normal pigmentation in tadpole stage 25; the tadpole in the middle derives from a pigmentless egg.

the largest is produced by the egg brooding hylid *Gastrotheca cornuta* (12 mm) (Duellman and Trueb, 1986).

The number and size of eggs produced by females are obviously connected with their fecundity. To compare fecundity among various species, one must consider different reproductive modes represented by anuran amphibians (Duellman and Trueb, 1986). There are general trends in a given reproductive mode that the positive correlation exists between clutch size and female body size, and between clutch size and ova size. However, there are some exceptions to these rules. Crump (1974) studied several species of tropical anurans and revealed that only about 40% of species showed a significant positive correlation between the size of eggs and the size of a clutch. The fecundity can be also measured as a ratio between the mass of ovaries (or number and size of mature oocytes or deposited eggs) and the size of females (mass or body length). These studies may be carried out to compare differences among species, or individual females within one species. Our own studies on *Rana temporaria*, *R. lessonae*, and *R. ridibunda* (Skierska *et al.*, 2007) show the positive correlation between clutch size and female body size, but no correlation between clutch size and ova size.

There are many controversial data on the relation between the egg size and the rate and duration of development of the resulting embryos (Duellman and Trueb, 1986). The most common generalization is that the

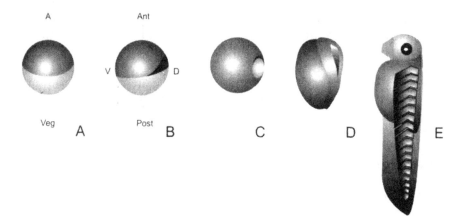

Fig. 24. The effect of egg polarity establishing the dorso-ventral and anterior-posterior axes of the embryo. A—Unfertilized oocyte is a polarized cell with the animal-vegetal axis. B—After fertilization and rotation, the former animal-vegetal axis (dotted line) is replaced by the antero-posterior axis of an embryo. The dorsal side is indicated by the grey crescent (black), and the ventral side is indicated by V (From Ogielska, 1999, with permission).

larger the ova, the longer the time needed to hatching or metamorphosis; that in turn is often explained as a result of the large amount of yolk, which causes the delay of cleavage furrows formation. However, at least the rate of early embryonic development depends not on the amount of yolk, but on the level of rDNA amplification, and thereby on the amount of rRNA and preribosomes stored during oogenesis and used later by the embryo. The best examples are species, which produce eggs of similar size and display various rate of development, such as *Eleutherodactylus coqui* (eggs 3.5 mm in diameter), *Flectonotus pygmaeus* (eggs 4.4 mm in diameter), and *Gastrotheca riobambae* (eggs 3.3 mm in diametr). Eggs of these species have similar diameter, but metamorphosis is achieved after 14-15 days, 25-31 days, and about 8 months, respectively (MacGregor and del Pino, 1982; del Pino *et al.*, 1982; Duellman and Trueb, 1986; del Pino, 1989a,b; Elinson and Fang, 1998). Certainly, the correlation between the rate of development and level of rRNA is also affected by temperature and by egg brooding in *G. riobambae* and *F. pygmaeus*.

Age of First Reproduction and Reproductive Life Span

Amphibians are long living animals, which have several breeding seasons during their life span. The age at which females become sexually mature (in most cases 2-3 years of life) is generally known, mostly from the data from captive animals. Data from field observations may often be estimated errorously in case when the age of individuals is judged by body size. It is a well-known phenomenon that anuran amphibians grow rapidly during juvenile period and after sexual maturity their growth almost stops; however such studies are still not numerous (for review, see Smirina, 1994). Also there are very scarce data on the relationship between the age of first reproduction and reproductive life span, and between the absolute longevity of individuals and their reproductive life span. In natural conditions such data can be obtained only from long-term studies on marked individuals, or with use of the skeletochronological method (for review see Castanet *et al.*, 1993; Smirina, 1994). Płytycz and Bigaj (1993, 2004) have studied a marked population of *Bombina variegata* since 1981, and 21 years later they recorded 11 out of 608 marked individuals, among them one pair in amplexus. It means that in this long-living species also very old specimens are active during breeding season. Unfortunately nothing is known about their fertility and fecundity in relation to age. Our own studies on the ranid frogs (*Rana temporaria, R. lessonae, R. ridibunda*) reveal that the age of breeding population ranges from 2 (males) to 8 (both sexes) years old individuals. There are some speculations that very old specimens probably do not take part in breeding, but such studies are lacking.

There are also differences of opinions concerning the measure of reproductive life span, which can be altered by inactive periods of aestivation or hibernation. One measure is the absolute age of a female and the alternative is the measure of the sum of active months. For example a species from the rainforest is active all year round, whereas a species from temperate zone hibernate 6 month a year. In such case the rainforest species is mature when is 1 year old, whereas the temerate zone species is mature when it is 2 years old, whereas the sum of active months is the same. Similar problem is with overwintering tadpoles and their siblings, which metamorphose in autumn (Khonsue *et al.*, 2001). Their absolute age is the same, their gonads are at the same stage of development, but their age estimated in relation to metamorphosis differs.

Regulation of Oogenesis in Anura

Hormonal Control of Oogenesis

One of the fundamental features of amphibian reproduction is its profound dependence on environmental conditions. Reproductive cycles are influenced by external factors, such as temperature, humidity, and to some extend also photoperiod. Because of this a great majority of amphibian species reproduces periodically and phases of their sexual activities alternate with inactive periods. On the other hand, in favorable climatic and nutritional conditions, the animals may reproduce continuously. Among the factors influencing amphibian reproduction, two are most crucial: access to water and food availability. During an annual cycle a female may go through one or more breedings separated by post-reproductive periods. The cyclic changes of reproductive activity are in turn coupled with well-defined changes in the ovaries. In most species two phases are distinguished during the breeding period: the pre-ovulatory phase, during which the processes of growth and maturation of follicles are accomplished, and a phase of ovulation when clutches of matured ova are released from the ovary to body cavity and then to oviducts. Around the time of ovulation females start to mate. During the post-reproductive period a new generation of ovarian follicles are recruited for growing (Lofts, 1974; Fernandez and Ramos, 2003; Jørgensen, 1973a,b,c, 1992).

The reproductive success of female amphibians depends on relations between environmental factors and hormones regulating reproductive functions (Fig. 26). Environmental influence is mediated by the central nervous system that cooperates with the hormonal system. Hormones regulating and coordinating reproductive functions of vertebrates act mainly through the hypothalamo-pituitary-gonadal axis and have the same chemical nature regardless the sex. Nevertheless, the endocrine machinery regulating ovarian function is more complicated than that of the testis (see chapter 2 'Spermatogenesis and Male Reproductive System in Amphibia—Anura' in this volume). The hypothalamo-pituitary-gonadal axis is hierarchic in its nature. The hypothalamus functions as the center that coordinates the activity of other elements of the axis. Hormone secretions are regulated by positive and negative feedback control. The positive feedback acts when an elevated level of one hormone increases secretion of another hormone, whereas a negative feedback takes place when an elevated level of one hormone suppresses the secretory activity of cells producing another hormone. Feedback interactions may operate between any of levels of the hierarchy described above, thereby making up so-called short or long loops. Hormones exert stimulatory or inhibitory effects through binding to their specific receptors located in cell membranes, cytoplasm, or nuclei of target cells. Steroids transported to the target organs through the blood circulation are associated with serum sex steroid binding proteins (SSBP), as was demonstrated in *Rana esculenta* (Paolucci and DiFiore, 1994; Paolucci *et al.*, 2000) and *Bufo arenarum* (Fernandez *et al.*, 1994).

Hormone levels and action fluctuate according to time and age of animals. As was described in the part devoted to ovary structure in this chapter, newly metamorphosed anuran ovaries contain oogonia and oocytes in their hormone-independent phase of differentiation. In juvenile and sexually mature females diplotene oocytes became sensitive to hormonal action (reviewed by Jørgensen, 1992). Neuroendocrine mechanisms that govern ovarian functions in amphibians are basically similar to those that regulate ovarian activity in other vertebrates. However, because anurans are primarily oviparous animals, only the follicular phase of the ovarian cycle occurs, after which sexually mature individuals empty their ovaries during ovulation. The luteal phase is characteristic of only a few viviparous and ovoviviparous species (for review, see Wake, 1993)

According to the chemical nature, the hormones belong to lipids, peptides, and monoamines. Lipid hormones are represented by steroids (estrogens, progesterone, and androgens) and eicosanoids (prosta-glandins). Peptide hormones are represented by gonadotropic glyco-proteins (luteinizing hormone LH and follicle-stimulating hormone FSH), polypeptides (prolactin PRL and growth hormone GH), and peptides (gonadotropin-releasing hormone GnRH). Monoamines are represented by melatonin MEL.

Hypothalamus

The organization and function of the hypothalamus is the same in both sexes of Anura. The secretory neurons of preoptic region and adjacent sites of hypothalamus are responsible for the secretion of gonadotropin releasing hormone (GnRH). The axons of these neurons extend down the median eminence to the pituitary and terminate on its capillaries forming the hypothalamo-median eminence tract (Sotowska-Brochocka, 1988; Muske and Moore, 1990; Miranda et al., 1998). By this route, neurohormones secreted by hypothalamic neurons reach the gonadotrophs in the glandular part of the pituitary and stimulate them to synthesis and secretion of gonadotropins: follicle-stimulating hormone (FSH) and lutenizing hormone (LH) (McCreery and Licht, 1983; McCreery et al., 1982; Licht et al., 1987; Stamper and Licht, 1993; Wang et al., 2001).

There are two distinct izoforms of GnRHs known in amhibians: mGnRH (GnRH-I) and cGnRH-II (King et al., 1994; Muske et al., 1994; Iela et al., 1996; Rastogi et al., 1998; Wang et al., 2001). The third izoform (sGnRH or GnRH-III) is still discussed (Cariello et al., 1989; Chieffi et al., 1991; D'Antonio et al., 1992; Fasano et al., 1990, 1993, 1995, (Licht et al., 1994; Collin et al., 1995; Yoo et al., 2000, Dubois et al., 2002), as was described in chapter 2 'Spermatogenesis and Male Reproductive System in Amphibia— Anura' in this volume (regulation of spermatogenesis). Although the mGnRH is the predominant form in the preoptic area and median eminence (Porter and Licht, 1986a; Miranda et al., 1998), it is known that both izoforms of amphibian GnRHs are capable of stimulating the pituitary for synthesis and secretion of gonadotropins (Licht et al., 1994; Licht and Porter, 1986; King and Millar, 1981). In the light of recent investigations it becomes clear that the influence of cGnRH-II on vertebrate reproduction had been underestimated. For instance, the analysis of structure and function of mammalian pituitary revealed that a large number of LH gonadotrops have cGnRH-II receptors, which reinforce GnRH-II role in regulation of gonadotropins secretion (Millar et al., 2004) and suggest the close cooperation of the two izoforms in stimulation of pituitary gonadotropin secretion. As in males, both types of immunoreactive GnRH are found not only in hypothalamus, but also in other parts of the brain (Iela et al., 1996; Miranda et al., 1998; Muske et al., 1994; Rastogi et al., 1998). The fact that cGnRH-II is expressed in different time and amount than mGnRH-I may suggest that they play different functions (Yuanyou and Haoran, 2000). Numerous studies show that cGnRH-II is expressed not only in brain, but also in spinal cord and sympathetic ganglia, as well as in non-neural tissues. There are many evidences that it is engaged in neuromodulation (Muske and More, 1990; Licht et al., 1994; Muske et al., 1994; D' Aniello et al., 1995; Collin et al., 1995; Troskie et al., 1997, Miranda et al., 1998; Rastogi et al., 1998), as well as in regulation of reproductive behavior and control of synthesis and secretion of steroids (Gobetti and Zerani, 1999; King et al., 1994, Collin et al., 1995; Yuanyou and Haoran, 2000; Millar, 2003).

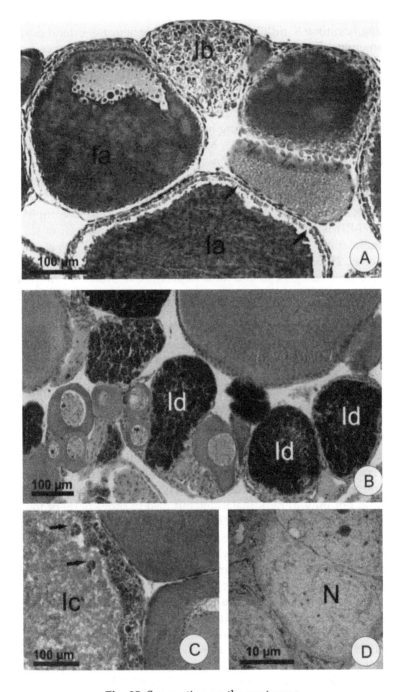

Fig. 25. See caption on the next page.

The secretory activity of hypothalamus depends on sexual maturity and phase of sexual cycle. The GnRH synthesis starts during larval life, but the number of GnRH immunoreactive neurons and the amount of secreted GnRH increase significantly after metamorphosis, as was reported for *Xenopus laevis* (King and Millar, 1981) and *R. cathesbeiana* (Crim, 1984; Whalen and Crim, 1985). The high level of synthesis and secretion of GnRH in sexually mature females fluctuates in species-specific manner with seasonal changes in reproductive activity. In *Bufo japonicus* examined in spring and autumn the immunoreactivity of the hypothalamo-median eminence tract was strong, whereas it was weak in toads examined in summer and during hibernation (Jokura and Urano, 1985). GnRH synthesized during hibernation is partly accumulated in the median eminence and used for intensive stimulation of pituitary for surge of gonadotropins after arousal, as was observed in *Rana temporaria* by Sotowska-Brochocka et al., (1992). GnRHs stimulate target organs through membrane bound receptors (GnRHRs) coupled with the G proteins (Millar, 2003; Millar et al., 2001, 2004; Guilgur et al., 2006). The GnRHRs vary in their structure and function (Troskie et al., 2000; Wang et al., 2001; Acharjee et al., 2002). Genes encoding GnRHRs have been cloned and the amino acid sequence has been determined for a broad range of vertebrate species including fish, amphibians, reptiles and mammals (Millar et al., 2001, 2004; Kah et al., 2004; Ikemoto et al., 2004; Guilgur et al., 2006). The structure and function of amphibian GnRHRs have been studied mainly in *Rana catesbeiana* (Wang et al., 2001), *Xenopus laevis* (Troskie et al., 2000), and *Rana dybowskii* (Seong et al., 2003). In the bullfrog *Rana catesbeiana* three types of receptors have been found, namely bf GnRHR-1, bf GnRHR-2, and bf GnRHR-3, which are encoded by different types of mRNAs. The bf GnRHR-1 is expressed mainly in the pituitary and plays an important role in integration of signals from the hypothalamus and the release of gonadotropins. The bf GnRHR-2 and bf GnRHR-3 are expressed in other parts of the brain. In *Xenopus laevis* two kinds of GnRH receptors are known: xlGnRHR-1 and xlGnRHR-2, whereas in *Rana dybowskii* three types are recorded: dyGnRHR-

---

Fig. 25. Degeneration and intrafollicular atresia in ovaries of Rana temporaria. Figures A-C show the most common type I divided into four stages Ia-Id. At stage Ia nucleoli fuse and the nucleus shrinks. Oocyte envelope (vitelline membrane) degenerates (arrows in A), the follicle cells transform into phagocytes and invade the ooplasm (arrows in C). At stage Ib the proliferating phagocytes fill the follicle and the phagocytes start to degenerate and accumulate dark pigment. At stage Ic the follicle cells hypertrophy and degenerate. The follicles were still filled with phagocyting cells, all of them containing dark pigment. At stage Id the condensed black pigment formes a mass within the follicle. D—Degenerating primary oogonium in a germ patch of adult female; N—nucleus.

1, -2, and -3. Despite their close similarity in the amino acid sequence, they exhibit significant differences in ligand recognition and the pathway of signal transduction (Seong et al., 2003).

Generally, the vertebrate receptors have been categorized into three families in respect of their sensitivity to tested ligands (which are the two GnRH izoforms) and differences in signal transduction pathways (Millar et al., 2004, Troski et al., 2000, Seong et al., 2003, Wang et al., 2001). According to this classification, bfGnRHR-2 of *Rana catesbeiana* and xlGnRH-1 of *Xenopus laevis* belong to the first family together with mammalian and fish GnRHR; receptor bfGnRhR-3 of *Rana catesbeiana* and xlGnRHR-2 of *Xenopus laevis* with human and other primates cGnRH-II form the second family, and bfGnRH-1 of *Rana catesbeiana* and some fish receptors form the third family (Millar et al., 2001, 2004).

The secretory activity of hypothalamus is regulated by a negative feedback mechanism from the pituitary and gonadal hormones. The release of GnRH from the hypothalamus is controlled by dihydrotestosterone (DHT) and estradiol (E) (McCreery and Licht, 1984a,b).

Besides the negative feedback mechanism from gonadal steroids and pituitary gonadotropins there is also an endogenous mechanism of GnRH regulation in the hypothalamus. As is known from experiments of Dierickx (1967a,b) and Sotowska-Brochocka (1983, 1988), a specialized dopaminergic area that inhibit a premature release of hypothalamic GnRH is localized in the ventral wall of the infundibular area of the hypothalamus. Experimantal lesions of this area destroy the inhibiting mechanism of the hypothalamus and cause the release of stored GnRH, which in turn stimulates the pituitary for LH secretion and causes preterm ovulation and spawning of preovulatory females *Rana temporaria* during hibernation.

Pituitary

The pituitary is a neuroepithelial organ, which develops in the amphibian embryo by a fusion of two different ectodermal derivatives. One part forms *pars nervosa*, whereas the other part forms solid glandular *adenohypophysis*. Pituitary gonadotropic cells (gonadotrophs) are distributed among other secretory cells in *pars distalis* of the *adenohypophysis* (Mikami, 1992). The synthesis and secretion of gonadotropins is stimulated by hypothalamic GnRHs (Licht and Papkoff, 1974; Papkoff et al., 1976; Takahashi and Hanaoka, 1981; McCreery and Licht, 1983a,b; Licht et al., 1987; Stamper and Licht, 1993; Polzonetti-Magni et al., 1998).

Gonadotropins (luteinizing hormone LH and follicle stimulating hormone FSH) belong to glycoproteins. They are heterodimers composed of two subunits: α and β. The α subunit is identical in both hormones, and very similar to the mammalian α subunit. The β chain, which is different in the two gonadotropins in its aminoacid sequence, ensures a specific biological activity and a unique character of each gonadotropin, and is mainly responsible for binding to the receptors (Mikami, 1992; Hayashi *et al.*, 1992a,b; Arai *et al.*, 1998). Gonadotropins, like many others proteins, are species specific. Small molecular differences in their structure or amino acid sequence may change their biological activities and cause that gonadotropin active in one species may be less potent or even inactive in other, even closely related species (Lofts, 1974; Licht and Crews, 1976; Fontaine, 1980; Licht *et al.*, 1983b).

The specific roles of LH and FSH of amphibians are not as well recognized as in mammals. Contrary to mammals, which pituitary has pulsative activity, the amphibian pituitary responds to continuous stimulation of GnRH (McCrerry and Licht, 1983a). Hypothalamic stimulation of pituitary fluctuates during the annual ovarian cycle. Gonadotropins synthesized in winter may be stored and used at the end of hibernation or after arousal. The levels of gonadotropins are generally low at the beginning of hibernation and start to rise progressively towards the end of the inactive period. Finally it is followed by a surge of gonadotropins, which results in ovulation. The concentration of both gonadotropins during the surge is elevated; however the LH level is several time higher than that of FSH. The surge of gonadotropins is accompanied by increase of testosterone and estradiol, and later on (in the ovulation phase) also progesterone (Itoh *et al.*, 1990; Polzonetti-Magni *et al.*, 1998; Kim *et al.*, 1998). Afterwards, concentration of gonadotropins drops to the basal level.

Gonadotropins act through receptors located on thecal and follicular cells of ovarian follicles (Ischii and Kubokawa, 1985; Polzonetti Magni *et al.*, 1998). Amphibian gonadotropic cells react with anti-FSH as well as with anti-LH antibodies. It indicates that both gonadotropins are most probably produced by the same secretory cell. However, some gonadothrops may secrete only one gonadotropin (for review see Fernandez and Ramos, 2003).

The oogonia and small non-vitellogenic oocytes in the ovaries of larval and newly metamorphosed females are gonadotropin independent (Fortune, 1983). Gonadotropin control begins when the newly recruited complement of small oocytes start synchronous vitellogenesis and lasts until the completion of their growth (Jørgensen, 1992). It is well known from early research that dissection of the pituitary (PDX) of sexually mature females causes degeneration of vitellogenic oocytes, whereas administration of exogenous gonadotropins to PDX females restores the function of their ovaries (Lofts, 1974; Jørgensen, 1975; Denver *et al.*, 2002).

Species may differ in their mode of pituitary control of a female annual cycle depending on the reproductive strategies employed (Itoh *et al.*, 1990). In *Rana esculenta* the highest value of FSH level at the beginning of the reproductive period is accompanied by a very high level of estradiol-17β and vitellogenin (VTG). On the contrary, the increase of plasma LH level, which occurs at the end of recovery period, remaines high during winter rest until the beginning of breeding period, after which it sharply decreases. Parallel to the changes in LH levels, the ovarian weight increases from the time of recovery until spawning (Polzonetti-Magni *et al.*, 1998). The plasma LH levels in *Rana dybowskii* and *R. nigromaculata* are much higher in females having uterine eggs or being in amplexus than in females with unovulated follicles, or in these that have just spawned. The absolute level of LH is also much higher than that of FSH. *Rana nigromaculata* has low plasma levels of both gonadotropins during hibernation, but ovulation and spawning are preceded by a surge of these hormones. Plasma FSH level increases after spawning, concomitantly with the growth of a new generation of recruited follicles (Kim *et al.*, 1998).

Besides the ovary, gonadotropin hormone receptor binding sites are located in the liver, which indicates their role in control of hepatic synthesis of yolk precursor VTG (Gobetti *et al.*, 1985; Kubokawa and Ischii, 1987; Carnevali and Mosconi, 1992). As it was shown in *R. esculenta*, plasma level of VTG increases during vitellogenesis, then it slightly decreases during hibernation, and increases again at the beginning of breeding period, and attains its peak value followed by a sharp decline soon after (Polzonetti-Magni *et al.*, 1998). *In vitro* study revealed that liver sensitivity to pituitary gonadotropins, as well as hepatic VTG synthesis, well correlate with seasonal changes of the reproductive cycle. Incubation of liver samples obtained from *R. esculenta* females at breeding period with *R. catesbeiana* gonadotropins (LH, FSH, or pituitary extract) induced an increase of hepatic VTG synthesis. When liver samples were taken from females being in the post reproductive period when physiologic synthesis of hepatic VTG was very low, the effect of gonadotropins was very weak. In liver samples taken during autumn ovarian recovery, it was possible to increase VTG synthesis only when they were incubated with LH and pituitary extract, whereas incubation with FSH had no effect (Polzonetti-Magni, 1998; Carnevali *et al.*, 1992). Polzonetti-Magni *et al.* (1998) found also distinct differences in ovarian response to pituitary gonadotropins. Ovarian follicles harvested during post-reproductive summer phase and cultured with the bullfrog FSH and LH, as well as with pituitary extract, significantly increased synthesis and secretion of estradiol 17β, whereas follicles collected during autumn recovery phase under the same experimental conditions increased the level of progesterone. Finally, incubation of follicles obtained from spawning *R. esculenta* females with bullfrog gonadotropins or pituitary extract has no effect on the level of steroids synthesized by gonads. The results obtained by Polzonetti-Magni *et al.* (1998) demonstrated that it is very difficult to distinguish between the results of FSH and LH actions, because both gonadotropins influence ovarian steroidogenesis in a similar way. During autumn recovery phase they stimulate follicles to secrete progesterone, whereas during post reproductive summer phase they stimulate estradiol 17β secretion.

Summarizing the effects of gonadotropins on ovarian function during the annual reproductive cycle, it appears that during long-lasting vitellogenic phase gonadotropins play a permissive rather than a regulative role. In contrast, ovulation must be preceded by the surge of gonadotropins secretion. The rupture of the follicles and ovulation defines the end of ovarian cycle. It is supposed that ovulation eliminates the probable paracrine inhibitory effect exerted by vitellogenic follicles on growth and differentiation of previtellogenic follicles. This makes possible the recruitment of a new wave of small follicles for vitellogenesis (Jørgensen, 1992; Fernandes and Ramos, 2003; Medina *et al.*, 2004).

The synthesis and secretion of gonadotropins is under negative feedback control by sex steroids acting on the hypothalamo-pituitary axis (Lin and Schuetz, 1985). From an early experiments on several anuran species: *Rana esculenta* (Rastogi and Chieffi, 1970a,b), *R. catesbeiana* (Licht *et al.*, 1983a,b; McCreery and Licht, 1983b, 1984a,b), and *Bufo bufo* (Vijayakumar *et al.*, 1971) it is known that gonadectomy causes hypertrophy of gonadotropic cells in pituitary, and that the hypertrophy can be reduced by administration of exogenous steroids. It has been demonstrated that steroid feedback control is exerted by androgens, especially by DHT (McCreery and Licht 1984a; Licht *et al.*, 1994), which can induce the pituitary to release gonadotropins. After prolonged stimulation of gonadectomized females the concentration of LH decreases, while FSH level remains high, probably due to different regulation of synthesis and secretion of the two gonadotropins (Stamper and Licht, 1991).

Gonadotropin synthesis is also influenced by prostaglandins. $PGF_{2\alpha}$ is known to stimulate gonadotropin synthesis. In the annual cycle the level of $PGF_{2\alpha}$ fluctuations is well correlated with circulation level of estradiol (Gobetti *et al.*, 1990).

Other polypeptide pituitary hormones that are known to influence amphibian reproduction are growth hormone (GH) and prolactin (PRL). The plasma levels of GH and PRL fluctuate significantly during annual reproductive cycle (Polzonetti-Magni *et al.*, 1995). The mode of annual fluctuation of the level of GH gives some evidence that it may stimulate the increase of ovarian weight. Mosconi *et al.* (1994) suggest that GH is involved in ovarian growth and regulation of vitellogenesis. In a series of experiments Carnevalli and Mosconi (1992) and Carnevalli *et al.* (1992,

1993, and 1995) gave the evidences that GH stimulates VTG mRNA transcription in both sexes *in vivo* and *in vitro,* in season- and dose-dependent manners. The direct stimulatory effect of GH on VTG synthesis has also been documented in experiments carried out in both intact and pituitary dissected females.

PRL has antigonadotropic effect. Acting on a paracrine way, it supresses the synthesis and secretion of gonadotropins. PRL-secreting cells are distributed mainly in the anterior part of the *distal lobe* of pituitary. They are frequently located in a close vicinity to gonadotropic cells (Polzonetti-Magni *et al.,* 1995). It was also suggested that PRL may increase the sensitivity of oocytes to gonadotropins and progesterone.

Ovary

Like in other vertebrates, two main and inseparable functions of anuran ovary are production of fertilizable ova and synthesis of hormones regulating reproductive functions. The cells of follicle wall produce a variety of hormones (steroids, prostaglandins, peptide hormones, and growth factors) that regulate and control the development of female gametes by creating favorable microenvironment for developing oocytes. Classical steroid receptors are able to function in dual path: in nucleus as transcription activators and outside the nucleus as signal transducers (Bayaa *et al.,* 2000).

Steroids: Steroid hormones, which play an important role in differentiation and function of female reproductive system, are the major secretory products of the ovary produced in response to gonadotropin stimulation. Cells that constitute the wall of ovarian follicles secrete three main types of steroids: estrogens, progestagens, and androgens. The members of each steroid type share the basal chemical structure and functional properties. The relative amount of each type of steroids varies during the reproductive cycle according to requirement of the developing oocytes (Medina *et al.,* 2004). Follicular and thecal cells have features typical of steroidogenic activity, such as well-developed smooth endoplasmic reticulum, tubular mitochondria, distinct Golgi apparatus, and lipid dropplets accumulated in the cytoplasm (Chieffi and Pierantoni, 1987). The mutual cooperation between thecal and follicular cells is essential for ovarian steroidogenesis, because both cell types vary in respect to the activity of specific enzymes involved in consecutive steps of steroidogenesis (Kwon *et al.,* 1991, 1993; Kwon and Ahn, 1994). Both follicular and thecal cells convert pregnenolone to progesteron. The interconversion of steroids from pregnenolone to progesterone depends on gonadotropins (Kwon *et al.,* 1993). Follicular cells, characterized by high activity of 17α hydroxylase and $C_{17,20}$ lyase, are able to convert progesterone to 17α-hydroxyprogesterone and androstenedione.

Androstenedione is transferred from follicular to thecal cells, where its further conversion to testosterone occurs in presence of high activity of 17β hydroxysteroid dehydrogenase. The final step of steroidogenesis, which is the aromatisation of testosterone to estradiol, also occurs in follicular cells and is independent of gonadotropins (Kwon *et al.*, 1993; Fernandes and Ramos, 2003). The activity of aromatase decreases with progression of vitellogenesis and less testosterone is aromatized to estrogens.

Estrogens (estrone—E_1, estradiol 17β—E_2, and estriol—E_3) maintain the physiological condition of female reproductive system and promote yolk synthesis and accumulation. The activity of 3β HSD and 17β HSD enzymes, necessary for transformation of estrone to estradiol and estriol, was detected in the follicular cells. Synthesis and secretion of estradiol 17β is stimulated by FSH (Polzonetti-Magni *et al.*, 1998). The main sources of estradiol 17β are the medium-sized vitellogenic follicles, as was demonstrated by Fortune (1983) in *Xenopus laevis*, and by Kwon *et al.* (1991, 1993) in *Rana nigromaculata*. According to these studies, the highest level of estradiol 17β is observed during vitellogenesis in early postovulatory period of the reproductive cycle (Girling *et al.*, 2002). The high concentration of estradiol inhibits oocyte maturation. Estrogens are the main inductors of hepatic synthesis of VTG. With progression of postovulatory period, growing follicles accumulate more and more yolk and ovarian weight increases significantly. At that time the follicles gradually change the mode of their secretion from estradiol to testosterone (Polzonetti-Magni *et al.*, 1998; Fernandez and Ramos, 2003). Change from estradiol to testosterone synthesis is connected with a decline of aromatase activity in follicular cells. A gradual decrease of estradiol 17β concentration occurs during the late postovulatory period attaining the lowest level before ovulation (Pierantoni *et al.*, 1987; Di Fiore *et al.*, 1998; Polzonetti-Magni *et al.*, 1998; Medina *et al.*, 2004). The low concentration of estradiol 17β enable nuclear maturation of fully grown oocytes (de Romero *et al.*, 1998; for references see also Medina *et al.*, 2004).

Progesterone, which belongs to progestagens, is synthesized in both thecal and follicular cells of fully grown ovarian follicles in response to LH surge that initiates oocyte maturation (Kwon and Ahn, 1994; Kwon *et al.*, 1993). Progesterone inhibits vitellogenesis and induces biochemical modification of cytoplasm and nucleus, which change oocyte metabolism and finally causes its maturation and ovulation. The level of progesterone fluctuates during the annual cycle. The highest serum concentration of progesterone occurs at the beginning of preovulatory phase of the reproductive cycle, sharply decreases after ovulation, remains low until the early postovulatory phase, and slowly increases during postovulatory phase (Fernandez and Ramos, 2003).

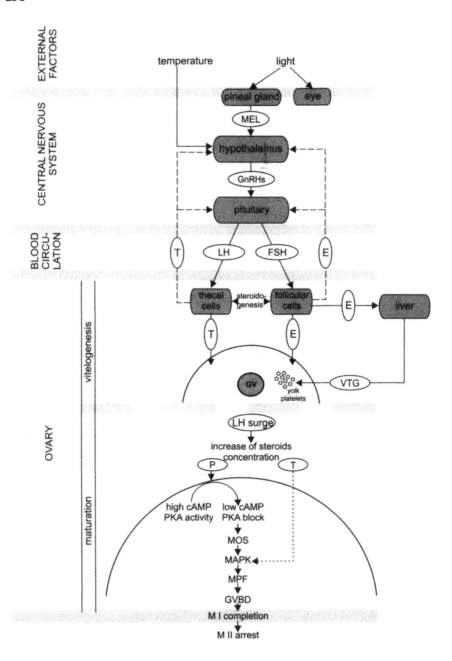

Fig. 26. See caption on the next page.

A characteristic feature of anuran females is the high concentration of androgens (testosterone T and dihydrotestosterone DHT) in the ovary and in the circulation (Fernandez *et al.*, 1994; Lutz *et al.*, 2001). In contrast to estrogens, serum concentration of androgens increases with ovarian weight (Fernandez and Ramos, 2003). During the annual reproductive cycle the concentration of T and DHT fluctuate in the same way, but the level of testosterone is always 4-5 times higher then that of DHT (Fernandez and Ramos, 2003). Similar to progesterone, both androgens reach the highest serum concentration during preovulatory phase, when follicles attain their maximal sizes. It suggests that androgens have a role in triggering oocyte nuclear maturation (Le Goascogne *et al.*, 1985; Lutz *et al.*, 2001, 2003; Ramos *et al.*, 2001). This suggestion has been confirmed by Ramos *et al.* (2001), who demonstrated nuclear maturation of oocytes after application of DHT that is a non-aromatizable androgen. There are many evidences that androgens are potent activators of oocyte maturation (Lutz *et al.*, 2001; Schmitt and Nebreda, 2002; Hammes, 2003, 2004). The highest levels of T and DHT occur at the same time when estradiol 17β drops to its lowest level, which might be a favorable condition required for oocytes maturation (Medina *et al.*, 2004). Levels of T and DTH decrease sharply at the beginning of ovulatory phase, remain low for the rest of the year, and start

---

Fig. 26. Schematic representation of main environmental, hormonal, and biochemical factors regulating growth and maturation of anuran oocytes. External factors stimulate the central nervous system; hypothalamus is no longer inhibited by melatonin and secretes GnRH, which stimulates secretion of pituitary gonadotropins: LH that stimulates thecal cells and FSH that stimulates follicular cells for steroid synthesis. During vitellogenesis and oocyte growth follicular cells secrete mainly estradiol 17 β that binds to hepatic receptors and stimulates synthesis and secretion of VTG. Blood transports VTG to oocyte surface, from where VTG enters the ooplasm by receptor-mediated endocytosis and is transformed into lipovitellin and fosvitin of yolk platelets. Fully grown oocyte is arrested in diplotene of meiotic prophase I. Maintenance of the arrest depends on elevated cAMP level and PKA. In response of LH surge, progesterone gives the signal for nuclear and cytoplasmic maturation of the oocyte. A decrease in cAMP level and block of PKA activity stimulates the cascade of secondary signals MOS and MAPK. The resumption of meiosis is additionally stimulated by testosterone, which also stimulates MAPK. The MOS-MAPK pathway is involved in activation of MPF that leads to meiosis resumption manifested by GVBD. Oocyte maturation is completed when meiosis is arrested in metaphase II. Negative feedback loops are indicated by dotted lines. E—estradiol 17 β; GnRH—gonadotropin releasing hormone; GVBD—germinal vesicle breakdown; MAPK—mitogen activated protein kinase; MPF—maturation promoting factor; P—progesterone; PKA—protein kinase A; T—testosterone; VTG—vitellogenin.

to increase slowly at the end of late post reproductive period (Fernandez and Ramos, 2003; Medina *et al.*, 2004).

Testosterone is the main source of estradiol 17β in the ovary (Kwon and Ahn, 1994; Fernandez and Ramos, 2003; Medina *et al.*, 2004). Aromatization of testosterone to estradiol occurs in follicular cells. It also occurs in the liver, where testosterone is aromatized to estrogen and used in VTG synthesis (DiFiore *et al.*, 1998). Besides that, testosterone is also aromatized to estradiol in oviducts (Kobayashi *et al.*, 1996) and brain (Guerriero *et al.*, 2000).

Prostaglandins: Anuran ovarian follicles synthesize prostaglandins *in vitro* in response to hypothalamic or pituitary stimulation (Schuetz, 1986; Gobetti and Zerani, 1992; Gobetti *et al.*, 1990). There is strong evidence to suggest that prostaglandins are involved in control of ovulation (Schuetz, 1986; Gobetti and Zerani, 1992; Chang *et al.*, 1995; Skoblina *et al.*, 1997). Skoblina *et al.* (1997) demonstrated in *Rana temporaria* that prostaglandin $PGF_{2\alpha}$ is capable to cause ovulation without pituitary stimulation. Ovaries taken from frogs in recovery stage, as well as oviducts collected during the reproductive stage, show an increase of $PGF_{2\alpha}$ level after incubation with mGnRH (mammalian GnRH) in culture medium. This suggests that $PGF_{2\alpha}$ may take part in control of eggs deposition (Gobetti and Zerani, 1990, 1992).

Various members of prostaglandin family are known to act in opposing manners to each other. As reported by Schuetz (1986), $PGF_{2\alpha}$ stimulates ovulation of *Rana pipiens* follicles cultured *in vitro* and exposed to progesterone or to frog pituitary homogenate (FPH), whereas PGE acts contrary to $PGF_{2\alpha}$. In *R. pipiens* $PGF_{2\alpha}$ is also able to trigger ovulation of immature ovarian oocytes. Plasma levels of $PGF_{2\alpha}$ fluctuate during annual reproductive cycle. $PGF_{2\alpha}$ was synthesized in frog ovarian follicles collected at hibernation and incubated with FPH or other ovulation-inducing factors (Chang *et al.*, 1995). In female *R. esculenta*, $PGF_{2\alpha}$ plasma level is positively correlated with estradiol fluctuations (Gobetti *et al.*, 1990). Gobetti and Zerani (1999) found that *R. esculenta* females in amplexus have higher brain aromatase level and plasma concentration of $PGF_{2\alpha}$, but lower PGE_2 level, than those which are not amplecting.

Peptide Hormones: Besides their role in steroid synthesis, pituitary gonadotropins also stimulate the ovary to synthesis of two polypeptides: activin and inhibin. They are members of TGF β superfamily of growth and differentiation factors. Both polypeptides are composed of subunits α and β. Inhibins are heterodimeric glycoproteins, which have one α and one of the two β subunits (β_A or β_B), forming inhibin A or inhibin B, respectively. Activin consists of two kinds of β subunits linked in $\beta_A\beta_A$ or $\beta_A\beta_B$ combinations (Ying, 1988). Both activin and inhibin are synthesized in follicular cells of the ovarian follicles. Besides the ovary they are also found in the pituitary gonadotropes (Uchiyama *et al.*, 1996, 2000). As generally in vertebrates, they are known to have multiple sources and are produced by a variety of cells. They act as endocrine, paracrine and autocrine regulators of various reproductive functions, and modulate or mediate the functions of other hormones, such as gonadotropins and steroids (Petrino *et al.*, 2007; Woodruff and Mather, 1995).

Both hormones are able to regulate pituitary gonadotropins release. In the bullfrog *Rana catesbeiana*, the action of activin B stimulates the secretion of pituitary FSH and LH in a dose-dependent manner, whereas inhibin B eliminates activin B-induced increse of gonadotropins without affecting their basal levels. Amphibians differ from mammals in respect of pituitary stimulation and biological potency of activin since in mammals activin selectively stimulate only FSH secretion (Ying *et al.*, 1988; Woodruff and Mather, 1995). Generally, activin stimulates differentiation of follicle cells induced by FSH and attenuate androgen rise induced by LH. Inhibin enhances LH-induced progesterone and androgen rise. Acting at a local autocrine/paracrine manner activin increases aromatase activity of follicular cells, thereby elevating the level of synthesized estrogens and inhibits androgen synthesis in thecal cells (Uchiyama *et al.*, 2000). Inhibin represses pituitary secretion of FSH through a negative feedback loop. Inhibin secretion is regulated by endocrine (FSH) and para/autocrine (activin) factors and the rising level of estrogens stimulates inhibin synthesis.

In vitro study of the action of inhibin on *Rana pipiens* oocyte maturation revealed that in the presence of FPH, which stimulates follicular cells multiplication, and progesterone secretion exogenous inhibin suppresses activity of FPH and progesterone in a dose-dependent manner. Inhibin also blocks GVBD induced by egzogenous progesterone in intact and denuded oocytes (Lin *et al.*, 1999). The results of above experiment demonstrated that inhibin acts on follicular cells (by blocking steroidogenesis) and on the oocytes (by suppressing their maturity). Inhibin affects early events of nuclear maturity, and its final effect depends of the duration of incubation of the follicles with inhibin. Generally, the activity of inhibin and activin in amphibians is poorly recognized. Based on the information about the roles of both polypeptides in mammals, it may be supposed that they may participate in a number of additional functions and similar regulatory pathways. Among others their possible role in recruitment and trans-formation of follicles from gonadotropins-independent to gonadotropins-dependent should be investigated.

Reproduction is regulated not only by hormones secreted by CNS or endocrine glands, which consist of components of the hypothalamo-pituitary-gonadal axis, but also other hormones that may participate directly or indirectly in the control of various reproductive processes. The two groups of such hormones, which are significant for reproduction, are insulin and insulin-like growth factors.

Insulin and Insulin-like Growth Factor I (IGF-I): Insulin is a polypeptide hormone produced by the endocrine pancreas. Insulin regulates many metabolic processes and influences most of internal organs, although its main function is a regulation of plasma glucose level. Insulin-like growth factors (IGFs), which are also polypeptides, are similar to insulin in their chemical structure and biological activities. IGFs are synthesized mainly in the liver and released to blood circulation. Both hormones exert the biological effect after binding to their specific receptors located in the cell membrane (El-Etr *et al.*, 1979, 1980). *In vitro* insulin and IGFs stimulate activities of specific protein kinases, which initiate series of events leading to activation of meiosis promoting factor (MPF), GVBD, and maturation of oocytes (Lessman and Schuetz, 1981; Maller and Koontz, 1981; Hainaut *et al.*, 1991; Debauche *et al.*, 1994; Liu *et al.*, 1995; Andersen *et al.*, 2003). There are conflicting opinions whether insulin and IGF-I may act independently (El-Etr *et al.*, 1979, 1980; Lessman and Schuetz, 1981), or exert a synergistic effect when associated with progesterone and testosterone action (Hirai *et al.*, 1983; Le Goascogne *et al.*, 1984, 1985).

Hormonal Regulation of Oocyte Growth and Maturation

Although oocyte growth and maturation are under multihormonal control, steroids are the main regulators that modify the metabolism of both oocytes and somatic cells of the ovary. Steroids initiate the transcription of specific genes into mRNA necessary for synthesis of various proteins, or act by a non-genomic signaling mechanisms. The levels and profiles of individual steroid hormones change during annual cycle according to demands of growing oocytes, such as vitellogenesis, maturation, meiosis resumption, and ovulation.

Vitellogenesis. In temperate zone anurans vitellogenesis takes place mainly in autumn when the ovary contains stage 3 and 4 medium-sized diplotene oocytes. The oocytes enter the phase of rapid growth in result of massive yolk accumulation. Under stimulation of gonadotropins follicular cells of growing oocytes secrete large amounts of estradiol 17β, which is the main regulator of vitellogenesis (Fortune, 1983). Throughout specific receptors in the liver, estradiol controls synthesis of vitellogenin and its transport with blood circulation to oocyte surface. Vitellogenin enters the oocyte by receptor-mediated endocytosis and is metabolized to lipovitellin and phosphitin, which are main components of yolk platelets. Liver synthesis and release of VTG is also under direct control of pituitary gonadotropins (Gobetti *et al.*, 1985). Other pituitary hormones (GH and PRL) influence the vitellogenesis indirectly. They participate in control of hepatic VTG synthesis by paracrine modulation of the activity of pituitary gonadotrops (Di Fiore *et al.*, 1998; Medina *et al.*, 2004).

Maturation and Meiosis Resumption: After vitellogenesis is completed oocytes attain their maximal size (stage 6) and the synthetic activity of the oocytes diminishes significantly. Fully grown oocytes are arrested in prophase I of meiosis. After the preovulatory surge of gonadotropins, fully-grown follicles synthesize progesterone, which trigger the resumption of meiosis and maturation of oocytes. It is well established that in amphibians, as in many vertebrates, the progesterone induces nuclear maturation, acting throughout specific membrane associated receptors, present in fully grown oocytes (Fernandez and Ramos, 2003; Liang *et al.*, 2007). During the preovulatory period, T, DHT, and P reach their highest concentrations, both in the ovary and in circulation. The role of androgens in re-initiaion of meiosis did not arouse adequate interest and was underestimated for a long time. During last years, there are many evidences that androgens, like progesteron, are able to physiological stimulations of nuclear maturation *in vivo*. There is no doubt, that they stimulate Cdc2 defosforylation and MAPK activation, however the way of their actions are still not clearly recognized (Lutz *et al.*, 2001, 2003; Ramos *et al.*, 2001; Fernandez and Ramos, 2003; Hammes, 2004; Rasar and Hammes, 2006; Liang *et al.*, 2007).

Meiosis may be also stimulated *in vitro* by exogenous steroids, polypeptides such as insulin or insuline-like growth factor (IGF), and other growth factors (Ferrell, 1999a,b; Schmitt and Nebreda, 2002, Liang *et al.*, 2007). After completion of metaphase I, oocytes are again arrested in meiotic metaphase II as mature gametes capable of ovulation and subsequent fertilization. Lutz *et al.* 2001 reported that after hCG (human chorionic gonadotropin) stimulation of *Xenopus laevis* females the concentration of progesterone in frog ovary and circulation was very low regardless of hCG dose, while androstenedione and testosterone were abundant. The level of T in female frogs injected with hCG is approximately 10 times higher than that of P.

Because of different (but complementary) enzymatic equipment, steroidogenesis requires close cooperation of the oocytes and cells of follicular walls. To test the ability of oocytes and follicular cells to synthesize steroids, Yang *et al.* (2003) separated oocytes from the follicle. In this way they gave an excellent evidence for close cooperation between oocytes and follicular cells. The follicular cells express activity of 3β-HSD and 17β-HSD, whereas the oocyte synthesizes CYP17; all the three enzymes are required for completion of androgen biosynthesis. CYP 17, a member of P 450 family, has the activity of both 17α hydroxylase and 17, 20 lyase. CYP 17 is the endogenous oocyte enzyme converting pregnenolone and progesterone to androgens, and stimulating oocyte maturation. Its presence in oocytes confirms that they are able to secrete factors, which in turn influence their own maturation (Hammes, 2004). There are two metabolic pathways for steroid synthesis: Δ4 (ketone) and Δ5 (with 3β hydroxysteroid dehydrogenase 3βHSD). It was shown that the synthesis of androgens follows almost entirely the Δ5 pathway (Young *et al.*, 2003). It bypasses progesterone and goes directly from pregnenolone to androgens. Pregnenolone synthesized in follicular cells is transferred to oocytes where it is converted to 17-hydroxypregnenolone, and then to dehydroepiandrosterone (DHEA) by a specific enzyme CYP 17 (Lutz *et al.*, 2001, 2003). DHEA in turn re-enters follicular cells, where it is metabolized to androstenedione (AD) by 3β hydroxysteroid dehydrogenase (3βHSD). AD may be transferred back to the oocyte and stimulate its maturation or is metabolized in follicular cells by 17β hydroxy steroid dehydrogenase (17βHSD) to more potent T, which moves back to the oocyte and triggers GVBD and release the oocyte from prophase G2-like arrest (Yang *et al.*, 2003). CYP 17 may also convert P to AD by catalyzing conversion of P to 17α OHP by 17 α hydroxylase activity, followed by conversion of 17OHP to androstenedion AD via 17, 20 lyase activity. Some amounts of pregnenolon are metabolized to P in follicular cells in the presence of 3βHSD (Δ4 pathway). According to these authors, the pathway plays marginal role in physiological induction of oocyte maturation, but when exogenous P is added, it causes oocytes maturation. Inactivation of CYP 17 does not inhibit the progesterone-mediated maturation. CYP 17, expressed only in germ cells, converts also P to AD that is transferred to follicular cells, where it is finally converted to T, as was described above. Although T is more potent than AD, both androgens are considered as physiological inducers of oocyte maturation (Lutz *et al.*, 2001; Hammes, 2004).

Steroid-stimulated maturation is unaffected by actinomycin D, a strong inhibitor of steroid-stimulated transcription, which means that it is transcription independent (Smith and Ecker, 1969, 1971; Masui and Markert, 1971; Cork and Robinson, 1994; Maller, 1998; Ferell, 1999a,b; Tian *et al.*, 2000; Lutz *et al.*, 2001, 2003; Rasar and Hammes, 2006). Thereby, it has been assumed that during oocyte maturation steroids act through separated class of surface-associated receptors (Maller, 1998; Tian *et al.*, 2000). Studies of Bayaa *et al.* (2000), Tian *et al.* (2000), and Bagowski *et al.* (2001) revealed that non-genomic, progesterone stimulated oocytes maturation is mediated by classical nuclear/cytoplasmic receptors. Simultaneously, Bagowski *et al.* (2001) found two izoforms of these receptors, from which one fraction, constituting about 5-10% of total progesterone receptors (PRs), appeared to be associated with oocyte cell membrane and could interact with the membrane signaling molecules to mediate P-induced maturation. Using specific biochemical and immuno-histochemical methods, Lutz *et al.* (2001, 2003) demonstrated that in *Xenopus laevis* ovary classical androgens receptors (ARs), acting outside of the oocyte nucleus, are used for transmission of androgen-induced signals that stimulate oocyte maturation.

Fully-grown oocytes may be arrested in the prophase of the first meiotic division for a long time (Fernandez and Ramos, 2003; Rasar and Hammes, 2006; Liang *et al.*, 2007). It is widely accepted that P is a hormone that induces maturation in lower vertebrates. The reinitiation of meiosis is preceded by complex metabolic changes in the ooplasm and germinal vesicle. Thus, oocyte maturation involves two processes, from which one concerns cytoplasmic, and the other nuclear maturation. Cytoplasmic maturation is stimulated by FSH and adrenaline and inhibited by noradrenaline and melatonine (MEL), acting individually or synergistically. Cytoplasmic maturation is associated with the synthesis of MPF (Matu-ration Promoting Factor), migration of the cortical granules to the periferal zone of oocyte cytoplasm, and activation of the Ca^{2+} dependent mechanism of phosphorylation of seryne/threonine proteins (Fernandez and Ramos, 2003). MPF stimulates the nuclear maturation, which is manifested by GVBD.

The obligatory step after P stimulation is inactivation of adenylate cyclase. This results in decrease of cellular cAMP level followed by protein kinase A (PKA) decrease (Matten *et al.*, 1994; Liang *et al.*, 2007). After the decrease of PKA, synthesis of specific new proteins required for subsequent steps of oocytes maturation is promoted. In *Xenopus laevis* one of specific molecules involved in initiation of oocyte maturation is germ cells specific serine-threonine (Ser/Thr) protein kinase mos (Mos), a product of the protooncogene *c- mos* (Sagata *et al.*, 1988, 1989; Sagata, 1997). The *mos* mRNA is detectable from early oogenesis until the time of gastrulation at nearly constant levels, but Mos protein is synthesized only at about the time of maturation before GVBD (Liang *et al.*, 2007). The Mos molecule stimulates MAPK kinase (Matten *et al.*, 1996) and is a potent activator of MAPK and Cdc2 pathways. Mos accumulation is essential for maturation of oocytes of many amphibian species. However, some authors demon-strated that, although Mos protein takes part in the promotion of the G_2/ M transition, it is not essential for meiosis resumption (Mood *et al.*, 2004). Meiosis in *Xenopus* can be resumed in Mos-independent way using alternatively Ras instead of Mos kinase (Liang *et al.*, 2007). Both kinases are able to activate MAPK and lead to MPF formation. There are many evidences that MAPK plays a pivotal role in the regulation of meiotic cell cycle progression in oocytes (Liang *et al.*, 2007). Among a large number of molecules participating in the regulation of the oocyte maturation on the MAPK pathway, the MPF is the most important molecule responsible for direct induction of GVBD. MPF is composed of a serine/threonine protein kinase p34 cdc2 and cyclin B (Masui and Markert, 1971; Nebreda and Ferby, 2000; Palmer and Nebreda, 2000). In G_2 arrested oocytes the process of formation and activation of MPF differs among amphibian species (Fernandez and Ramos, 2003). In *Xenopus laevis* oocytes MPF is present in the inactive form as a pre-MPF composed of Cdc2 kinase and cyclin B, inactivated by inhibitory phosphorylation of Cdc2 kinase on threonine 14 (Thr 14) and thyrosine 15 (Tyr 15). Activation of pre MPF through thyrosine 15/threonine 14 dephosphorylation initiate oocyte maturation. In other amphibian species (*Rana japonica, Bufo japonicus*) growing oocytes contain only Cdc 2, whereas cyclin B is synthesized in ooplasm after P signal for maturation (for review see: Fernandez and Ramos, 2003). After association of cyclin B and Cdc2 it is activated through threonine 161 phosphorylation. Activated MPF triggers the transition of the oocyte from prophase I to the metaphase I. After completion of meiosis I oocytes are arrested in the metaphase of the second meiotic division that is mediated by cytostatic factor (CSF) (Masui and Markert, 1971). From this arrest they are released by fertilization.

Summarizing, it should be pointed out that despite decades of intensive investigations, the details of biochemical identity of some factors engaged in oocyte maturation, and/or the signaling pathway still remain poorly recognized (Eyers *et al.*, 2005; Rasar and Hammes, 2006; Liang *et al.*, 2007).

Pineal Complex and Environmental Cues

Contrary to other endocrine organs, pineal complex is directly influenced by photoperiod and temperature (Jørgensen, 1992; Alonso-Bedate *et al.*, 1990; Alonso-Gomez *et al.*, 1992; Delgado and Vivien Roels, 1989; Fernandez and Ramos, 2003; Chieffi and Minucci, 2004). The pineal complex of anurans consists of the frontal organ (parietal eye) lying under the head skin and the proper pineal gland (epiphysis) located within the braincase. Epiphysis forms small, flattened sac-like structures, evaginated from the diencephalic roof, and connected to it by a stalk. The pineal lumen communicates with the third ventricle of the brain. Both parts of the pineal complex are connected by the pineal nerve, which forms the pineal tract penetrating the diencephalon roof (Sato and Wake, 1992). Both parts are photoreceptive, and epiphysis functions also as a neuroendocrine gland (Mayer *et al.*, 1997). Pineal gland is known to play an important role in mediating photoperiodic effects on reproduction of many vertebrate species, especially mammals and birds (Vanecek, 1999; Hazlerigg, 2001). Although amphibian reproduction is under control of environmental factors, the influence of photoperiod seems to be of much lower importance. The primary hormone of the pineal gland is melatonin MEL (5-methoxy-N-acetyl-tryptamine). The synthesis and secretion of MEL, as well as its contents in serum, is controlled by endogenous circadian rhythms and shows seasonal variations (Klein and Weller, 1970a,b). Melatonin synthesis is higher during the night than during the day, because light inhibits the activity of serotonin N-acetyltransferase (NAT), the enzyme responsible for MEL processing (Jørgensen, 1992). In some species the night elevation of MEL level lasts longer under short then under long photoperiod (Delgado and Vivien Roels, 1989). Besides the pineal gland, MEL is also synthesized in eye neural retina (Delgado and Vivien Roels, 1989; Rohde *et al.*, 1985). Retinal MEL is the main source of this hormone in amphibians (Isorna *et al.*, 2004a). Both pineal and retinal MELs are released to blood circulation (Delgado and Vivien Roels, 1989).

The major target organ for MEL is the pituitary (Wiechmann *et al.*, 2003). Melatonin modulates pituitary activity and release of some of its hormones, although details of its influence are still not clear. Melatonin exerts its physiological action upon its target tissues through expression of G protein-coupled receptors (Wiechman *et al.*, 2003; Isorna *et al.*, 2004a; Dubocovich, 1988; Reppert *et al.*, 1996). Wiechman *et al.* (2003) found that MEL receptors in the pituitary of *Xenopus laevis* differ in their expression and distribution in relation to other pituitary hormone-producing cells. Type Mel_{1a} receptors were found on cells of *pars intermedia*, and Mel_{1c} receptors in *pars distalis*. Both types together were present only in *pars nervosa* of the pituitary. This may suggest that the receptors mediate specifically the response of various pituitary cells to MEL. The authors did not find expression of the MEL receptors on pituitary gonadotrops, which suggest that there is no direct influence of MEL on gonadotropin-secreting cells. However it appears that the MEL and arginine-vasotocin (AVT) receptors are co-localized in cells of *pars distalis*, and give them the possibility to influence AVT release. Besides the control of hydromineral balance, AVT is known to stimulate and mediate the action of gonadal steroids on reproductive behavior in sex-specific manner (Boyd, 1992; Diakow, 1978). Modulating the AVT release, MEL may influence reproductive processes. Chowdhury *et al.* (2007) found in the bullfrog that MEL induces the expression of hypothalamic neuropeptides (frog GH releasing peptides: fGRP and fGRP-RPs) stimulating release of GH and PRL from the pituitary. They demonstrate the expression of MEL receptors (Mel_{1b}) on neurons of suprachiasmatic nucleus and median eminence of diencephalon expressing the fGPR and fGRP-RPs neuropeptides. The expression of these neuropeptides is regulated by photoperiod; it decreaseas after orbital enucleation and pinealectomy, and increases after MEL administration. Besides of pituitary and hypothalamus, which both originate from diencephalon, MEL receptors were also present in telen- mesen- and myelencephalon (Tavolaro *et al.*, 1995,). Such widespread distribution of MEL binding sites in the brain suggest that it may act at different levels and influence many organs, making possible synchronization of reproduction with changing environment (Isorna *et al.*, 2004a).

The photoperiodic control of amphibian reproduction is still poorly understand and has to be clarified. The results of studies concerning photoperiodic influence are frequently controversial. Jørgensen (1992) in his review quoted the early experiments on *Rana pipiens*, *Xenopus laevis*, and *Bufo fowleri* in which short photoperiod or even complete darkness have no significant effect on gonadal cycles and vitellogenesis. Saidapur and Hoque (1995) working on *Rana tigrina* came to the same conclusions. According to de Vlaming *et al.* (1974) MEL-injected *Hyla cinerea* females have smaller ovaries and lower GSI (Gonosomatic Index) than non-injected females. On the contrary Alonso-Bedate *et al.* (1990) reported that the application of MEL (100 µg), to *Rana perezi* in winter did not influence significantly neither the GSI of females, nor histology of their ovaries and plasma concentration of estradiol and testosterone. However they noticed that MEL-injected females had lower number of previtellogenic follicles when compared with control animals. Jørgensen (1992) reported results of exposure of *Bufo bufo* to artificially shortened photoperiod in the spring. This treatment caused the lost of the entire newly recruited complement of follicles at early stages of vitellogenesis in about 30% of experimental females and highly increased proportion of atretic follicles in the ovaries of all experimental females. After exposure to short photoperiod more than 30% of experimental females showed massive degeneration of vitellogenic oocytes soon after their recruitment. The negative effect of artificially changed photoperiod was also evident in *Rana catesbeiana* (Horseman *et al.*, 1978). Administration of MEL negatively affected vitellogenesis in *Rana cyanophlyctis* (Kupwade and Saidapur, 1986), decreased the number of previtellogenic follicles in *Rana perezi* (Alonso-Bedate *et al.*, 1990), and inhibited compensatory hypertrophy of the ovary after unilateral ovariectomy in *Rana tigrina* (Hoque *et al.*, 1993). Since there are several reports that MEL exerts direct inhibitory effect on the ovary and influences various phases of oocyte development, it was suggested that MEL receptors can be located at both the gametogenic and somatic cells of the ovary (Kupwade and Saidapur, 1986; de Atenor *et al.*, 1994, Udaykumar and Joshi, 1997). It is suggested that besides MEL, the pineal gland may synthesize and secrete other biologically active substances, because the injection of pineal extract exerts stronger inhibition of reproductive function than pure MEL (Skene *et al.*, 1991). It should be mentioned that experimental procedures concerning changes of MEL levels or/and manipulation of light—dark regime may alter not only synthesis and secretion of MEL itself, but also expression, density, and affinity of its receptors (Isorna *et al.*, 2004b). Continuous exposure to light inhibits, and to darkness enhances, MEL synthesis, alters melatonin binding in the retina, whereas seasonal and temperature changes do not affect MEL binding (Wiechmann and Smith, 2001; Isorna *et al.*, 2004b).

Besides the photoreceptive function, the pineal complex serves as neuroendocrine transducer of temperature signals (Fernandez and Ramos, 2003; Alonso-Bedate *et al.*, 1990; Mayer *et al.*, 1997). Temperature is known to be the most important external factor regulating sexual activity in amphibians. Synthesis and transduction of MEL signal is also temperature-dependent (Isorna *et al.*, 2005). Low environmental temperature may decrease the MEL synthesis and diminish its diurnal rhythm (Delgado and Vivien Roels, 1989). In *Rana tigrina*, regardless of the photoperiod, the high temperature during quiescent phase stimulates the growth of higher number of previtellogenic oocytes, whereby in recrudescent phase it induces production of greater number of eggs, comparing to the control females maintained at ambient temperature (Saidapur and Hoque, 1995). Pineal complex and eyes can also contribute, at least in part, to reaction of *Rana perezi* females on temperature stimulation (Alonso-Gomez *et al.*, 1990).

The influence of photoperiod and temperature on amphibian reproduction is often analyzed together, although the effect of temperature may change or mask the effect of photoperiod and make the exact valuation of photoperiodic influence difficult.

Acknowledgements

We would like to thank Piotr Kierzkowski for preparing and arranging drawings and photos, Agnieszka Kotusz for photos in Fugures 3A; 4A; 5A; 6A; 7A; 8; 13A,B; 15B,D; Renata Augustyńska for photos 1A-D; 21A, D, and 25A,B; and Beata Rozenblut for photos 1E,F; 7B,C; 17A,B; 21C,D, and 25A.

References

Acharjee S, K Maiti, JM Soh, WB Im, JY Seong, HB Kwon. 2002. Differential desensitization and internalization of three different bullfrog gonadotropin-releasing hormone receptors. Mol. Cells. 31: 14 (1): 101-107.

Al Mukhtar KAK, AC Webb. 1971. An ultrastructural study of primordial germ cells, oogonia and early oocytes in *Xenopus laevis*. J. Embryol. Exp. Morphol. 26: -217.

Al-Anzi B, DE Chandler. 1998. A sperm chemoattractant is released from *Xenopus* egg jelly during spawning. Dev. Biol. 198: 366-375.

Alonso-Bedate M, R Carballada, MJ Delgado. 1990. Effects of melatonin on gonadal steroids and glucose plasma levels in frogs (*Rana perezi* and *Rana temporaria*). J. Pineal Res. 8: 79-89.

Alonso-Gómez AL, M Tejera, M Alonso-Bedate, MJ Delgado. 1990. Response to pinealectomy and blinding in vitellogenic female frogs (*Rana perezi*) subjected to high temperature in autumn. Can. J. Physiol. Pharmacol. 68: 94-98.

Andersen CB, H Sakaue, T Nedachi, KS Kovacina, C Clayberger, Conti, RA Roth. 2003. Protein kinase B/Akt is essential for the insulin- but not progesterone-stimulated resumption of meiosis in *Xenopus oocytes*. Biochem J. 15: 369 (Pt 2): 227-38.

Andreucetti P, C Campanella. 1980. Origin and cytochemistry of the animal dimple granules in *Discoglossus pictus* (Anura) eggs. J. Embryol. Exp. Morph. 56: 239-252.

Andreucetti P, C Campanella. 1982. Regional difference in the pattern of vitellogenesis in the painted frog *Discoglossus pictus*. Tissue and Cell. 14: 681-690.

Arai Y, K Kubokawa, S Ishii. 1998. Cloning of cDNAs for the Pituitary Glycoprotein Hormone Alpha Subunit Precursor Molecules in Three Amphibian Species, *Bufo japonicus*, *Rana catesbeiana*, and *Cynops* pyrrhogaster. Gen. Comp. Endocrinol. 112: 46-53.

Arranz SE, IE Albertali, MO Cabada. 1997. *Bufo arenarum* egg jelly coat: purification and characterization of two highly glycosylated proteins. Biochem. J. 323: 307-312.

Augustyńska R, B Rozenblut, M Ogielska. 2007. Intrafollicular atresia in amphibian ovary (Anura: Ranidae). J. Morphol. 268: 1047.

Bagowski CP, JW Myers, JE Ferrell Jr. 2001. The classical progesterone receptor associates with p42 MAPK and is involved in phosphatidylinositol 3—kinase signaling in *Xenopus* oocytes. J. Biol. Chem. 276 (40): 37708-37714.

Balinsky BI. 1965. An Introduction to Embryology. The Sounders Company Philadelphia and London.

Barisone GA, IE Albertali, M Sanchez, MO Cabada. 2003. The envelopes of amphibian oocytes: physiological modifications in *Bufo arenarum*. Reproductive Biol. and Endocrinol. I: 18. (http://www.RBEj.com/content/1/1/18).

Barth LG. 1953. Embryology. The Dryden Press Inc.

Basu SL, A Mondal. 1960. Effect of estradiol implants on the gonads and other accessory genital organs of the common toads, *Bufo melanosticus* Schneid. J. Animal Morphol. Physiol. 7: 150-155.

Bayaa M, RA Booth, Y Sheng, XJ Liu. 2000. The classical progesterone receptor mediates *Xenopus* oocyte maturation through a nongenomic mechanism. Proc. Nat. Acad. Sci. 97 (23): 12607-12612.

Beçak ML, KF Pizzocaro. 1980. Chromatin circles in amphibian previtellogenic oocytes. Experientia 36: 164-166.

Bergink EW, RA Wallace. 1974. Precursor-product relationship between amphibian vitellogenin and the yolk proteins lipovitellin and phosvitin. J. Biol. Chem. 294: 2897-2903.

Berrill NJ, G Karp. 1976. Development. McGraw-Hill Book Company.

Bhaduri JL, SL Basu. 1957. A study of the urogenital system of Salientia. Part I.

Ranidae and Hyperolidae of Africa. Annales du Musee Royal du Congo Belge. 55: 9-34.

Billet FS, E Adam. 1976. The structure of the mitochondrial cloud of *Xenopus laevis.* J. Embryol. Exp. Morph. 33: 697-710.

Billeter E, CB Jørgensesn. 1976. Ovarian development of young toads *Bufo bufo bufo* (L.): Effects of unilateral ovariectomy, hypophysectomy, treatment with gonadotropins (hCG), growth hormon, and prolactin, and importance of body growth. Gen. Comp. Endorcinol. 29: 521-544.

Birelli GB, PB Anderson, PC Jost, OH Griffith, LJ Banaszak, J Seeling. 1982. Lipid environments in the yolk lipoprotein system. A spin-labelling study of the lipovitellin-phosvitin complex from *Xenopus laevis.* Biochemistry 21: 2444-2452.

Blüm V. 1986. Vertebrate reproduction. The Formation of Gametes. pp. 75-110. Springer Verlag.

Bonnell BS, D Reinhart, DE Chandler. 1998. *Xenopus laevis* egg jelly coats consists of small diffusible proteins bound to a complex system of structurally stable networks composed of high-molecular-weight glycoconjugates. Dev. Biol. 174: 32-42.

Bounoure L, R Aubry, ML Huck. 1954. Nouvelles recherches experimentales sur les origines de la lignee reproductive chez la grenouille rousse. J. Embryol. Exp. Morphol. 2: 245-263.

Boyd SK. 1992. Sexual differences in hormonal control of release calls in bullfrogs. Horm. Behav. 26: 522-535.

Brizzi R, G Delfino, S Jantra. 2003. An overview of breeding glands. In BGM Jamieson [ed.] 2003. Reproductive Biology and Phylogeny of Anura. Science Publisher, Inc. NH, USA, pp 253-317.

Browder LW, CA Erickson, WR Jeffrey. 1991. Developmental Biology. Saunders College Publishing. Third Edition.

Brown FD, EM del Pino, G Krhone. 2002. Bidder's organ in the toad *Bufo marinus*: Effects of orchidectomy on the morphology and expression of lamina-associated polypeptide2. Develop. Growth Differ. 44: 527-535.

Cabada MO, ME Manes, MI Gomez. 1989. Spermatolysins in *Bufo arenarum*: their activity on oocyte surface. J. Exp. Zool. 249: 229-234.

Callan HG. 1963. The nature of lampbrush chromosomes. Int. Rev. Cytol. 15: 1-34.

Callen JC, N Dennebouy, JC Mounoulou. 1980. Development of the mitochondrial mass and accumulation of mtDNA in previtellogenic stages of *Xenopus laevis* oocytes. J. Cell. Sci. 41: 307-320.

Callen JC, JC Dennebouy, JC Mounoulou. 1986. Early onset of a large pool of previtellogenic oocytes and cyclic escape by vitellogenesis: The pattern of ovarian activity of *Xenopus laevis* females and its physiological consequences. Reprod. Nutr. Develop. 26: 13-30.

Callen JC. 1986. Differential pinocytosis and the origin of the animal-vegetal polarity in *Xenopus laevis* oocytes of growing or post-spawning adult females. Biol. Cell 57: 207-220.

Callery EM, H Fang, RP Elinson. 2001. Frogs without polliwogs: Evolution of anuran direct development. BioEssays 23: 233-241.

Campanella C, R Carotenuto, V Infante, G Maturi, U Atripaldi. 1997. Sperm-egg interaction in the painted frog *Discoglossus pictus*. Mol. Reprod. Dev. 47: 323-333.

Cariello L, G Romano, A Spagnuolo, L Zanetti, S Fasano, S Minucci, L Di Matteo, R Pierantoni, G Chieffi. 1989. Molecular forms of immunoreactive gonadotropin-releasing hormone in hypothalamus and testis of the frog, *Rana esculenta*. Gen. Comp. Endocrinol. 75: 343-348.

Carnevali O, G Mosconi, K Yamamoto, T Kobayashi, S Kikuyama, AM Polzonetti-Magni. 1993. In-vitro effects of mammalian and amphibian prolactins on hepatic vitellogenin synthesis in *Rana esculenta*. J. Endocrinol. 137(3): 383-389.

Carnevali O, G Mosconi. 1992. In vitro induction of vitellogenin synthesis in *Rana esculenta*: role of the pituitary. Gen. Comp. Endocrinol. 86(3): 352-8.

Carnevali O, G Mosconi. 1992. In vitro induction of vitellogenin synthesis in *Rana esculenta*: Role of the pituitary. Gen. Comp. Endocrinol. 86(3): 352-8.

Carnevali O, MG Sabbieti, G Mosconi, AM Polzonetti-Magni. 1995. Multihormonal control of vitellogenin mRNA expression in the liver of frog, *Rana esculenta*. Mol. Cell. Endocrinol. 30: 114(1-2): 19-25.

Carroll EJ Jr, SH Wei, GM Nagel, R Ruibal. 1991. Structure and macromolecular composition of the egg and embryo jelly coats of the anuran *Lepidobatrachus laevis*. Develop. Growth Differ. 33: 37-43.

Castanet J, H Francillon-Vieillot, FJ Meunier, A De Ricqles. 1993. Bone and individual aging. In: Bone Growth,(Ed.) BH Hall, Boca Raton, Ann Arbor, London, Tokyo. CRC Press Inc. Vol. 7: 245-283.

Chan AP, M Kloc, L Etkin. 1999. *fatvg* encodes a new localized RNA that uses a 25-nucleotide element (FVLE1) to localize to the vegetal cortex of *Xenopus* oocytes. Development 126: 4943-4953.

Chan AP, M Kloc, S Bilinski, L Etkin. 2001. The vegetally localized mRNA *fatvg* is associated with germ plasm in early embryo and is later expressed in the fat body. Mech. Dev. 100: 137-140.

Chang KJ, JW Kim, J Lee, WB Im, HB Kwon, AW Schuetz. 1995. Prostaglandin production and ovulation during exposure of amphibian ovarian follicles to gonadotropin or phorbol ester in vitro. Gen. Comp. Endocr. 100 (3): 257-266.

Charbonneau M, RD Grey, RJ Baskin, D Thomas. 1986. A freeze-fracture study of the cortex of *Xenopus laevis* eggs. Develop. Growth Differ. 28: 75-84.

Charbonneau M, RD Grey. 1984. The onset of activation responsiveness during maturation coincides with the formation of the cortical endoplasmic reticulum in oocytes of *Xenopus laevis*. Dev. Biol. 102: 90-97.

Cheng TH. 1932. The germ line history of *Rana cantabrigensis* BAIRD. I Germ cell origin and gonad formation. Z. Zellforsch. Mikrosk. Anat. 16: 495-541.

Chieffi G, R Pierantoni, S Fasano. 1991. Immunoreactive GnRH in hypothalamic and extrahypothalamic areas. Int. Rev. Cytol. 127: 1-55.

Chieffi G, R Pierantoni. 1987. Regulation of ovarian steroidogenesis. In: Hormones and Reproduction in Fishes, Amphibians, and Reptiles, (Eds) DO Norris and RE Jones. Plenum Press, New York, pp. 117-144.

Chieffi P, S Minucci. 2004. Environmental influence of tesicular MAP kinase (ERK1) activity in the frog *Rana esculenta*. J. Exp. Biol. 207: 2209-2213.

Chowdhury VS, K Yamamoto, I Saeki, I Hasunuma, T Shimura, K Tsutsui. 2007.

Melatonin stimulates the release of growth hormone and prolactin by a possible induction of the expression of frog Growth hormone-releasing peptide and its related peptide- 2 in the amphibian hypothalamus. Endocrinology 1493: 962-970.

Coggins LW. 1973. An ultrastructural and radioautographic study of early oogenesis in the toad *Xenopus laevis*. J. Cell. Sci. 12: 71-93.

Collin F, N Chartrel, A Fasolo, JM Conlon, F Vandesande, H Vaudry. 1995. Distribution of two molecular forms of gonadotropin-releasing hormone (GnRH) in the central nervous system of the frog *Rana ridibunda*. Brain Res. 703(1-2): 111-128.

Coppin A, D Florea, E Maes, D Cogálniceanu, G Strecker. 2003. Comparative study of carbohydrate chains release from the oviductal mucins of the two very closely related amphibian species *Bombina bombina* and *Bombina variegata*. Biochimie 85: 53-64.

Coppin A, E Maes, C Flahaut, B Codeville, G Strecker. 1999. Acquisition of species-specific O-linked carbohydrate chains from oviductal mucins in *Rana arvalis*. A case study. Eur. J. Biochem. 266: 370-382.

Coppin A, E Maes, G Morelle, G Strecker. 1999. Structural analysis of 13 neutral oligosaccharide-alditols released by reductive β-elimination from oviductal mucins of *Rana temporaria*. Eur. J. Biochem. 266: 94-104.

Cork RJ, KR Robinson. 1994. Second messenger signalling during hormone-induced *Xenopus* oocyte maturation. Zygote. 2(4): 289-299.

Crim JW. 1984. Immunocytochemistry of luteinizing hormone-releasing hormone in brains of bullfrogs (*Rana catesbeiana*) during spontaneous metamorphosis. J. Exp. Zool. 229 (2): 327-337.

Crump ML. 1974. Reproductive strategies in a tropical Auran community. The Universtity of Kansas Lawrence, USA.

D'Aniello B, C Pinelli, MM Di Fiore, L Tela, JA King, RK Rastogi. 1995. Development and distribution of gonadotropin-releasing hormone neuronal systems in the frog (*Rana esculenta*) brain: Immunohistochemical analysis. Brain Res. Dev. Brain Res. 89 (2): 281-288.

Danilchik MV, JM Denegre. 1991. Deep cytoplasmic rearrangements during early development in *Xenopus laevis*. Development 111: 845-856.

Danilchik MV, JC Gerhart, 1987. Differentiation of the animal-vegetal axis in *Xenopus laevis* oocytes. I Polarized intracellular translocation of platelets establishes the yolk gradient. Dev. Biol. 122: 101-112.

D'Antonio M, S Fasano, R de Leeuw, R Pierantoni. 1992. Effects of gonadotropin-releasing hormone variants on plasma and testicular androgen levels in intact and hypophysectomized male frogs, *Rana esculenta*. J. Exp. Zool. 261: 34-39.

De Atenor MS, IR de Romero, E Brauckmann, A Pisano, AH Legname. 1994. Effects of the pineal gland and melatonin on the metabolism of oocytes in vitro and on ovulation in *Bufo arenarum*. J. Exp. Zool. 268: 436-441.

De Romero IR, MB de Atenor, AH Legname. 1998. Nuclear maturation inhibitors in *Bufo arenarum* oocytes. Biocell. 22: 27-34.

de Vlaming VL, M Sage, CB Charlton. 1974. The effects of melatonin treatment on gonadosomatic index in the teleost fundulus similis, and the tree frog, *Hyla cinerea*. Gen. Comp. Endocrinol. 22: 433-438.

Debauche P, B Baras, P Devos. 1994. Insulin but not progesterone promotes the biosynthesis of glycogen in *Xenopus laevis* oocytes: Implications on the control of glycogen synthase by phosphorylation, dephosphorylation. J. Exp. Zool. 269: 1-11.

del Pino EM. 1973. Interactions between gametes and environment in the toad *Xenopus laevis* (Daudin) and their relationship to fertilization. J. Exp. Zool. 185: 121-132.

del Pino EM, H Steinbeisser, A Hofmann, C Dreyer, M Campos MF Trendelenburg. 1986. Oogenesis in the egg-brooding frog *Gastrotheca riobambae* produces large oocytes with fewer nucleoli and low RNA content in comparison to *Xenopus laevis*. Differentiation 32: 24-33.

del Pino EM. 1989a. Modification of oogenesis and development in marsupial frogs. Development 107: 169-187.

del Pino EM. 1989b. Marsupial frogs. Sci. Am., May: 76-84.

Delgado MJ, B Vivien-Roels. 1989. Effect of environmental temperature and photoperiod on the melatonin levels in the pineal, lateral eye, and plasma of the frog, *Rana perezi*: Importance of ocular melatonin. Gen. Comp. Endocrinol. 75: 46-53.

Denis H, M Wegnez. 1977. Biochemical research on oogenesis. Oocytes of *Xenopus laevis* synthesise, but do not accumulate 5S RNA of somatic type. Dev. Biol. 85: 212-217.

Denis-Donini S, C Campanella. 1977. Ultrastructural and lectin binding changes during the formation of the animal dimple in oocytes of *Discoglossus pictus*. Dev. Biol. 61: 140-152.

Denver RJ, KA Glennemeier, G Boorse. 2002. Endocrinology of Complex Life Cycles: Amphibians In: Hormones, Brain and Behavior, (Eds) D Pfaff, A Arnold, A Edgen, S Fahrback, R Rubin, Vol. II. Non-mammalian hormone Behavior systems. pp. 469-513.

Di Fiore M, L Assisi, V Botte. 1998 Aromatase and testosterone receptor in the liver of the female green frog, *Rana esculenta* Life Sci. 62 (21): 1949-1958.

Dierickx K. 1967a. The gonadotropic centre of the tuber cinereum hypothalami and ovulation. Z. Zellforsch. Mikrosk. Anat. 77: 188-203.

Dierickx K. 1967b. The function of the hypophysis without preoptic neurosecretory control. Z. Zellforsch. Mikrosk. Anat. 78: 114-130.

Dubocovich ML. 1988. Pharmacology and the function of melatonin receptors. FASEB J., 2,2765-2773.

Dubois EA, MA Zandbergen, J Peute, HJ Goos. 2002. Evolutionary development of three gonadotropin-releasing hormone (GnRH) systems in vertebrates. Brain Res. Bull. 57: 413-418.

Duellman WE, L Trueb, 1986. Biology of Amphibia. McGraw-Hill Book Company, USA. pp. 670.

Dumont JN, AR Brummett. 1978. Oogenesis in *Xenopus laevis* (Daudin). I. Relationship between developing oocytes and their investing follicular tissues. J. Morphol. 155: 73-98.

Dumont JN, JJ Eppig. 1971. A method for the production of pigmentless eggs in *Xenopus laevis*. J. Exp. Zool. 178: 307-312.

Dumont JN, RA Wallace. 1968. The synthesis, transportand uptake of yolk proteins in *Xenopus laevis*. J. Cell. Biol. 39: 37a-38a.

Dumont JN. 1972. Oogenesis in *Xenopus laevis* (Daudin). I. Stages of oocyte development in laboratory maintained animals. J. Morphol. 136: 153-180.

Dumont JN. 1978. Oogenesis in *Xenopus laevis*. VI. The route of injected tracer transport in the follicle and developing oocyte. J. Exp. Zool. 204: 193-218.

El-Etr M, S Schorderet-Slatkine, EE BaulieuE. 1979. Meiotic maturation in *Xenopus laevis* initiated by insulin. Science 205: 1397-1399.

El-Etr M, S Schorderet-Slatkine, EE Baulieu. 1980. The role of zinc and follicle cells in insulin initiated meiotic information in *Xenopus laevis* oocytes. Science 210: 928-930.

Elinson RP, Fang H. Sdary coverage of the yolk by the body wall in the direct developing frog *Eleutherodactylus coqui*: An unusual process for amphibian embryos. Dev. Genes Evol. 208: 457-466.

Elinson RP. 1971a. Fertilization of partially jellied and jellyless oocytes of the frog *Rana pipiens*. J. Exp. Zool. 176: 415-428.

Elinson RP. 1971b. Sperm lytic activity and its relation to fertilization in the frog *Rana pipiens*. J. Exp. Zool. 177: 207-218.

Elinson RP. 1973. Fertilization of frog body cavity eggs enhanced by treatments affecting the vitelline coat. J. Exp. Zool. 183: 291-302.

Elinson RP. 1974. A block to cross-fertilization located in the egg jelly of the frog *Rana clamitans*. J. Embryol. Exp. Morphol. 32: 325-335.

Eppig JJ. 1970. Melanogenesis in Amphibians. III. The buoyant density of oocyte melanosomes from the eyes of PTU-treated larvae. J. Exp. Zool. 175: 476-476.

Eyers PA, J Liu, NR Hayashi, AL Lewellyn, J Gautier, JL Maller. 2005. Regulation of the G(2)/M transition in *Xenopus* oocytes by the cAMP—dependent protein kinase. J. Biol. Chem. 280 (26): 24339–24346.

Falconi R, S Petrini, A Quaglia, F Zaccanti. 2001. Fine structure of undifferentiated gonads in *Rana dalmatina* tadpoles. Ital. J. Zool. 68: 15-21.

Fang H, RP Elinson. 1999. Evolution alteration in anterior patterning: otx2 expression in the direct developing frog *Eleutherodactylus coqui*. Dev. Biol. 205: 233-239.

Farias CF, SP Carvalho-e-Silva, L de Brito-Gitirana. 2002. Bidder's organ of *Bufo ictericus*: A light and electron microscopy analysis. Micron. 33: 673-679.

Fasano S, M D'Antonio, P Chieffi, G Cobellis, R Pierantoni. 1995. Chicken GnRH-II and salmon GnRH effects on plasma and testicular androgen concentrations in the male frog, *Rana esculenta*, during the annual reproductive cycle. Comp. Biochem. Physiol. C Pharmacol. Toxicol. Endocrinol. 112: 79-86.

Fasano S, R de Leeuw, R Pierantoni, G Chieffi, PG van Oordt. 1990. Characterization of gonadotropin-releasing hormone (GnRH) binding sites in the pituitary and testis of the frog, *Rana esculenta*. Biochem. Biophys. Res. Commun. 168: 923-932.

Fasano S, HJ Goos, C Janssen, R Pierantoni. 1993. Two GnRHs fluctuate in correlation with androgen levels in the male frog Rana esculenta. J. Exp. Zool. 266: 277 283.

Fernandez SN, ZC Mansilla-Fritacre, DC Miceli. 1994. Characterization and properties of steroid binding protein in *Bufo arenarum* serum. Mol. Reprod. Develop. 38: 364-372.

Fernandez SN, DC Miceli, CM Whitacre. 1997. Ultrastructural studies of the effect of steroid hormones on pars recta secretions in *Bufo arenarum*. J. Morphol. 231: 1-10.

Fernandez SN, I Ramos. 2003. Endocrinology of reproduction. In: Reproductive Biology and Phylogeny of Anura, (Ed.) BGM Jamieson pp. 73-117.

Ferrel JE Jr. 1999a. Building a cellular switch: More lessons from a good egg. BioEssays 21: 866-870.

Ferrel JE Jr. 1999b. *Xenopus* oocyte maturation: new lessons from a good egg. BioEssays 21: 833-842.

Florea D, E Maes, G Strecker. 1997. Primary structure of seven sulfated oligo-saccharide-alditols release by reductive β-elimination from oviductal mucins of *Rana temporaria*. Characterization of the sequence $HSO_3(3)GlcA(\beta1-3)$ Gal. Carbohydr. Res. 302: 179-189.

Florea D, E Maes, M Haddad, G Strecker. 2002. Structural analysis of the oligo-saccharide alditols released from the jelly coat of *Rana dalmatina* eggs by reductive β-elimination. Biochemie. 84: 611-624.

Follet BK, TJ Nicholls, MR Redshaw. 1968: The vitellogenic response in the south african clawed toad (*Xenopus laevis* Doudain). J. Cell. Physiol. Suppl. 1 1: 91-102.

Fontaine YA. 1980. Pituitary gonadotropic hormones: comparative biochemistry and biology: Specificity and evolution. Reprod. Nutr. Dev. 20: 381-418.

Fortune JE. 1983. Steroid production by *Xenopus* ovarian follicles at different developmental stages. Dev. Biol. 99: 502-509.

Franchi LL, AM Mandi, S Zuckerman. 1962. The development of the ovary and the process of oogenesis. In: The Ovary. (Ed.) S Zuckerman, Vol. 1, 1-88, Academic Press.

Gall JG, HG Callan. 1989. The sphere organelle contains small nuclear ribonucleo-proteins. Proc. Natl Acad. Sci. USA 60: 6635-6639.

Gall JG. 2000. Cajal bodies: the first 100 years. Annu. Rev. Cell. Dev. Biol. 16: 273-300.

Gard DL, BJ Cha, MM Schroeder. 1995. Confocal immunofluorescence microscopy of microtubules, microtubule-associated proteins, and microtubule-organizing centers during amphibian oogenesis and early development. In: Current Topics in Developmental Biology, (Ed.) DG Capco, Academic Press. Vol. 31: 383-431.

Gardiner DM, RD Grey. 1983. Membrane junctions in *Xenopus* eggs: their distribution suggests a role in calcium regulation. J. Cell. Biol. 96: 1159-1163.

Gerhart J, S Black, S Scharf, R Gimlich, JP Vincent, M Danilchik, B Rowning, J Roberts. 1986. Amphibian early development. BioScience 36: 541-549.

Gerhart J, JP Vincent, S Scharf, S Black, R Gimlich, M Danilchik. 1984. Localization and induction in early development of *Xenopus*. Phil. Trans. R. Soc. Lond. B. 307: 319-330.

Gilbert SF. 2000. Developmental Biology. Sinauer Associates, Inc., Publishers Sunderland, Massachusetts. Sixth Edition.

Gobbetti, A., A Polzonetti-Magni, M Zerani, O Carnevali, V Botte. 1985. Vitello-genin hormonal control in the green frog, *Rana esculenta*. Interplay between estradiol and pituitary hormones. Comp. Biochem. Physiol. A. 82: 855-8.

Gobbetti A, M Zerani, O Carnevali, V Botte. 1990. Prostaglandin F2 alpha in female water frog, *Rana esculenta*: Plasma levels during the annual cycle and effects of exogenous PGF2 alpha on circulating sex hormones. Gen. Comp. Endocrinol. 80: 175-80.

Gobbetti A, M Zerani. 1992. A possible involvement of prostaglandin F2 alpha (PGF2alpha) in *Rana esculenta* ovulation: Effects of mammalian gonadotropin-releasing hormone on *in vitro* PGF2 alpha and 17 beta estradiol production from ovary and oviduct. Gen. Comp. Endocrinol. 87: 163-170.

Gobetti A, M Zerani. 1999. Hormonal and cellular brain mechanisms regulating the amplexus of male and female water frog (*Rana esculenta*). J. Neuroendocr. 11: 589-596.

Goldenberg M, RP Elinson. 1980. Anima/vegetal differences in cortical granule exocytosis during activation of the frog egg. Develop. Growth Differ. 22: 345-356.

Gosner LK. 1960. A simplified table for staging anuran embryos and larvae with notes on identification. Herpetologica 16: 513-543.

Gramapurohit NP, BA Shanbhag, SK Saidapur. 2000. Pattern of gonadal sex differentiation, development and onset of steroidogenesis in the frog, *Rana curtipes*. Gen. Comp. Endocrin. 119: 256-264.

Grant P. 1953. Phosphate metabolism during oogenesis in *Rana temporaria*. J. Exp. Zool. 124: 513-543.

Grey RD, DP Wolf, JL Hedrick. 1974. Formation and structure of the fertilization envelope in *Xenopus laevis*. Dev. Biol. 36: 44-61.

Grey RD, PK Working, JL Hedrick. 1976. Evidence that fertilization envelope blocks sperm entry in eggs of *Xenopus laevis*: Interaction of sperm with isolated envelope. Dev. Biol. 54: 52-60.

Gruzova MN, VN Parfenov. 1973. The karyosphere in late oocytes of frogs. Monnitore Zool. Ital. 7: 225-242.

Gruzova MN, VN Parfenov. 1977. Ultrastructure of late oocyte nuclei in *Rana temporaria*. J. Cell. Sci. 28: 1-13.

Guerardel Y, O Kol E Maes, T Lefebvre, B Boilly, M Davril, G Strecker. 2000. O-glycan variability of egg-jelly mucins from *Xenopus laevis*: Characterization of four phenotypes that differ by the terminal glycosylation of their mucins. Biochem. J. 352: 449-463.

Guerriero G, CHE Roselli, M Paolucci, V Botte, G Ciarcia. 2000. Estrogen receptors and aromatase activity in the hypothalamus of the female frog *Rana esculenta*. Fluctuations throughout the reproductive cycle. Brain Res. 880: 92-101.

Guilgur LG, NP Moncaut, AV Canário, GM Somoza. 2006. Evolution of GnRH ligands and receptors in Gnathostomata. Comp. Biochem. Physiol. A Mol. Integr. Physiol. 144: 272-83.

Guraya S. 1969. Histochemical study of follicular atresia in the amphibian ovary. Acta Biol. Acad. Sci. Hung. 20: 43-56.

Guraya S. 1979. Recent advances in the morphology, cytochemistry and function of the Balbiani's vitelline body in animal oocytes. Int. Rev. Cytol. 59: 249-321.

Hainaut P, S Giorgetti, A Kowalski, R Ballotti, E Van Obberghen. 1991. Antibodies to phosphotyrosine injected in *Xenopus laevis* oocytes modulate maturation induced by insulin/IGF-I. Exp. Cell. Res. 195: 129-36.

Hammes SR. 2003. The further redefining of steroid mediated signaling. PNAS 100: 2168-2170.

Hammes SR. 2004. Steroids and oocyte maturation—A new look at an old story. Mol. Endocr. 18 (4): 769-775.

Hanocq F, M Kirsch-Volders, J Hanocq-Quertier. 1972. Characterization of yolk DNA from Xenopus laevis oocytes ovutaled in vitro. Proc. Acad. Nat. Sci. (Washington) 69: 1322-1326.

Hara K, P Tydeman. 1979. Cinematographic observation of an "activation wave" (AW) on the locally inseminated egg of Xenopus laevis. Wilhelm Roux's Arch. 186: 91-94.

Hardisty MW. 1966. The number of vertebrate primordial germ cells. Biol. Rev. 42: 265-287.

Hardy DM, JL Hendrick. 1992. Oviductin. Purifiation and properties of the oviductal protease that processes the molecular weight 43000 glycoprotein of Xenopus laevis egg envelope. Biochemistry 31: 4466-4472.

Hausen P, YH Wang, C Dreyer, R Stick. 1985. Distribution of nuclear proteins during maturation of the Xenopus oocyte. J. Embryol. Exp. Morphol. 89: 17-34.

Hayashi H, T Hayashi, Y Hanaoka. 1992a. Amphibian lutropin and follitropin from the bullfrog Rana catesbeiana. Complete amino acid sequence of the alpha subunit. Eur. J. Bio. 203: 185-191.

Hayashi H, T Hayashi, Y Hanaoka. 1992b. The complete aminoacid squence of the follitropin beta subunits of the bullfrog Rana catesbeiana. Comp. Endocrinol. 88: 144-150.

Hazlerigg DG. 2001. What is the role of melatonin within the anterior pituitary. J. Endocrinol. 170: 493-501.

Heasman J, J Quarmby, CC Wylie. 1984. Mitochondrial cloud of Xenopus laevis; the source of germinal granule material. Dev. Biol. 105: 458-469.

Hirai S, C Coascogne, EE Baulieu. 1983. Induction of germinal vesicle breakdown in Xenopus laevis oocytes. Response of denuded oocytes to progesteron and insulin. Dev. Biol. 100: 214-221.

Hiyoshi M, K Takamune, K Mita, H Kubo, Y Sugimoto, C Katagiri. 2002. Oviductin, the oviductal protease that mediates gamete interaction by affecting the vitelline coat in Bufo japonicus: Its molecular cloning and analyses of expression and post-translational activation. Dev. Biol. 243: 176-184.

Hoque B, SK Saidapur, K Pancharatna. 1993. Melatonin supresses compensatory hypertrophy and follicular development in the unilaterally ovariectomized bullfrog Rana tigrina: defailed analysis of follicular kinetics. Zool. Anz. 230: 85-93.

Horn EC. 1962. Extranucleolar histone in the amphibian oocyte. Proc. Acad. Nat. Sci. (Washington) 48: 257-265.

Horrel A, J Shuttleworth, A Colman. 1987. Transcript levels and translation control of hsp70 synthesis in Xenopus oocytes. Genes Dev. 1: 433-444.

Horseman ND, CA Smith, DD Culley. 1978. Effects of age and photoperiod on ovary size and condition in bullfrogs (Rana catesbeiana Shaw) (Amphibia, Anura, Ranidae): Herpertol. 12: 287-290.

Hsü C-Y, KL Li, MH Lü, HM Liang. 1982. The presence of intramitochondrial yolk-crystals in oocytes of hypophysectomized bullfrog tadpoles. Develop. Growth Differ. 24: 319-325.

Hyioshi M, K Takamune, K Mita, H Kubo, Y Sugimoto, C Katagiri. 2002. Oviductin, the oviductal protease that mediates gamete interaction by affecting the vitelline coat in *Bufo japonicus*: Its molecular cloning and analyses of expression and posttranslational activation. Dev. Biol. 243: 176-184.

Iela L, JF Powell, NM Sherwood, B D'Aniello, RK Rastogi, JT Bagnara. 1996. Reproduction in the Mexican leaf frog, *Pachymedusa dacnicolor*. VI. Presence and distribution of multiple GnRH forms in the brain. Gen. Comp. Endocrinol. 103 (3): 235-243.

Ijiri KI, N Egami. 1975. Mitotic activity of germ cells during normal development of *Xenopus laevis* tadpoles. J. Embryol. Exp. Morphol. 34: 687-694.

Ikemoto T, M Enomoto, MK Park. 2004. Identification and characterization of a reptilian GnRH receptor from the leopard gecko. Mol. Cell. Endocrinol. 214: 137-147.

Imoh H, M Okamoto, G Eguchi. 1983. Accumulation of annulate lamellae in the subcortical layer during progesterone-induced oocytes maturation in *Xenopus laevis*. Dev. Growth Differ. 25: 1-10.

Ischii S., K Kubokawa. 1985. Adaptation of vertebrate gonadotropin receptors to environmental temperature. In: Current Trends in Comparative Endocrinology (Ed.) B Lofts, WN Holmes, Hong Kong University Press, Hong Kong, pp. 751-754.

Ishihara K, J Hosono, H Kanatani, C Katagiri. 1984. Toad egg-jelly as a source of divalent cations essential for fertilization. Dev. Biol. 105: 435-442.

Isorna E, MJ Delgado, AI Guijarro, MJ Delgado, M Alonso-Bedate, A Alonso-Gomez. 2004a.Characterization of melatonin binding sites in the brain and retina of the frog *Rana perezi*. Gen. Comp. Endocr. 135: 259-267.

Isorna E, MJ Delgado, AI Guijarro, MA Lopez-Patino, M Alonso-Bedate, A Alonso-Gomez. 2004b. 2-[125I] Melatonin binding sites in the central nervous system and neural retina of the frog *Rana perezi*: Regulation by light and temperature. Gen. Comp. Endocr. 139: 95-102.

Isorna E, A Guijarro, MA Lopez-Patino, MJ Delgado, M Alonso-Bedate, A Alonso-Gomez. 2005. Effect of temperature on 2-[125I] iodomelatonin binding to melatonin receptors in the neural retina of the frog *Rana perezi*. J. Pineal Res. 38: 176-181.

Itoh M, M Inoue, S Ischii. 1990. Annual cycle of pituitary and plasma gonadotropins and sex steroids in a wild population of the toad *Bufo japonicus*. Gen. Comp. Endocr. 109: 13-23.

Iwao Y, C Katagiri. 1982. Properties of the vitelline coat lysin from toad sperm. J. Exp. Zool. 219: 87-95.

Iwasawa H, K Yamaguchi. 1984. Ultrastructural study of gonadal development in *Xenopus laevis*. Zool. Sci. (Japan), 1: 591-600.

Jokura Y, A Urano. 1985. An immunohistochemical study of seasonal changes in luteinizing hormone-releasing hormone and vasotocin in the forebrain and the neurohypophysis of the toad, *Bufo japonicus*. Gen. Comp. Endocrinol. 59: 238-245.

Jørgensen CB. 1973c. Mechanisms regulating ovarian function in amphibians (toads). In: The Development and Maturation of the Ovary and its Functions.

(Ed.) Hannah Peters, Excerpta Medica Amsterdam. International Congress Series, 675: 133-151.

Jørgensen CB. 1973a. Pattern of recruitment of oocytes to second growth phase in normal toads and in hypophysectomized toads *Bufo bufo bufo* (L.) treated with gonadotropin (HCG). Gen. Comp. Endocrinol. 21: 152-159.

Jørgensen CB. 1973b. Mechanisms regulating ovarian function in amphibians (toads). In: The development and maturation of the ovary and its functions. Ed: Hannah Peters, Excerpta Medica Amsterdam. Int. Cong. Series. 675: 133-151.

Jørgensen CB. 1974. Mechanisms regulating ovarian cycle in the toad *Bufo bufo bufo* (L.). Role of presence of second growth phase oocytes in controlling recruitment from pool of first growth phase oocytes. Gen. Comp. Endocrinol. 23: 170-177.

Jørgensen CB. 1975. Factors controlling the annual ovarian cycle in the toad *Bufo bufo bufo* (L.). Gen. Comp. Endocrinol. 25(3): 264-273.

Jørgensen CB. 1981. Ovarian cycle in a temperate zone frog, *Rana temporaria*, with special reference to factors determining number and size of eggs. J. Zool. Lond. 195: 449-458.

Jørgensen CB. 1984. Ovarian functional patterns in Baltic and Mediterranean populations of a temperate zone anuran, the toad *Bufo viridis*. Oikos 43: 309-321.

Jørgensen CB. 1986. Effect of fat body excision in female *Bufo bufo* on the ipsilateral ovary, with a discussion of fat body-gonad relationship. Acta. Zool. 67: 5-10.

Jørgensen CB. 1992. Growth and reproduction. In: Environmental Physiology of the Amphibians, (Ed.) ME Feder ME, WW Burggren. pp. 439-466.

Jurand A, GS Selman. 1964. Yolk utilization in the notochord of newt as studied by electron microscopy. J. Embryol. Exp. Morphol. 12: 43-50.

Kah O, C Lethimonier, JJ Lareyre. 2004. Gonadotrophin-releasing hormone (GnRH) in the animal kingdom. J. Soc. Biol. 198: 53-60.

Karasaki S. 1963a. Studies on amphibian yolk. 1. The ultrastructure of the yolk platelet. J. Cell. Biol. 18: 135-151.

Karasaki S. 1963b. Studies on amphibian yolk. 5. Electron microscopic observations on the utilization of yolk platelets during embryogenesis. J. Ultrastr. Res. 9: 225-247.

Katagiri C, Y Iwao, N Yoshizaki. 1982. Participation of oviductal pars recta secrtetions in inducing the acrosome reaction and release of vitelline coat lysine in fertilizing toad sperm. Dev. Biol. 94: 1-10.

Katagiri C, N Yoshizaki, M Kotani, H Kubo. 1999. Analyses of oviductal pars recta-induced fertilizability of coelomic eggs in *Xenopus laevis*. Dev. Biol. 210: 269-276.

Katagiri C. 1965. The fertilizability of coelomic and oviductal eggs of the toad *Bufo buf o formosus*. J. Fac. Sci. Hokkaido Univ. Ser. VI Zool. 15: 633-643.

Katagiri C. 1974. A high frequency of fertilization in premature and mature coelomic toad eggs after enzymic removal of vitelline membrane. J. Embryol. Exp. Morphol. 31: 573-587.

Katagiri C. 1975. Properties of the hatching enzyme from frog embryos. J. Exp. Zool. 193: 109-118.

Katagiri C. 1987. Role of oviductal secretions in mediating fusion in anuran amphibians. Zool. Sci. 4: 1-14.

Kemp N. 1953. Synthesis of yolk in oocytes of *Rana pipiens* after induced ovulation. J. Morphol. 92: 487-511.

Kemp NE, NL Istock. 1967. Cortical changes in growing oocytes and in fertilized or pricked eggs of *Rana pipiens*. J. Cell. Biol. 34: 111-122.

Kessel RG, LR Ganion. 1980. Cytodifferentiation in the *Rana pipiens* oocyte. VI. The origin and morphogenesis of primary yolk precursors compklexes. J. Submicrosc. Cytol. 12: 647-654.

Kessel RG. 1968. Annulate lamellae. J. Ultrastruct. Res. Suppl. 10: 5-82.

Kessel RG. 1992. Annulate lamellae: A last frontier in cellular organs. Int. Rev. Cytol. 133: 43-120.

Khonsue W, M Matsui, T Hirai, Y Misawa. 2001. Age determination of wrinkled frog, *Rana rugosa* with special reference to high variation in postmetamorphic body size (Amphibia, Ranidae). Zool. Sci. 18: 605-612.

Kim JW, WB Im, HH Choi, S Ischii, HB Kwon. 1998. Seasonal fluctuation in pituitary gland and plasma level of gonadotropic hormones in *Rana*. Gen. Comp. Endocrinol. 109: 13-23.

King JA, RP Millar. 1981. TRH, GH-RIH, and LH-RH in metamorphosing *Xenopus laevis*, Gen. Comp. Endocrinol. 44: 20-27.

King JA, AA Steneveld, RP Millar. 1994. Differential regional distribution of gonadotropin-releasing hormones in amphibian (clawed toad, *Xenopus laevis*) brain. Regul. Pept. 50: 277-289.

King ML, Y Zhou, M Babunenko. 1999. Polarizing genetic information in the egg: RNA localization in the frog oocyte. BioEssays 21: 546-557.

King ML. 1995. mRNA localization during frog oogenesis. In: Localized RNAs, (Ed.) HD Lipshitz. Springer. pp. 137-148.

Klag JJ, GA Ubbels. 1975. Regional morphological and cytochemical differentiation in the fertilized egg of *Discoglossus pictus* (Anura). Defferentiation 3: 15-20.

Klein DC, J Weller. 1970a. Input and output signals in a model neural system: The regulation of melatonin production in the pineal gland. In Vitro 6(3): 197-204.

Klein DC, JL Weller. 1970b. Indole metabolism in the pineal gland: A circadian rhythm in N-acetyltransferase. Science 169(950): 1093-1095.

Kloc M, LD Etkin. 1995. Genetic pathways involved in the localization of RNA in *Xenopus* oocytes. In: Localized RNAs, (Ed.) HD Lipshitz, Springer. pp. 149-156.

Kloc M, S Bilinski, APY Chan, LH Allen, NR Zearfoss, LD Etkin. 2001. RNA localization and germ cell determination in *Xenopus*. Int. Rev. Cytol. 203: 63-91.

Kloc M, S Bilinski, MT Dougherty, EM Brey, LD Etkin. 2004a. Formation, architecture and polarity of female germline cyst in *Xenopus*. Dev. Biol. 266: 43-61.

Kloc M, S Bilinski, DE Etkin. 2004b. The Balbiani body and germ cell determinants: 150 years later. Cur. Topics in Dev. Biol. 59: 1-31.

Kloc M, C Larabell, APY Chan, LD Etkin. 1998. Contribution of METRO pathway localized molecules to the organization of the germ cell lineage. Mech. Dev. 75: 81-93.

Klymkovsky MW. 1995. Intermediate filament organization, reorganization, and function in the clawed frog, *Xenopus*. In: Current Topics in Developmental Biology, (Ed.) DG Capco, Academic Press, Vol. 31, pp. 455-486.

Klymkowsky MW, A Karnovsky. 1994. Morphogenesis and the cytoskeleton: studies of the *Xenopus* embryo. Dev. Biol. 165: 372-384.

Kobayaschi F, SJ Zimniski, KN Smalley. 1996. Characterization of oviductal aromatase in the northern leopard frog *Rana pipiens*. Comp. Bioch. Physiol. B 113: 653-657.

Kordylewski L. 1983. Experimental evidence for the accumulation of egg pigment in the brain cavities of *Xenopus laevis*. J. Exp. Zool. 227: 93-97.

Kordylewski L. 1984. Egg pigment is accumulated in the tadpole's brain. Experientia 40: 277-279.

Kress A, UM Spornitz. 1972. Ultrastructural studies of oogenesis in some European amphibians. 1. *Rana esculenta* and *Rana temporaria*. Z. Zellforsch. Mikr. Anatom. 128: 438-456.

Kress A, UM Spornitz. 1974. Paracrystalline inclusions in mitochondria of frog oocytes. Experientia (Basel) 30: 786-788.

Kress A. 1982. Ultrastructural indications for autosynthetic proteinaceous yolk formation in amphibian oocytes. Experientia 38: 761-771.

Kubokawa K, S Ischii. 1987. Native gonadotropin receptors in amphibian liver. Gen. Comp. Endocrinol. 68: 260-270.

Kupwade VA, SK Saidapur. 1986. Effect of melatonin on oocyte growth and recruitment, hypophyseal gonadotrophs, and oviduct of the frog *Rana cyanophlyctis* maintained under natural photoperiod during the prebreeding phase. Gen. Comp. Endocrinol. 64: 284-292.

Kwon HB, MJ Choi, RS Ahn, YD Yoon. 1991. Steroid production by amphibian (*Rana nigromaculata*) ovarian follicles at different developmental stages. J. Exp. Zool. 260: 66-73.

Kwon HB, RS Ahn, WK Lee, WB Im, Lee CC, Kim K. 1993. Changes in the activities of steroidogenic enzymes during the development of ovarian follicles in *Rana nigromaculata*. Gen. Comp. Endocrinol. 92: 225-232.

Kwon HB, RS Ahn. 1994. Relative roles of theca and granulosa cells in ovarian follicular steroidogenesis in the amphibian, *Rana nigromaculata*. Gen. Comp. Endocrinol. 94(2): 207-214.

Lanzavecchia G. 1965. Structure and demolition of yolk in *Rana esculenta* L. J. Ultrastr. Res. 12: 147-159.

Larabell CA, DE Chandler. 1988. Freeze-fracture analysis of structural reorganization during meiotic maturation in oocytes of *Xenopus laevis*. Cell and Tissue Res. 251: 129-136.

Larabell CA, DE Chandler. 1989. The coelomic envelope of *Xenopus laevis* eggs: A quick freeze, deep-etch analysis. Dev. Biol. 131: 126-135.

Larabell CA. 1995. Cortical cytoskeleton of the *Xenopus* oocyte, egg, and early embryo. In: Current Topics in Developmental Biology, (Ed.) DG Capco, Academic Press, Vol. 31: 433-453.

Lau YT, JK Reynhout, SB Horowitz. 1984. Regional water changes during oocyte meiotic maturation: Evidence of ooplasmic segregation. Dev. Biol. 104: 106-110.

Le Goascogne CL, S Hirai, EE Baulieu. 1984. Induction of germinal vesicle breakdown in *Xenopus laevis* oocytes: Synergistic action of progesterone and insulin. J. Endocr. 101: 7-12

Le Goascogne C, N Sananes, M Gouezou, EE Baulieu. 1985. Testosterone-induced meiotic maturation of *Xenopus laevis* oocytes: Evidence for an early effect in the synergistic action of insulin. Dev. Biol. 109: 9-14.

Lessman CA, AW Schuetz. 1981. Role of follicle wall in meiosis reinitiation induced by insulin in *Rana pipiens* oocytes. Am. J. Physiol. 241: E51-56.

Liang ChG, YQ Su, HY Fan, H Schatten, QY Sun. 2007 Mechanisms regulating oocyte meiotic resumption: Roles of mitogen-activated protein kinase. Mol. Endocrinol. 21: 2037-2055.

Licht P, D Crews. 1976. Gonadotropin stimulation of *in vitro* progesterone production in reptilian and amphibian ovaries. Gen. Comp. Endocrinol. 29: 141-151.

Licht P, BR McCreery, R Barnes. 1983b. Relation between acute pituitary responsiveness to gonadotropin releasing hormone (GnRH) and the ovarian cycle in the bullfrog, *Rana catesbeiana*. Gen. Comp. Endocrinol. 51: 148-153.

Licht P, BR McCreery, H Papkoff. 1983a. Effects of gonadectomy on polymorphism in stored and circulating gonadotropins in the bullfrog, *Rana catesbeiana*. II. Gel filtration chromatography. Biol. Reprod. 29: 646-657.

Licht P, H Papkoff. 1974. Separation of two distinct gonadotropins from the pituitary gland of the bullfrog *Rana catesbeiana*. Endocrinology. 94: 1587-1594.

Licht P, D Porter, RP Millar. 1987. Specificity of amphibian and reptilian pituitaries for various forms of gonadotropin-releasing hormones in vitro. Gen. Comp. Endocrinol. 66: 248-255.

Licht P, PS Tsai, J Sotowska-Brochocka. 1994. The nature and distribution of gonadotropin-releasing hormones in brains and plasma of ranid frogs. Gen. Comp. Endocrinol. 94: 186-198.

Licht P, DA Porter. 1986. Role of gonadotropin-releasing hormone in regulation of gonadotropin secretion from amphibian and reptilian pituitaries. In: Hormones and Reproduction in Fishes, Amphibians, and Reptiles, (Eds) DO Norris, RE Jones, Plenum Press, New York, pp. 61-85.

Lin YP, WA Schuetz. 1985. Intrafollicular action of estrogen in regulating pituitary—induced ovarian progesterone synthesis and oocyte maturation in *Rana pipiens*: Temporal relationship and locus of action. Gen. Comp. Endocrinol. 58: 421-435.

Lin Y-WP, T Petrino, AM Landin, S Franco, I Simeus. 1999. Inhibitory action of the gonadopeptide inhibin on amphibian (*Rana pipiens*) steroidogenesis and oocyte maturation. J. Exp. Zool. 284: 232-240.

Lindsay LL, JL Hedrick. 1998. Treatment of *Xenopus laevis* coelomic eggs with trypsin mimics pars recta oviductal transit by selectively hydrolyzing envelope glycoprotein gp43, increasing sperm binding to the envelope, and rendering eggs fertilizable. J. Exp. Zool. 281: 132-138.

Lindsay LL, MJ Wieduwilt, JL Hedrick. 1999. Oviductin, the *Xenopus laevis* oviductal protease that processes egg envelope glycoprotein gp43, increases sperm binding to envelopes, and is transient as part of an unusual mosaic protei composed of two preotease and several CUB domains. Biol. Reprod. 60: 980-995.

Liu XJ, A Sorisky, L Zhu, T Pawson. 1995. Molecular cloning of an amphibian insulin receptor substrate 1—like cDNA and involvement of phosphaty-dylinositol 3—kinase in insulin-induced *Xenopus* oocyte maturation. Mol. Cell. Biol. 15 (7): 3563-3570.

Lofts B. 1974. Reproduction. In: Physiology of the Amphibia, (Ed.) B Loft, Academic Press, Vol. 2 pp. 107-200.

Lopez K. 1989. Sex differentiation and early gonadal development in *Bombina orientalis* (Anura: Discoglossidae). J. Morphol. 199: 299-311.

Ludwig H. 1847. Über die Eibildung in Tierreiche. Arb. Physiol. Lab. Würzburg. 1: 287-510.

Lutz LB, LM Cole, MK Gupta, KW Kwist, RJ Auchus, SR Hammes. 2001. Evidence that androgens are the primary steroids produced by *Xenopus laevis* ovaries and may signal through the classical androgen receptor to promote oocyte maturation. Proc. Nat. Acad. Sci. 98: 13728-13733.

Lutz LB, M Jamnongjit, WH Yang, D Jahani, A Gill, SR Hammes. 2003. Selective modulation of genomic and nongenomic androgen responses by androgen receptor ligands. Mol. Endocrinol. 17: 1106-1116.

MacGregor HC, EM del Pino. 1982. Ribosomal gene amplification in multi-nucleolate oocytes of the egg-brooding hylid frog *Flectonotus pygmaeus*. Chromosoma 85: 475-488.

MacGregor HC, J Kezer. 1970. Gene amplification in oocytes wuith eight germinal vesicles from the tailed frog *Ascaphus truei* Stejneger. Chromosoma 29: 189-206.

MacGregor HC. 1980. Recent developments in the study of lampbrush chromo-somes. Heredity 44: 3-35.

MacGregor HC. 1982. Ways of amplifying ribosomal genes. In: The Nucleolus, (Ed.) EG Jordan, CA Cullis. Soc. Exp. Boil. Seminar, Cambridge New York University Press. Ser. 15: 129-152.

Maes E, D Florea, A Coppin, G Strecker. 1999. Structural analysis of 20 oligo-saccharide-alditol released from the jelly coat of *Rana palustris* eggs by reductive β-elimination. Characterization of the polymerized sequence [Gal(β1-3)GalNAc(α1-4)n]. Eur. J. Biochem. 264: 301-313.

Mais C, B McStay, U Scheer. 2002. On the formation of amplified nucleoli during early *Xenopus* oogenesis. J. Structural Biol. 140: 214-226.

Maller JL, JW Koontz. 1981. A study the induction of cell division in amphibian oocytes by insulin. Dev. Biol. 85: 309-316.

Maller JL. 1998. Recurring themes in oocyte maturation. Biol. Cell. 90: 453-460.

Mariano MI, MG de Martin, A Pisano. 1984. Morphological modifications of oocyte vitelline envelope from *Bufo arenarum* during different functional states. Develop. Growth Differ. 26: 33-42.

Massover WH. 1971b. Nascent yolk platelets of anuran amphibian oocytes. J. Ultrastr. Res. 37: 574-591.

Massover WH. 1971a. Intramitochondrial yolk-crystals of frog oocytes. I. Formation of yolk-crystals by mitochondria during bullfrog oogenesis. J. Cell. Biol. 48: 266-279.

Masui Y, HJ Clarke. 1979. Oocyte maturation. Int. Rev. Cytol. 37: 185-282.

Masui Y, CL Markert. 1971. Cytoplasmic control of nuclear behavior during meiotic maturation of frog oocytes. J. Exp. Zool. 177: 129-145.

Matten W, I Daar, GF Vande Woude. 1994. Protein kinase A acts at multiple points to inhibit *Xenopus* oocyte maturation. Mol. Cell. Biol. 14: 4419-4426.

Matten WT, TD Copeland, NG Ahn, GF Vande Woude. 1996. Positive feedback between MAP kinase and Mos during *Xenopus* oocyte maturation. Dev. Biol. 179: 485-492.

Maturi G, V Infante, R Carotenuto, R Focarelli, M Caputo, C Campanellsa. 1998. Specific glycoconjugates are present at the oolemma of the fertilization site in the egg of *Discoglossus pictus* (Anurans) and bind spermatozoa in an in vitro assay. Dev. Biol. 204: 210-223.

Mayer I, C Bornestaf, B Borg. 1997. Melatonin in non-mammalian Vertebrates: physiological role in reproduction? Comp. Biochem. Physiol. 118A, 3: 515-531.

McCreery B, P Licht. 1983a. Induced ovulation and changes in pituitary responsiveness to continuous infusdion of gonadotropin-releasing hormone during the ovarian cycle in the bullfrog, *Rana catesbeiana*. Biol. Repr. 29: 863-871.

McCreery BR, P Licht. 1983b. Effects of gonadectomy on polymorphism in stored and circulating gonadotropins in the bullfrog, *Rana catesbeiana*. I. Clearance profiles. Biol. Reprod. 29(3): 637-645.

McCreery BR, P Licht. 1984a. The role of androgen in the development of sexual differences in pituitary responsiveness to gonadotropin releasing hormone (GnRH) agonist in the bullfrog, *Rana catesbeiana*. Gen. Comp. Endocrinol. 54: 350-359.

McCreery BR, P Licht. 1984b. Effects of gonadectomy and sex steroids on pituitary gonadotrophin release and response to gonadotrophin-releasing hormone (GnRH) agonist in the bullfrog, *Rana catesbeiana*. Gen. Comp. Endocrinol. 54: 283-296.

McCreery BR, P Licht, R Barnes, JE River, WW Vale. 1982. Actions of agonistic and antagonistic analogs of gonadotropin releasing hormone (GnRH) in bullfrog *Rana catesbeiana*. Gen. Comp. Endocrinol. 46: 511-520.

Medina MF, I Ramos, CA Crespo, S Gonzales-Calvar, SN Fernandez. 2004. Changes in serum sex steroid levels throughout the reproductive cycle of *Bufo arenarum* females. Gen. Comp. Endocrinol. 136: 143-151.

Merriam RW, RA Sauterer, R Christensen. 1983. A subcortical, pigment-containing structure in *Xenopus* eggs with contractile properties. Dev. Biol. 95: 439-446.

Miceli DC, SN Fernandez. 1982. Properties of an oviductal protein involved in amphibian oocyte fertilization. J. Exp. Zool. 221: 357-364.

Mignotte F, M Tourte, JC Mounoulou. 1987. Segregation of mitochondria in the cytoplasm of *Xenopus* vitellogenic oocytes. Biol. Cell 60: 97-102.

Mikami SI. 1992. Hypophysis. In: Atlas of endocrine organs. Vertebrates and Invertebrates, (Eds) A Matsumoto, S Ischii, Springer-Verlag, pp. 39-60.

Millar R, S Lowe, D Conklin, A Pawson, S Maudsley, B Troskie, T Ott, M Millar, G Lincoln, R Sellar, B Faurholm, G Scobie, R Kuestner, E Terasawa, A Katz. 2001. A novel mammalian receptor for the evolutionarily conserved type II GnRH. Proc. Natl Acad. Sci. USA. 98: 9636-9641.

Millar RP, ZL Lu, AJ Pawson, CA Flanagan, K Morgan, SR Maudsley. 2004. Gonadotropin-releasing hormone receptors. Endocr. Rev. 25: 235-275.

Millar RP. 2003. GnRH II and type II GnRH receptors. Trends Endocrinol. Metab. 14: 35-43.

Miranda LA, DA Paz, JM Affanni, GM Somoza. 1998. Identification and neuro-anatomical distribution of immunoreactivity for mammalian gonadotropin-releasing hormone (mGnRH) in the brain and neural hypophyseal lobe of the toad *Bufo arenarum*. Cell Tissue Res. 293: 419-425.

Mitchell ELD, RS Hill. 1986. The occurrence of lampbrush chromosomes in early diplotene oocytes of *Xenopus laevis*. J. Cell. Sci. 83: 213-221.

Mizell S. 1964. Seasonal differences in spermatogenesis and oogenesis in *Rana pipiens*. Nature (London) 202: 875-876.

Mood K, YS Bong, HS Lee, A Ishimura, IO Daar. 2004. Contribution of JNK, Mek, Mos and PI-3K signaling to GVBD in *Xenopus* oocytes. Cell Signal. 16: 631-642.

Morelle W, MO Cabada, G Strecker. 1998. Structural analysis olgosaccharide-alditols released by reductive β-elimination from the jelly coats of the anuran *Bufo arenarum*. Eur. J. Biochem. 252: 253-260.

Morelle W, R Guétant, G Strecker. 1998. Structural analysis of oligosaccharide-alditols released by reductive β-elimination from oviductal mucins of *Rana dalmatina*. Carbohydr. Res. 306: 435-443.

Morelle W, G Strecker. 1997. Structural analysis of hexa to dodecaoligosaccharide-a-alditols released by reductive β-elimination from oviductal mucins of *Bufo bufo*. Glycobiology 7: 1129-1151.

Morelle W, G Strecker. 1998. Structural analysis of a new series oligosaccharide-alditols released by reductive β-elimination from oviductal mucins of *Rana urticularia*. Biochem.J. 330: 469-478.

Moriguchi Y, A Tanimura, H Iwasawa. 1991. Annual changes in the Bidder's organ of the toad *Bufo japonicus formosus*: histological observation. Sci. Rep. Niigata Univ. Ser. D Biol. 28: 11-17.

Mosconi G, K Yamamoto, O Carnevali, M Nabissi, A Polzonetti-Magni, S Kikuyama. 1994. Seasonal changes in plasma growth hormone and prolactin concentrations of the frog *Rana esculenta*. Gen. Comp. Endocrinol. 93: 380-387.

Mouraud R, W Morelle, A Nevau, G Strecker. 2001. Diversity of O-linked glycosylation patterns between species. Characterization of 25 carbohydrate chains from oviductal mucins of *Rana ridibunda*. Eur. J. Biochem. 268: 1990-2003.

Muske LE, FL Moore. 1990. Ontogeny of immunoreactive gonadotropin-releasing hormone neuronal systems in amphibians. Brain Res. 534(1-2): 177-187.

Muske LE, JA King, FL Moore, RP Millar. 1994. Gonadotropin-releasing hormones in microdissected brain regions of an amphibian: concentration and anato-mical distribution of immunoreactive mammalian GnRH and chicken GnRH II. Reg. Pept. 54: 373-384.

Nagahama Y. 1987. Endocrine control of oocyte maturation. In: Hormones and Reproduction in Fishes, Amphibians, and Reptiles. (Eds) DO Norris, RE Jones. Plenum press, New York, London, pp. 171-202.

Nebreda AR, I Ferby. 2000. Regulation of the meiotic cell cycle in oocytes. Curr. Opin. Cell. Biol. 12: 666-675.

Ogielska M, R Augustyńska, A Kotusz, J Ihnatowicz. 2007. Definite number of ova for a lifespan is established during juvenile period of anuran amphi-bians. J. Morphol. 268: 1112-1113.

Ogielska M, A Kotusz. 2004. Pattern and rate of ovary differentiation with reference to somatic development in anuran amphibians. J. Morphol. 259: 41-54.

Ogielska M, E Wagner. 1990. Oogenesis and development of the ovary in European green frog, Rana ridibunda (Pallas). I. Tadpole stages until metamorphosis. Zool. J. Anat. 120: 211-221.

Ogielska M, E Wagner. 1993. Oogenesis and ovary development in natural hybridogenetic water frog, Rana esculenta L. 1. Tadpole stages until metamorphosis. Zool. J. Physiol. 97: 349-368.

Ogielska M. 1990. The fate of intramitochondrial paracrystalline inclusion bodies in germ line cells of water frogs (Amphibia, Anura). Experientia 46: 98-101.

Ogielska M. 1999. Wczesny rozwój Płazów: Indukcja mezodermy i tworzenie się osi ciała (Early amphibian development: Mesoderm induction and formation of body axes). Przegl. Zool. XLIII: 33-52.

Ogielska-Nowak M. 1985. Development of embryos deriving from pigmentless eggs of the natural hybrid, Rana esculenta L. (Amphibia, Anura). Zool. Pol. 32: 223-228.

Ohlendorf DH, RF Wrenn, LJ Banaszak. 1978. Three-dimentional structure of lipovitellin-phosvitin complex from amphibian oocytes. Nature (London) 272: 28-32.

Ohno S, S Karasaki, K Takata. 1964. Histo- and cytochemical studies on the superficial layer of yolk platelets in the Triturus embryo. Exp. Cell. Res. 33: 310-318.

Olson JH, DE Chandler. 1999. Xenopus laevis egg jelly contains small proteins that are essential to fertilization. Dev. Biol. 210: 401-410.

Olson JH, X Xiand, T Ziegert, A Kittelson, A Rawls, AL Bieber, DE Chandler. Allurin, a 21-kDa sperm chemoattractant from Xenopus egg jelly, is related to mammalian sperm-binding proteins. Proc. Natl Acad. Sci. USA 98: 11205-11210.

Opresko LK, HS WileyS. 1987a. Receptor-mediated endocytosis in Xenopus oocytes. I. Characterization of the vitellogenin receptor system. J. Biol. Chem. 262: 4109-4155.

Opresko LK, HS Wiley. 1987b. Receptor-mediated endocytosis in Xenopus oocytes. II. Evidence for two novel mechanisms of hormone regulation. J. Biol. Chem. 262: 4116-4123.

Palmer A, AR Nebreda. 2000. The activation of MAP kinase and p34cdc2/cyclin B during the meiotic maturation of Xenopus oocytes. Prog. Cell Cycle Res. 4:131-143.

Pancak-Roessler MK, DO Norris. 1991. The effects of orchidectomy and gonado-tropis on steroidogenesis and oogenesis in Bidder's organs of the toad Bufo woodhousii. J. Exp. Zool. 260: 323-336.

Paolucci M, MM Di Fiore. 1994. Sex steroid binding proteins in the plasma of the green fog Rana esculenta: Changes during the reproductive cycle and dependence on pituitary gland and gonads. Gen. Comp. Endocrinol. 96: 401-411.

Paolucci M, G Guerriero, G Ciarcia. 2000. Effect of 17β –estradiol and testosterone treatment on sex steroid binding proteins in the female of the green frog Rana esculenta. Zool. Sci. 17: 797-803.

Papkoff H, SW Farmer, P Licht. 1976. Isolation and characterization of luteinizing hormone from amphibian (*Rana catesbeiana*) pituitaries. Life Sci. 18(2): 245-250.

Parfenov VN, MN Gruzova. 1984. The peculiarities of the karyosphere formation in *Rana ridibunda*. Electron microscope data. Citologia. 26: 165-172.

Pepling ME, M de Cuevas, AC Spradling. 1999. Germline cysts: A conserved phase of germ cell development? Trends in Cell. Biol. 9: 257-262.

Perry JS, IW Rowlands. 1962. The ovarian cycle in vertebtates. In: The Ovary, (Ed.) S Zuckerman, Academic Press, New York, pp. 279-309.

Petrino TR, G Toussaint, YW Lin. 2007. Role of inhibin and activin in the modulation of gonadotropin- and steroid-induced oocyte maturation in the teleost *Fundulus heteroclitus*. Reprod. Biol. Endocrinol. 5: 5-21.

Phillips S, M Cotton, F Laengle-Rouault, G Schaffner, ML Birnstiel. 1992. Amphibian oocytes and sphere organelles: Are the U snRNA genes amplified? Chromosoma 101: 549-556.

Pierantoni R, B Varriale, S Fasano, S Minucci, L Di Matteo, G Chieffi. 1987. Seasonal plasma and intraovarian sex steroid profiles, and influence of temperature on gonadotropin stimulation of in vitro estradiol-17 beta and progesterone production, in *Rana esculenta* (Amphibia: Anura). Gen. Comp. Endocrinol. 67: 163-168.

Pinto MR, L Santella, G Casazza, F Rosati, A Monroy. 1985. The differentiation of the vitelline envelope of *Xenopus* oocyte. Develop. Growth Differ. 27: 189-200.

Płytycz B, J Bigaj. 1993. Studies on the growth and longevity of the yellow-bellied toad, *Bombina variegata*, in natural environment. Amphibia-Reptilia 14: 35-44.

Płytycz B, J Bigaj. 2004. Długowieczność kumaków górskich, *Bombina variegata*. (Longevity of the yellow-bellied toads, Bombina variegata). In: Biologia gadów i płazów—ochrona herpetofauny, (Ed.) W Zamachowski. Wydawnictwo Naukowe Akademii Pedagogicznej Kraków, pp. 71-73.

Płytycz B, J Mika, J Bigaj. 1995. Age-dependent changes in thymuses in the European common frog, *Rana temporaria*. J. Exp. Zool. 273: 451-460.

Polzonetti-Magni A, O Carnevali, K Yamamoto, S Kirujama. 1995. Growth hormone and prolactin in amphibian reproduction. Zool. Sci. 12: 683-694.

Polzonetti-Magni AM, G Mosconi, O Carnevali, K Yamamoto, Y Hanaoka, S Kikuyama. 1998. Gonadotropins and reproductive function in the anuran amphibian, *Rana esculenta*. Biol. Reprod. 58: 88-93.

Porter DA, P Licht. 1986. Effects of temperature and mode of delivery on responses to gonadotropin releasing hormone by superfused frog pituitaries. Gen. Comp. Endocrinol. 63: 236-244.

Prasadmurthy YS, SK Saidapur. 1987. Role of fat bodies in oocyte growth and recruiment in the frog *Rana cyanophlyctis* (Sch.). J. Exp. Zool. 243: 153-162.

Quill TA, JL Hedrick. 1996. The fertilization layer mediated block to polyspermy in *Xenopus laevis*: Isolation of the cortical granule lectin ligand. Arch. Bioch. Biophys. 333: 326-332.

Raff RA. 1996. The shape of life: Genes, development and the evolution of animal form. University of Chicago Press.

Ramos I, S Cisint, C Crespo, MF Medina, S Fernandez. 2001. Nuclear maturation inducers in *Bufo arenarum* oocytes. Zygote 9: 353-359

Räsänen K, M Pahkala, A Laurila, J Merilä. 2003. Does jelly envelope protect the common frog *Rana temporaria* embryos from UV-B radiation? Herpetologica 59: 293-300.

Rasar MA, SR Hammes. 2006. The physiology of the Xenopus laevis ovary. Meth. Molec. Biol. 322: 17-30.

Rastogi RK, G Chieffi. 1970a. A cytological study of the pars distalis of the pituitary gland of normal, gonadectomized and gonadectomized, steroid hormone treated green frog, *Rana esculenta* L. Gen. Comp. Endocrinol. 15: 247-263.

Rastogi RK, G Chieffi. 1970b. Cytological changes in the pars distalis of pituitary of the green frog *Rana esculenta* L., during the reproductive cycle. Z. Zellforsch. 111: 505-518.

Rastogi RK, DL Meyer, C Pinelli, M Fiorentino, B D'aniello. 1998. Comparative analysis of GnRH neuronal systems in the amphibian brain. Gen. Comp. Endocrinol. 112: 330-45.

Raven CP. 1961. Oogenesis. The storage of developmental information. International Series of Monographs on Pure and Applied Biology. Vol. 10. Pergamon Press.

Redshaw MR, BK Follet. 1971. The crystalline yolk-platelet proteins and their soluble plasma precursor in an amphibian, *Xenopus laevis*. Biochem. J. 124: 759-766.

Redshaw MR. 1972. The hormonal control of the amphibian ovary. Am. Zool. 12: 289-306.

Reppert SM, DR Weaver, C Godson. 1996. Melatonin receptors step into the light: Cloning and classification of subtypes. Trends Pharmacol. Sci. 17: 100-102.

Richter HP. 1987. Membranes during yolk-platelet development in oocytes of the toad *Bufo marinus*. Roux's Arch. Dev. Biol. 196: 367-371.

Robertson N. 1979. The carbohydrate content of isolated yolk platelets from early developmental stages of *Xenopus laevis*. Cell. Diff. 8: 173-186.

Rohde BH, MA McLaughlin, LY Chiou. 1985. Existence and role of endogenous ocular melatonin. J. Ocul. Pharmacol. 1: 235-243.

Romek M, A Krzysztofowicz. 2001. Utilization of yolk platelets during early embryonic development of *Rana temporaria* and *Bufo bufo*. Folia Histochem. Cytobiol. 39: 283-291.

Rugh R. 1951. The frog. Its Reproduction and Development. The Blakiston Company, Philadelphia, Toronto.

Ruiz I, A Altaba, DA Melton. 1990. Axial patterning and the establishment of the polarity in the frog embryo. Trends Genet. 6: 57-64.

Ryabova LV, SG Vassetzky. 1997. A two-component cytoskeletal system of *Xenopus laevis* egg cortex: concept of its contractility. Int. J. Dev. Biol. 41: 843-851.

Sagata N, I Daar, M Oskarsson, SD Showalter, GF Vande Woude. 1989. The product of the mos proto-oncogene as a candidate "initiator" for oocyte maturation. Science. 245: 643-646.

Sagata N, M Oskarsson, T Copeland, J Brumbaugh, GF Vande Woude. 1988. Function of c-mos proto-oncogene product in meiotic maturation in *Xenopus* oocytes. Nature (London) 335(6190): 519-525.

Sagata N. 1997. What does Mos do in oocytes and somatic cells? Bioessays 19(1): 13-21.

Saidapur SK, NP Gramapurohit, BA Shanbhag. 2001. Effect of sex steroids on gonad differentiation and sex reversal in the frog, *Rana curtipes*. Gen. Comp. Endocrinol. 124: 115-123.

Saidapur SK, B Hoque. 1995. Effect of photoperiod and temperature on ovarian cycle of the frog *Rana tigrina* (Daud.) J. Biosci. 20 (3): 445-452.

Sánchez SS, EI Vilecco. 2003. Oogenesis. Reproductive Biology and Phylogeny of Anura, (Ed.) BGM Jamieson, Science Publisher, Inc. NH, USA, pp. 27-71.

Sato T, K Wake. 1992. Pineal gland. In: Areas of Endocrine Organs, (Eds) A Matsumoto, S Ischii, Springer-Verlag, Berlin, Heidelberg, New York.

Scheer U. 1973. Nuclear pore flow rate of ribosomal RNA and chain growth rate of its precursor during oogenesis of *Xenopus laevis*. Dev. Biol. 30: 13-28.

Schmitt A, AR Nebreda. 2002. Signalling pathways in oocyte meiotic maturation. J. Cell. Sci 115: 2457-2459.

Schneider WJ. 1996. Vitellogenin receptors: oocyte-specific members of the low-density lipoprotein receptor supergene family. Int. Rev. Cytol. 166: 103-137.

Schuetz AW, 1986. Hormonal dissociation of ovulation and maturation of oocytes: ovulation of immature amphibian oocytes by prostaglandin. Gam. Res. 15: 99-113.

Semik D, W Kilarski. 1998. Scanning electron microscopy of the egg surface of the common toad, *Bufo bufo*, following fertilization and during first cleavage. Folia Histochem. Cytobiol. 36: 29-34.

Seong JY, L Wang, DY Oh, O Yun, K Maiti, JH Li, JM Soh, HS Choi, K Kim, H Vaudry, HB Kwon. 2003. Ala/Thr(201) in extracellular loop 2 and Leu/Phe (290) in transmembrane domain 6 of type 1 frog gonadotropin-releasing hormone receptor confer differential ligand sensitivity and signal transduction. Endocrinology. 144: 454-466.

Sever DM, WC Hamlett, R Slabach, B Stephenson, PA Verrell. 2003. Internal fertilization in the Anura with special refce to mating and female sperm storage in *Ascaphus*. In:. Reproductive biology and phylogeny of Anura, (Ed.) BGM Jamieson, Science Publisher, Inc. NH, USA, pp. 319-341.

Sever DM, EC Moriarty, LC Rania, WC Hamlett. 2001. Sperm storage in the oviduct of the internal fertilizing frog *Ascaphus truei*. J. Morphol. 248: 1-21.

Skene DJ, B Vivien-Roels, P Pevet. 1991. Day and nighttime concentrations of 5-methoxytryptophol and melatonin in the retina and pineal gland from different classes of vertebrates. Gen. Comp. Endocrinol. 84(3): 405-411.

Skierska K, K Pierzchot, M Socha, M Ogielska. 2007. Age of males and females in amplecting pairs, and number and size of eggs in the grass frog *Rana temporaria*. Abstracts of the 14th European Congress of Herpetology, Porto Portugal 2007: 301.

Skoblina MN, OT Kondrat'eva, GP Nikiforova, I Huhtaniemi. 1997. The role of eicosanoids and progesterone in the ovulation of the oocytes in the common frog. Ontogenez 28: 211-216.

Sladeèek F. 1964. The development of "white eggs" mutants of *Rana temporaria* L. in normal conditions and parabiotic and chiearic combinations with pigmented embryos. Folia Biol. 10: 23-29.

Smith CL. 1955. Reproduction of female amphibia. Mem. Soc. Endocrinol. 4: 39-56.

Smith DL. 1986. Regulation of translation during amphibian oogenesis and oocyte maturation. Gametogenesis and the Early Embryo, (Ed.) Alan R. Liss, Inc., New York, pp. 131-150.

Smith JC, R White. 2003. Patterning the Xenopus embryo. In: Patterning in Vertebrate Development, (Ed.) Ch. Tickle, Oxford University Press, pp. 24-47.

Smith LD, RE Ecker. 1969. Cytoplasmic regulation in early events of amphibian development. Proc. Can. Cancer Conf. 8: 103-129.

Smith LD, RE Ecker. 1971. The interaction of steroids with Rana pipiens Oocytes in the induction of maturation. Dev. Biol. 25 (2): 232-247.

Sotowska-Brochocka J. 1983. Role of the hypothalamus in the control of ovulation in Rana temporaria L. Acta Physiol. Pol. 34 (5-6): 621-624.

Sotowska-Brochocka J. 1988. The stimulatory and inhibitory role of the hypothalamus in the regulation of ovulation in grass frog, Rana temporaria L. Gen. Comp. Endocrinol. 70: 83-90.

Sotowska-Brochocka J, P Licht. 1992. Effect of infundibular lesions on GnRH and LH release in the frog, Rana temporaria, during hibernation. Gen. Comp. Endocrinol. 85: 43-54.

Sotowska-Brochocka J, L Martńyska, P Licht. 1992. Changes of LH level in the pituitary gland and plasma in hibernating frogs, Rana temporaria. Gen. Comp. Endocrinol. 87: 286-291.

Spornitz UM. 1972. Some properties of crystalline inclusion bodies in oocytes of Rana temporaria and Rana esculenta. Experientia 28: 66-67.

Stamper DL, P Licht. 1991. Time-dependent changes in gonadotropin synthesis following gonadectomy in the frog, Rana pipiens. Biol. Reprod. 44: 798-805.

Stamper DL, P Licht. 1993. Further studies on the influence of GnRH on the biosynthesis of gonadotropins in female frogs (Rana pipiens). Gen. Comp. Endocrinol. 92: 104-112.

Stebbins RC, NW Cohen. 1995. A Natural History of Amphibians, Princeton Univ Press, Princeton.

Szymura JM, NH Barton. 1986. Genetic analysis of a hybrid zone between the fire-bellied toads, Bombina bombina and B. variegata near Cracow in southern Poland. Evolution 45: 237-261.

Szymura JM. 1993. Analysis of hybrid Zones with Bombina. In: Hybrid Zones and the Evolutionary Process, (Ed.) RG Harrison, Oxford University Press, New York, pp. 261-289.

Takahashi H, Y Hanaoka. 1981. Isolation and characterization of multiple components of basic gonadotropin from bullfrog (Rana catesbeiana) pituitary gland. J. Biochem. 90(5): 1333-1340.

Takamune K, N Yoshizaki, C Katagiri. 1986. Oviductal pars recta-induced degradation of vitelline coat proteins in relation to acquisition of fertilizability of toad eggs. Gam. Res. 14: 215-224.

Tandler CJ, JL la Torre. 1967. An acid polysaccharide in yolk platelets of Bufo arenarum oocytes. Exp. Cell. Res. 45: 491-494.

Tanimura A, H Iwasawa. 1986. Development of gonad and Bidder's organ in

Bufo japonicus formosus: Histological observation. Sci. Rep. Nilgata Univ. Ser. D. 23:11-21.

Tanimura A, H Iwasawa. 1987. Germ cell kinetics in gonadal development in the toad *Bufo japonicus formosus*. Zool. Sci. 4: 657-664.

Tanimura A, H Iwasawa. 1989. Origin of somatic cells and histogenesis in the primordial gonad of the Japanese tree frog, *Rhacophorus arboreus*. Anat. Embryol. 180: 165-173.

Tanimura A, H Iwasawa. 1992. Ultrastructural observations of the ovary and Bidder's organ in young toad, *Bufo japonicus formosus*. Sci. Rep. Niigata Univ. D. Biol. 29: 27-33.

Tata JR. 1976. The expression of the vitellogenin gene. Cell 9: 1-14.

Tata JR, BS Baker, JV Deeley. 1980. Vitellogenin as a multigene family. J. Biol. Chem. 255: 6721-6726.

Tavolaro R, M Canonaco, MF Franzoni. 1995. Comparison of melatonin-binding sites in the brain of two amphibians: An autoradiographic study. Cell Tissue Res. 279(3): 613-617.

Thiébaud CH. 1979a. Quantitive determination of amplified rDNA and its distribution during oogenesis in *Xenopus laevis*. Chromosoma 73: 37-44.

Thiébaud CH. 1979b. The intra-nucleolar localization of amplified rDNA in *Xenopus laevis*. Chromosoma 73: 29-36.

Tian J, H Gong, GH Thomsen, WJ Lennarz. 1997. *Xenopus laevis* sperm-egg adhesion is regulated by modifications in the sperm receptor and the egg vitelline envelope. Dev. Biol. 187: 143-153.

Tian J, S Kim, E Heilig, JV Ruderman. 2000. Identification of XPR-1, a progesterone receptor required for *Xenopus* oocyte activation. Proc. Nat. Acad. Sci. 97 (26): 14358-14363.

Tokarz RR. 1978. Oognial proliferation, oogenesis and folliculogenesis in non-mammalian vertebrates. In: The Vertebrate Ovary, Comparative Biology and Evolution, (Ed.) RE Jones. Plenum Press, pp. 145-279.

Tourte M, C Besse, JC Mounoulou, 1991. Cytochemical evidence of an organized microtubular cytoskeleton in *Xenopus lae vis* oocytes: Invovement in the segregation of mitochondrial populations. Mol. Reprod. Dev. 30: 353-359.

Tourte M, F Mignotte, JC Mounoulou. 1981. Organization and replication activity of the mitochondrial mass of oogonia and and previtellogenic oocytes in *Xenopus laevis*. Dev. Growth Differ. 23: 9-21.

Tourte M, F Mignotte, JC Mounoulou. 1984. Heterogenous distribution and replication activity of mitochondria in *Xenopus laevis* oocytes. Eur. J. Cell. Biol. 34: 171-178.

Troskie B, JA King, RP Millar, YY Peng, J Kim, H Figueras, N Illing. 1997. Chicken GnRH II-like peptides and a GnRH receptor selective for chicken GnRH II in amphibian sympathetic ganglia. Neuroendocrinology 65(6): 396-402.

Troskie BE, JP Hapgood, RP Millar, N Illing. 2000. Complementary deoxyribo-nucleic acid cloning, gene expression, and ligand selectivity of a novel gonadotropin-releasing hormone receptor expressed in the pituitary and midbrain of *Xenopus laevis*. Endocrinology 141(5): 1764-71.

Tyler MJ. 2003. The gross anatomy of the reproductive system, In: Reproductive Biology and Phylogeny of Anura, (Ed.) BGM Jamieson, Science Publisher, Inc. NH, USA, pp. 19-26.

Uchiyama H, S Komazaki, M Asashima, S Kikuyama. 1996. Occurrence of immunoreactive activin/inhibin beta (B) in gonadotrophs, thyrotrophs, and somatotrophs of the *Xenopus* pituitary. Gen. Comp. Endocrinol. 102: 1-10.

Uchiyama K, A Koola, S Komazaki, N Oyama S Kikuyama. 2000. Occurence of immunofreactive activin/inhibin β in thyrotropes and gonadotropes in the bullfrog pituitary: Possible paracrine/autocrine effects of activin β on gonadotropin secretion. Gen. Comp. Endocrinol. 118: 68-76.

Udaykumar K, BN Joshi. 1997. Effect of exposure to continuous light and melatonin on ovarian follicular kinetics in the skipper frog, *Rana cyanophlyctis*. Biol. Signals 6(2): 62-66.

Ueda Y, N Yoshizaki, Y Iwao. 2002. Acrosome reaction in sperm of the frog, *Xenopus laevis*: Its detection and induction by oviductal pars recta secretion. Dev. Biol. 243: 55-64.

Van den Hoef MHF, WJAG Dictus, WJ Hage, JG Bluemink. 1984. The ultrastructural organization of gap junctions between follicle cells and the oocyte in *Xenopus laevis*. Eur. J. Cell. Biol. 33: 242-247.

Vanecek J. 1999. Inhibitory effect of melatonin on GnRH-induced LH release. Rev. Reprod. 4: 67-72.

Viertel B, S Richter. 1992. Anatomy: viscera and endocrines. In: Tadpoles: The Biology of Anuran Larvae, (Eds) RW McDiarmid and R Altig, The University of Chicago Press, pp. 92-148.

Vijayakumar S, CB Jorgensen, K Kjaer. 1971. Regulation of ovarian cycle in the toad *Bufo bufo bufo* (L.): Effects of autografting pars distalis of the hypophysis, of extirpating gonadotropic hypothalamic region, and of partial ovariectomy. Gen. Comp. Endocrinol. 17(3): 432-443.

Wagner E, M Ogielska. 1990. Oogenesis and development of the ovary in European green frog, *Rana ridibunda* (Pallas). 2. Juvenile stages until adults. Zool. Jb. Anat. 120: 223-231.

Wagner E, M Ogielska. 1993. Oogenesis and ovary development in natural hybridogenetic water frog, *Rana esculenta* L. 2. After metamorphosis until adults. Zool. Jb. Physiol. 97: 349-368.

Wake MH. 1978. The reproductive biology of *Eleutherodactylus jasperi* (Amphibia, Anura, Leptodactylidae), with comments on the evolution of live-bearing system. J. Herpetol. 12: 121-133.

Wake MH. 1980. The reproductive biology of *Nectophrynoides malcolmi* (Amphibia: Bufonidae) with comments on the evaluation of reproductive modes in the genus *Nectophrynoides*. Copeia 1980 2: 193-209.

Wake MH. 1993. Evolution of oviductal gestation in amphibians. J. Exp. Zool. 266: 394-413.

Wallace RA, JN Dumont. 1968. The induced synthesis and transport of the yolk proteins and their accumulation by the oocytes in *Xenopus laevis*. J. Cell. Physiol. 72 Suppl. 1: 73-89.

Wallace RA, DW Jared. 1968. Studies on amphibian yolk. 7. Serum phosphoprotein synthesis by vitellogenic femàles and estrpgen-treated males of ·*Xenopus laevis*. Can. J. Biochem. 46: 953-959.

Wallace RA, DW Jared. 1969. Studies on amphibian yolk. 8. The estrogen-induced hepatic synthesis of a serum lipophosphoprotein and its selective uptake

270

by the ovary and transformation into yolk platelet proteins in *Xenopus laevis*. Dev. Biol. 19: 498-526.

Wallace RA, JM Nickol, JM Ti Ho, DW Jared. 1972. Studies on amphibian yolk. 10. The relative roles of authosynthesic and heterosynthetic processes during yolk protein assembly by the isolated oocytes. Dev. Biol. 29: 255-272.

Wallace RA, K Selman. 1990. Ultrastructural aspects of oogenesis and oocyte growth in fish and amphibians. J. Electr. Micr. Tech. 16: 175-201.

Wallace RA. 1963a. Studies on amphibian yolk. 3. Resolution of yolk platelet component. Biochem. Biophys. Acta 74: 495-504.

Wallace RA. 1963b. Studies on amphibian yolk. 4. An analysis of the main body component of yolk platelets. Biochem. Biophys. Acta 74: 505-518.

Wang JJ, CY Hsü. 1974. Ultrastructural changes in oocytes and follicle cells during tadpole development. Bull. Inst. Zool. Academia Sinica 13: 75-86.

Wang JJ, CY Hsü, HM Liang. 1980. Fine structural changes in tadpole oocytes after heat treatment. Acta Embryol. Morphol. Exper. 1: 43-57.

Wang L, J Bogerd, HS Choi, JY Seong, JM Soh, SY Chun, M Blomenrohr, BE Troskie, RP Millar, WH Yu, SM McCann, HB Kwon. 2001. Three distinct types of GnRH receptor characterized in the bullfrog. Proc. Nat. Acad. Sci. USA. 98(1): 361-366.

Ward RT. 1962a. The origin of protein and fatty yolk in *Rana pipiens*. 1. Phase microscopy. J. Cell. Biol. 14: 303-308.

Ward RT. 1962b. The origin of protein and fatty yolk in *Rana pipiens*. 2. Electron microscopical and cytochemical observations of young and mature oocytes. J. Cell. Biol. 14: 309-341.

Ward RT. 1978a. The origin of protein and fatty yolk in *Rana pipiens*. 3. Intramitochondrial and primary vesicular yolk formation in frog oocytes. Tissue and Cell 10: 515-524.

Ward RT. 1978b. The origin of protein and fatty yolk in *Rana pipiens*. 4. Secondary vesicular yolk formation in frog oocytes. Tissue and Cell 10: 525-434.

Ward RT, L Opresko, RA Wallace. 1985. A comparison of the proteins of yolk platelets and intramitochondrial crystals in the oocytes of the bullfrog. Dev. Biol. 112: 59-65.

Ward RT, E Ward. 1968. The origin and growth of cortical granules in the oocytes of *Rana pipiens*. J. Microscopie 7: 1021-1030.

Weeks DL, C Bailey, E Bullock, J Dagle, R Gururajan, J Linnen, F Longo. 1995. mRNA localized to the animal hemisphere of *Xenopus laevis* oocytes and early embryos, and the proteins that they encode. In: Localized RNAs, (Ed.) HD Lipshitz. Springer. pp. 173-183.

Whalen R, JW Crim. 1985. Immunocytochemistry of luteinizing hormone-releasing hormone during spontaneous and thyroxine-induced metamorphosis of bullfrogs. J. Exp. Zool. 234: 131-144.

Wiechmann AF, AR Smith. 2001. Melatonin receptor RNA is expressed in photoreceptors and displays a diurnal rhythm in Xenopus retina. Mol. Brain Res. 91: 104-111.

Wiechmann AF, MJ Vrieze, CR Wirsig-Wiechmann. 2003. Differential distribution of melatonin receptors in the pituitary gland of *Xenopus laevis*. Anat. Embryol. 206 (4): 291-299.

Wiley HS, RA Wallace. 1981. The structure of vitellogenin. J. Biol. Chem. 256: 8626-8643.

Winik BC, MF Alcaide, CA Crespo, MF Medina, I Ramos, SN Fernandez. 1999. Ultrastructural changes in the oviductal mucosa throughout the sexual cycle in *Bufo arenarum*. J. Morphol. 239: 61-73.

Wischnitzer S. 1966. The ultrastructure of the cytoplasm of the developing amphibian egg. In: Advances in Morphogenesis. (Eds) M Abercombie and J Brachet, Vol. 5, Academic Press. pp. 131-179.

Witschi E. 1929. Studies on sex differentiation and sex determination in amphibians. I. Development and sexual differentiation of the gonads of *Rana silvatica*. J. Exp. Zool. 52: 135-265.

Witschi E. 1956. Development of Vertebtares. Saunders Company. pp. 24-33.

Wolf DP. 1974. The cortical granule reaction in living eggs of the toad, *Xenopus laevis*. Dev. Biol. 36: 62-71.

Woodruff TK, JP Mather. 1995. Inhibin, activin and the female reproductive axis. Annu. Rev. Physiol. 57: 219-244.

Xavier F. 1977. An exceptional reproductive strategy in Anura: *Nectophrynoides occidentalis* Angel (Bufonidae), an example of adaptation to terrestrial life by viviparity. In: Major Patterns in Vertebrate Evolution, (Eds) MK Hecht, PC Goody and BM Hecht, Plenum Press, New York, pp. 545-442.

Yamaguchi S, JL Hedrick, C Katagiri. 1989. The synthesis and localization of envelope glycoproteins in oocyte of *Xenopus laevis* using immunocyto-chemical methods. Develop. Growth Differ. 31: 85-94.

Yamasaki H, C Katagiri. 1991. Egg exudate-induced reduction of sperm lysin sensitivity in the vitelline coat after fertilization of *Bufo japonicus* and its participation in polyspermy block. J. Exp. Zool. 258: 404-413.

Yang WH, LB Lutz, SR Hammes. 2003. *Xenopus laevis* ovarian CYP17 is a highly potent enzyme expressed exclusively in ooctes. J. Biol. Chem. 278 (11): 9552-9559.

Ying SY. 1988. Inhibins, activins, and follistatins: gonadal proteins modulating the secretion of follicle-stimulating hormone. Endocr. Rev. 9: 267-293.

Yisraeli JK, F Oberman, SP Schwartz, L Havin, Z Elisha. 1995. Vg1 mRNA localization in *Xenopus* oocytes: A paradigm for interion decorating. In: Localized RNAs, (Ed.) HD Lipshitz, Springer Verlag, pp. 157-171.

Yoo MS, HM Kang, HS Choi, JW Kim, BE Troskie, RP Millar, HB Kwon. 2000. Molecular cloning, distribution and pharmacological characterization of a novel gonadotropin-releasing hormone ([Trp8] GnRH) in frog brain. Mol. Cell. Endocrinol. 164: 197-204.

Yoshizaki N, CH Katagiri. 1981. Oviductal contribution to alternation of the vitelline coat in the frog, *Rana japonica*. An electron microscope study. Develop. Growth Diff. 23: 495-506.

Yoshizaki N, CH Katagiri. 1982. Acrosome reaction in sperm of the toad, *Bufo bufo japonicus*. Gam. Res. 6: 343-352.

Yoshizaki N, CH Katagiri. 1984. Necessity of oviductal pars recta for the formation of the fertilization layer in *Xenopus laevis*. Zool. Sci. 1: 255-264.

Yoshizaki N. 1986. Properties of the cortical granule lectin isolated from *Xenopus* eggs. Develop. Growth Differ. 28: 275-283.

Yoshizaki N. 1989. Immunoelectron microscopic demonstration of cortical granule lectins in coelomic, unfertilized and fertilized eggs of *Xenopus laevis*. Develop. Growth Differ. 31: 325-330.

Yuanyou L, L Haoran. 2000. Differences in mGnRH and cGnRH-II contents in pituitaries and discrete brain areas of *Rana rugulosa* W. according to age and stage of maturity. Comp. Biochem. Physiol. C Toxicol. Pharmacol. 125(2): 179-188.

Zbarsky IB, LS Filatova. 1979. A study of karyosphere in the *Rana temporaria* oocytes. Citologia 10: 502-506.

6

Oogenesis and Female Reproductive System in Amphibia—Urodela

Mari Carmen Uribe Aranzábal

A female reproductive system of urodeles consists of two ovaries, two oviducts, spermathecae and cloacal glands. The ovaries are located in the abdominal cavity and connected to the body wall by a peritoneal tissue, the mesovarium. The ovaries lie symmetrically on either side of the midline of the body, in parallel position to the kidneys, the Müllerian ducts (oviducts), the Wolffian ducts (urinary ducts), and the fat bodies (Hope *et al.*, 1963; Lofts, 1984; Wake, 1985; Duellman and Trueb, 1986; Norris, 1997) (Figs. 1, 2A and 3).

The female reproductive system of urodeles shows seasonal cyclic changes based on climatic patterns. Similar to those seen in other vertebrates, the seasonal variation of environmental conditions such as light, temperature, rainfall and humidity, influence the function of the ovaries via the central nervous system and the hypophyseal hormones that stimulate oogenesis and control the reproductive process. An adequate supply of nutrients is also an essential factor for the reproductive process, particularly in females where the vitellogenesis and maturation of the oocytes require abundant energy reserves (Lofts, 1984; Duellman and Trueb, 1986; Licht and Porter, 1987). Temperate zone urodeles and several tropical species show cyclic changes, attaining the maximum activity during autumn and winter, when females produce various clutches of eggs and males produce numerous spermatophores (Adams, 1940; Bruce, 1975; Whittier and Crews, 1987; Lofts, 1984; 1987; Uribe *et al.*, 1991; Sharon *et al.*, 1997; Wake and Dickie, 1998). Whittier and Crews (1987) commented the reproductive cycle, as in *Pseudotriton montanus*, considering that oocyte maturation was completed in autumn, ovulation was delayed over early

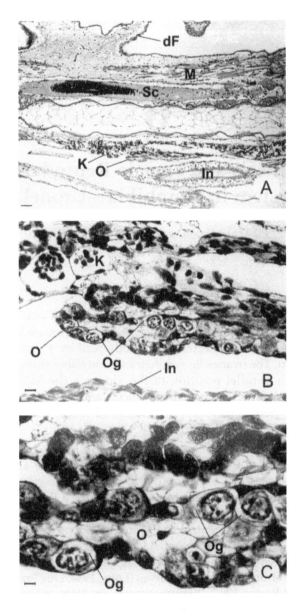

Fig. 1. *Ambystoma dumerilii*. Longitudinal sections of the abdominal region of a larva three weeks after hatching. A-C—Sequence of images in three progressive magnifications. Early ovary (O) at the ventral side of the kidney (K). The ovary is dorsal position to the intestine (In). The spinal cord (Sc), dorsal muscles (M) and dorsal fin (dF) are seen. Note the position of the oogonia (Og) at the periphery of the ovarian cortex. Masson's trichrome staining. Bars—(A) 0.3 mm, (B) 20 μm, (C) 10 μm.

winter, and hatching occurred in spring, under favorable conditions.

There are some aseasonal reproducers, such as cave-dwelling salamanders, *e.g.*, *Euproctus asper* (Lofts, 1984; Wake and Dickie, 1998). In other tropical species, as plethodontid salamanders, the reproduction is associated with the rainy season, *e.g.*, *Bolitoglossa rostrata* lay eggs early in the dry season and hatching occurs in the wet season (Whittier and Crews, 1987). The maturation of the oocytes in some cases requires more than one year, as it is described in several temperate and tropical female plethodontid salamanders that reproduce biennially (Bruce, 1975; Whittier and Crews, 1987).

Structure of Ovary

The structure and cyclic changes of the ovaries in urodeles are determinated by major functions, as in most other animals, including: 1) the process of oogenesis, when oogonia develop into mature oocytes; 2) ovulation, and 3) the hormonal activity associated with the secretion of female sex hormones essential for reproduction.

The ovaries are elongated and irregular in shape. Their size changes seasonally according to the phase of reproductive cycle. The previtello-genic ovaries are thin, but during the breeding season the ovaries grow in size occupying most of the abdominal cavity. At that time the ovaries contain not only previtellogenic oocytes, but also numerous vitellogenic oocytes. The ovaries during previtellogenesis are light-yellow in color, whereas during vitellogenesis in species that produce pigmented eggs they show mixed color of light-yellow corresponding to the vegetal hemisphere of the oocytes, and dark brown corresponding to the pigmented animal hemisphere of the oocytes. During early summer, primary oocytes grow slowly and the germ-line cells are represented by oogonia and small previtellogenic diplotene oocytes in follicles. Consequently, the ovarian weight is low at this time. Yolk deposition starts during late summer and autumn when many oocytes contain various amounts of yolk, according to the stage of vitellogenesis. Then, in autumn and winter, the ovaries attain the maximum size, containing an abundance of late vitellogenic oocytes. After ovulation, the follicular cells transform into *corpora lutea* and become lipoidal cells surrounded by the theca.

Histological examinations of ovaries of urodeles (Adams, 1940; Hope *et al.*, 1963; Joly and Picheral, 1972; Lofts, 1974, 1984; Dodd, 1977; Guraya, 1978; Wallace, 1978; Wallace and Selman, 1990; Bement and Capco, 1990; Sharon *et al.*, 1997; Uribe, 2001, 2003) is composed of several ovarian sacs containing a lumen as a central cavity filled with lymph (Figs. 2A, 11A). The wall of the ovary is limited externally by cuboidal epithelium (Fig. 2),

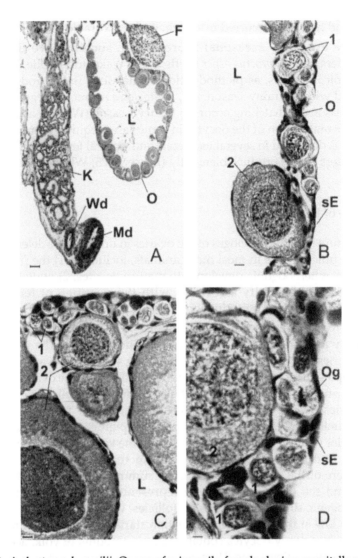

Fig. 2. *Ambystoma dumerilii.* Ovary of a juvenile female during previtellogenesis. A-D—Sequence of images in three progressive magnifications. Anatomical relationships of the ovary in the abdominal cavity. Ovary (O), kidney (K), fat body (F), Wolffian duct (Wd), Müllerian duct (Md). The saccular structure of the ovary is evident, with the central lumen (L), surrounded by the ovarian wall. The ovarian wall contains oocytes at different stages of development. Observe oogonia (Og) during mitotic metaphase, abundant early meiotic oocytes in previtellogenic stage 1 (1) and previtellogenesis 2 (2) with a large nucleus and lampbrush diplotene chromosomes. Squamous epithelium (sE) limits the lumen (L) of the ovary. Hematoxylin and eosin (H+E) staining. Bars—(A) 0.1 mm, (B, C) 20 µm, (D) 10 µm.

originating from the coelomic mesothelium. During early embryogenesis, the coelomic mesothelium is colonized by the primordial germ cells giving rise to the gonadal anlage containing germ and somatic cells (Fig. 1). Thus, the cortex of an early ovary consists of oogonia, early oocytes formed when oogonia enter meiosis, bordered by somatic external epithelial cells that covers the gonad, and internal ones that lines the lumen (Figs. 2B,C,D) supported by basal lamina. Subjacent to the basal lamina, the stroma shows vascularized loose connective tissue containing collagen fibers, fibroblasts, and various types of cells of the immunological system (macrophages and limphocytes), melanocytes, lymphatic vessels, and nerves. The saccular structure of the ovary provides the space for the progressively growing oocytes, which are supported by a peduncle at the external epithelium and project into the lumen of the ovary (Figs. 2, 11A) (Dodd, 1977; Lofts, 1984). The cortex of the ovary during oogenesis contains follicles at different stages of development, including degenerating (atretic) follicles (Hope et al., 1963; Lofts, 1984; Uribe, 2001, 2003).

Female Reproductive Tract

During ovulation, when follicles rupture, the ovary releases mature oocytes into the body cavity, where they are propelled to the cephalic end of the oviduct, the infundibulum, by the cilia of the coelomic epithelium. Then the oocytes enter the oviducts and continue their transit to the caudal portions of the oviducts. The urodeles show interesting features of reproduction that define several morphological and physiological characteristcs of the female reproductive tract. These are: 1) the majority of species are oviparous, but there are also larviparous and viviparous species that belong to the family Salamandridae, such as *Mertensiella luschani*, *Salamandra atra*, *S. salamandra*, and *S. s. infraimmaculata* (Wake, 1993; Greven and Guex, 1994; Sharon et al., 1997; Greven, 1998, 2003). 2) Even though oviparity is the most frequent type of reproduction in urodeles, internal fertilization occurs in a high percentage of species (90%) (Lofts, 1984; Duellman and Trueb, 1986). 3) The internal fertilization is correlated with specific adaptations in males. The males of most species of urodeles produce spermatophores secreted by the cloacal gland complex; these are gelatinous bodies that contain spermatozoa (Zalisko et al., 1984; Joly et al., 1994). The spermatophores are deposited either directly into the female oviducts by cloacal apposition, as is the case of *Salamandra atra* and *Amphiuma tridactylum*, or males first deposit the spermatophores externally, and subsequently a female picks up the spermatophore with her cloacal labia and introduces it to the oviduct where fertilization occurs, as is observed in *S. salamandra* or *Pleurodeles waltlii* (Lofts, 1984; Wake, 1993). 4) The internal fertilization supported by the presence of the spermatheca in females; the spermatheca is composed of a cluster of the cloacal glands that store spermatozoa (Trauth, 1983, 1984; Sever, 1994, 2003; Brizzi et al., 1995). 5) The internal fertilization occurs in the oviduct and usually takes place in the caudal part of the oviduct or in the cloacal cavity immediately before oviposition (Duellman and Trueb, 1986; Joly et al., 1994; Wake and Dickie, 1998). 6) According to Wake and Dickie (1998), the ancestral salamander condition was the external fertilization. The species of urodeles that show external fertilization belong to the families Hynobiidae, Cryptobranchidae, and Sirenidae (Lofts, 1984; Duellman and Trueb, 1986; Sever et al., 1996), in which oviposition and extrusion of spermatozoa or spermatophores occur simultaneously and fertilization is external. 7) Oviductal features change according to the reproductive cycle. During non-breeding seasons, the oviducts are less folded, narrower and show scarce secretory activity, whereas during the breeding season they become larger, thicker, and with an intense secretory activity (Adams, 1940; Brandon, 1970; Duellman and Trueb, 1986; Norris, 1997).

Oviducts

The oviducts are paired and elongated organs, extended on each side of the midline of the body cavity, attached to the dorsal body wall by the mesotubaria. The oviducts originate from the Müllerian ducts under the influence of the ovarian estrogens during ontogenesis. The Müllerian ducts develop as paired invaginations of coelomic genital ridge epithelium near the pronephric ducts (Figs. 2A, 3). These invaginations grow caudally, until they reach the cloaca, to which they connect (Adams, 1940; Potemkina, 1967; Sever et al., 1996; Wake and Dickie, 1998; Greven, 2003).

According to the terminology proposed by Sever et al. (1996), Wake and Dickie (1998), and Greven (2003), the oviducts of urodeles are divided into four morphological and functionally different regions: 1) the anterior infundibulum or ostium (Fig. 4); 2) the *pars recta* or atrium (Figs. 5, 6); 3) the *pars convoluta* or ampulla, or middle glandular convoluted region (Figs. 7, 8, 9); and, 4) the ovisac or posterior aglandular region (Fig. 10). The infundibulum is formed by the expanded cephalic end of the oviduct described as a funnel-like opening that receives the eggs after ovulation. The atrium is a straight region, narrower than the infundibulum. The ampulla is the largest region of the oviduct, showing large convolutions. Two types of convolutions have been described in urodeles: the *Ambystoma*-type showing large and thick loops taking almost the entire width of the body cavity, and the *Triturus*-type showing thinner and shorter loops (Greven, 2002, 2003). The ovisac is short and straight. Both ovisacs approach in the caudal portion joining the cloaca.

The histology of the oviduct were studied in several species of

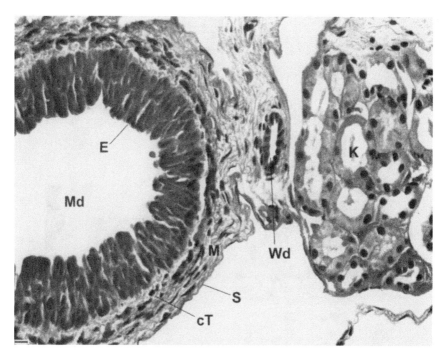

Fig. 3. *Ambystoma mexicanum.* Cross section of ducts of a juvenile female during previtellogenesis. Wolffian duct (Wd) and Müllerian duct (Md) adjacent to the kidney (K). Folded epithelium (E), connective tissue (cT), muscle fibers (M), and serosa (S) are seen. H+E staining. Bar 10 μm.

urodeles, such as *Triturus torosus* (McCurdy, 1931); *Triturus pirrhogaster* (Kambara, 1956); *Hydromantes genei* (Vilter and Thorn, 1967); *Salamandra atra, S.salamandra* (Vilter, 1967; Greven, 1977, 1998; Greven and Guex, 1994); *Pleurodeles waltlii* (Jego, 1974); *Ambystoma tigrinum* (Guillette *et al.*, 1985); *Eurycea cirrigera* (Sever, 1988); *Ambystoma mexicanum* (Uribe *et al.*, 1989); *Siren intermedia* (Sever *et al.*, 1996); *Ambystoma dumerilii* (Uribe, 2001). Wake and Dickie (1998) and Greven (2003) summirize the features of the oviduct of several species of urodeles.

The walls of the urodelan oviducts are composed of three tissue layers (Mc Curdy, 1931; Kambara, 1956; Vilter, 1967; Uribe *et al.*, 1989; Sever *et al.*, 1996; Greven, 1998, 2003; Uribe, 2001). The inner layer is the mucosa, which is composed of luminal simple columnar epithelium, glands and vascularized connective tissue; the middle layer is the myometrium formed by smooth muscle fibers; and the outer layer is the serosa composed of thin connective tissue and the squamous mesothelial epithelium.

Infundibulum. The infundibulum (Fig. 4) is formed by a thin wall that possesses a highly folded mucosa. The mucosal folds are irregular in shape

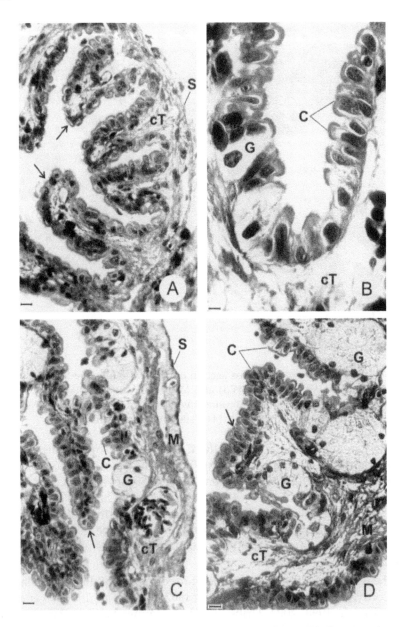

Fig. 4. Oviduct of an adult female *Ambystoma dumerilii*. (A-D) Cross sections of various levels of the infundibulum in cephalo-caudal axis. The wall shows large and irregular folds (arrows limited by columnar epithelium containing abundant ciliated cells (C) and scarce gland cells (G) which increase in number caudally. Connective tissue (cT), muscle fibers (M) and serosa (S). Masson's trichrome. Bars—(A, C, D) 10 μm, (B) 20 μm.

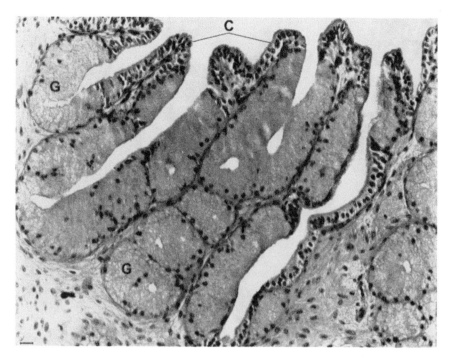

Fig. 5. Oviduct of an adult female of *Ambystoma mexicanum*. Cross section of the *pars recta*. The wall shows thicker folds and more numerous gland cells (G) compared to those seen in the previous region. The columnar epithelium contains also ciliated cells (C). H+E. Bar 10 μm.

and size, and project into the lumen. The mucosa of the infundibulum is limited by a cubic or columnar epithelium with ciliated and secretory cells and the epithelium is separated from the stroma by a basal lamina. The stroma is formed by loose connective tissue that contains blood vessels. There is a thin myometrium formed by smooth muscle fibers. The myometrium consists of one or two layers of muscle cells in circular arrangement. The serosa surrounds the periphery of the wall of the infundibulum.

Pars recta. The *pars recta* (Figs. 5, 6) is characterized as a region of transition between the aglandular infundibulum and the highly glandular ampulla. The wall of the *pars recta* is folded, as in the infundibulum, but the folds are shorter and thicker and the epithelium shows some glandular cells, occuring singly or in groups, and interspersed among the cubic epithelial cells. The glandular cells, increasing in number progressivelly along this region, characterize the transition to the next oviducal region, the ampulla. The secretory cells become larger than the non-secretory cells of the epithelium, their nuclei are basal, and the cytoplasm shows abundant

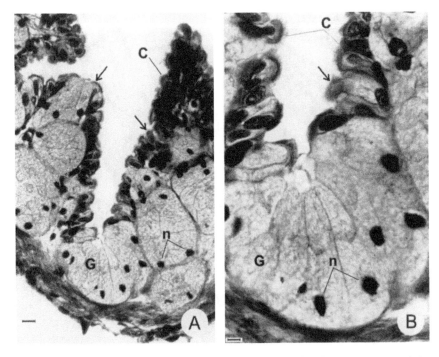

Fig. 6. Oviduct of an adult female *Ambystoma dumerilii*. Cross sections of the *pars recta*. Thick folds contain numerous and large gland cells (G), their nuclei (n) are basal. The ciliated columnar epithelium (C) cells are smaller compared to the glandular cells. Masson's trichrome. Bars—(A, B) 20 μm.

lightly stained globules. The myometrium and the serosa are similar to those seen in the previous region.

Pars convoluta. This region, also called the ampulla or middle glandular convoluted region, (Figs. 7, 8, and 9) shows progressively increasing number of exocrine tubular glands that form the thick wall of the oviduct. The luminal epithelium contains scarce ciliated cells scattered among numerous glandular cells. The glandular cells are large, prismatic in shape, and have basal nuclei and light vacuolated cytoplasm. The glands of the *pars convoluta* are greatly hypertrophied during the reproductive season and surrounded by thin layers of vascularized connective tissue. The myometrium and the serosa are similar to those seen in the previous regions. The size and quantity of glands diminishes in caudal direction. The glands open into the lumen, secreting the abundant components of the jelly-coats that surround the eggs during their transit along this region. A progesterone-like steroid synthesized by the ovary stimulates the oviductal glands to produce the jelly-coats (Saidapur and Nadkarni, 1974). The jelly-coats could be composed of several sublayers, and each layer has

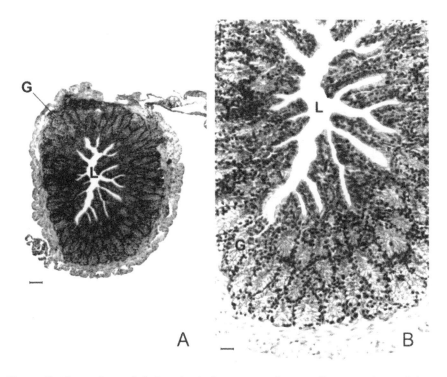

Fig. 7. Oviduct of an adult female *Ambystoma mexicanum*. Cross sections of the *pars convoluta* during previtellogenesis. A, B—The glands (G) form the thicker wall of the oviduct. Lumen (L). H+E staining. Bars—(A) 0.3 mm, (B) 0.1 mm.

specific carbohydate components that accumulate during the passage through different regions of the oviduct (Wake and Dickie, 1998; Watanabe and Onitake, 2003). The glands produce glycosylated proteins, acid and neutral mucopolysaccharides, sulphated mucopolysaccharides and neutral carbohydrates (Greven, 2003).

Greven (1998) reported that the spermatozoa did not fertilize eggs lacking jelly-coats. Watanabe and Onitake (2003) revealed that in the mechanics of internal fertilization in urodeles, the egg-spermatozoa interaction is induced in the jelly-coat, suggesting that the signal for the initiation of spermatozoa motility is present in the egg jelly. In case of the internal fertilization, hundreds of spermatozoa are deposited onto the surface of the jelly-coats, and they pass across them. After oviposition, when the eggs are released into the environment, the jelly-coats have other important functions, such as anchorage to the substrate by its stickiness, thermal insulation (Salthe, 1963), and protection against mechanical damage, predators (Ward and Sexton, 1981), and prevention of dessiccation by acting as a water reserve (Heatwole, 1961). The ampulla can be sub-

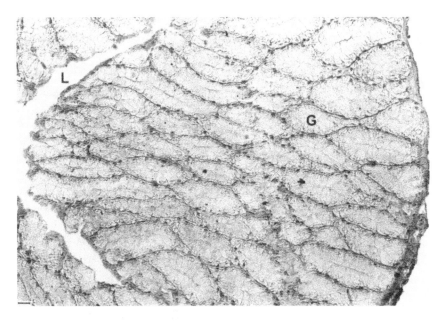

Fig 8. Oviduct of an adult female *Ambystoma dumerilii*. Cross sections of the *pars convoluta* during vitellogenesis. Note the hypertrophy of the glands (G) compared with the previous stage. The thickness of the wall attains the maximum size. Lumen (L). H+E staining. Bar 0.1 mm.

divided into different number of functional portions, depending on species. These subdivisions correspond to different type of secretions that form the jelly-coats (summarized by Greven, 1998; Wake and Dickie, 1998). Some authors subdivided the ampulla into two (Kambara, 1956), three (Vilter, 1967), or four (Adams, 1940; Humphries and Hughes, 1959), portions based on the nature of secretory products.

Ovisac. The ovisac (Fig. 10) is the most caudal region of the oviduct. The ovisac forms a short and straight aglandular region. The mucosa is limited by columnar epithelium with ciliated cells. The cilia are large and abundant. In contrast to the previous regions, the ovisac contains thick layers of connective tissue and myometrium. The myometrium is arranged into the inner circular and the outer longitudinal layers. The increase in the number of muscle fibers in this region supports oviposition. Neuro-hypophysial peptides, such as arginine, vasotocin and mesotocin, stimulate muscular contractions during oviposition (Heller *et al.*, 1970; Guillette *et al.*, 1985). Scattered, irregular stellate-shaped melanocytes are seen at the periphery of this region. They possess long and thin cytoplasmic processes that extend into the connective tissue. The cytoplasm of these cells contains dark-brown fine granules of melanin.

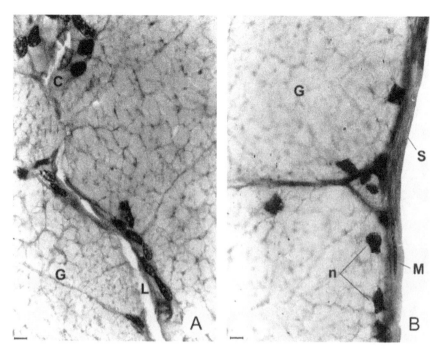

Fig. 9. Oviduct of an adult female *Ambystoma dumerilii*. Details of the regions similar to those indicated in Figure 8. A—Apical region of the wall. Epithelial ciliated cells (C), and reduced lumen (L). B—Periphery of the wall. Hypertrophied gland cells (G), with basal nuclei (n). Thin muscle (M) and serosa (S) layers are situated at the periphery. H+E staining. Bar 10 μm.

Oviducts of both oviparous and viviparous (including larviparous) species have similar structure, but some modifications are observed in viviparous species. All regions of the oviduct of viviparous species is composed of simple epithelium, connective tissue, scarce muscle cells and a squamous peritoneal epithelium, which show unique adaptations to the presence and retention of developing embryos. The oviduct of viviparous species as *Salamandra atra, S. salamandra*, and *Mertensiella luschani* were described by Vilter (1967); Greven and Guex (1994); Guex and Greven (1994); Greven (1998, 2003). The *pars convoluta* of viviparous species produces the egg jelly, which also is present in oviparous species. However, in viviparous species this oviductal region has fewer convolutions, as well as fewer gland cells than those seen in oviparous species. These authors mentioned that the *pars convoluta* is differentiated into four parts according to the appearence of epithelial cells; these parts are *pars convoluta* I, II, III, and the uterus where the embryo develops. The parts I, II, and III are glandular and contain tubular glands. The secretions produced by these

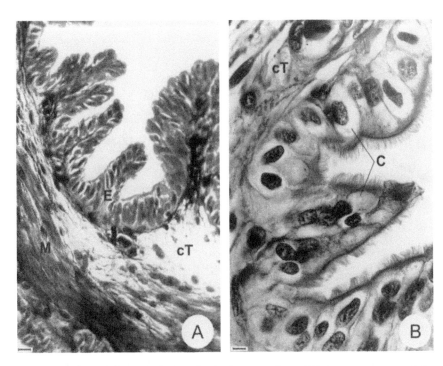

Fig. 10. Oviduct of an adult female *Ambystoma dumerilii*. Cross sections of the ovisac A—The wall contains columnar epithelium (E), thick layers of connective tissue (cT) and muscle (M). The muscle layers show inner circular and outer longitudinal arrangement. B—The epithelium is formed by columnar ciliated cells (C). Masson's trichrome. Bars—(A) 10 μm, (B) 20 μm.

three parts of the *pars convoluta* differ histologically and histochemically. The fourth part of this region, the uterus, has characteristic anterior part called the *zona trophica*. During gestation, the *zona trophica* of the uterus contains irregularly shaped giant epithelial cells, and thicker and more compact connective tissue layer. The uterus lacks glands, but epithelial cells produce carbohydrate-rich material. The uterine secretion may contribute to the fluid surrounding the young, but does not seem to support nutrition. Sources of nutrition during gestation in the uterus could be provided solely by yolk of the egg accumulated during oogenesis (lecithotrophy), as is the case of *S. salamandra,* or additionally by ingestion of various materials provided by the mother (matrotrophy), as is the case of *S. atra*. The examples of the maternal support involve the remaining unfertilized eggs (oophagy), ingestion of siblings (adelphophagy), and ingestions of uterine secretions of epithelial cells derived from a specialized uterine trophic zone (epitheliophagy) (Wake, 1985; Guex and Chen, 1986; Joly *et al.*, 1994; Greven, 1998). Guex and Chen (1986) suggested that the *zona trophica* is stimulated during gestation by the presence of the embryo. When the yolk mass is exhausted, the embryo is nourished with epithelial cells of the *zona trophica* until the end of pregnancy. Regeneration and detachment of the uterine epithelial cells occur simultaneously during gestation in the *zona trophica*. The density of blood vessels in the connective tissue increases in the uterine portion (Greven and Guex, 1994; Greven, 2002). The viviparity observed in the genus *Salamandra* displays various degrees of young development, which permits giving birth to aquatic larvae, as is the case of the subspecies *S.s. fastuosa*, or metamorphosed offspring on land as is the case of *S. atra*. Giving birth to more advanced offspring seems to be advantageous in the protection of the offspring (Joly *et al.*, 1994).

Cloacal Glands

The presence of exocrine cloacal glands is a common anatomical feature in urodelan females. Functional morphology, phylogeny, and cyclic changes of the cloacal glands, including the spermathecae, have been widely investigated by Sever (see reviews 1991, 1994, 2002, 2003). Three types of cloacal glands have been described, 1) the spermathecae (see below); 2) the ventral glands, and 3) the dorsal glands (Trauth *et al.*, 1994). These glands are composed of the tubulo-acinar structures that connect to the cloaca by a tube. Glandular morphology and secretory activity of cloacal glands change during the reproductive cycle, during which the maximal development corresponds to the peak of reproductive activity (Trauth *et al.*, 1994).

The ventral glands secrete onto the medial portion of the cloacal chamber. The structure of these glands corresponds to simple, tubulo-alveolar, exocrine glands that produce merocrine secretion, and the secretion is accumulated as globules in the lumen. The tubuloalveolar portions are limited by cubic epithelium that contains abundant secetory material during the breeding season. Trauth *et al.* (1994) mentioned that the ventral glands of *Ambystoma talpoideum* secrete throughout the year, but are hypertrophied and more active during the breeding season. These authors hypothesized that ventral glands may have great significance and may be involved in territorial marking or individual recognition throughout the year. Sever (1988), who studied the ventral glands of *Eurycea cirrigera*, added that these glands were sources of a potent pheromone to attract males during the mating period, suggesting that this may be the ancestral function of the cloacal glands (Sever, 1988). Trauth *et al.* 1994; Norris, 1997 agreed with Sever (1988) suggesting that the ventral glands produce pheromones.

The dorsal glands, which are best developed in ambystomatids and rudimentary in plethodontids, are described as small number of short tubes connected to a single duct. The structure of these glands is similar to that described in the ventral glands, corresponding to simple, tubuloalveolar, exocrine glands that produce merocrine secretion, but differ from these by less expanded alveolar regions and by the possible presence of some spermatozoa (Trauth *et al.*, 1994). Dorsal glands secrete cephally and/or caudally to the spermathecae.

Spermathecae

Most of the salamander species have sperm-storage glandular structures called spermathecae, which open dorso-laterally into the cloacal lumen. The spermathecae are not considered as parts of the oviduct (*sensu* Müllerian duct) because they are cloacal derivatives (Wake and Dickie, 1998). Sever (1994) described two types of spermathecae: simple containing tubuloalveolar glands (found in most species excluding members of the Plethodontidae family) and complex containing alveolar glands (characteristic of the Plethodontidae). The simple spermathecae are formed by several acinar portions, as in *Amphiuma phloleter*, the complex spermathecae are formed by a common tube that branches into tubules that expand into distal bulbs forming acinar, as in *Desmognathus ocoee*. The acinar epithelium is formed by cubic or columnar cells with irregular apical borders. Secretions of the spermathecae are also of the merocrine type (Trauth *et al.*, 1994). Adjacent to the basal lamina is a sheet of myoepithelial cells. The ducts are lined by ciliated columnar epithelium, active in the transportation of spermatozoa into and out of the spermatheca (Trauth, 1983; Sever, 1991). During the breeding season, the lumen of the ducts contains spermatozoa. Trauth *et al.* (1994) observed in *Ambystoma talpoideum* the presence of spermatozoa in the spermathecae of all females examined during the breeding season (from November to March), whereas no females contained spermatozoa during the non-breeding season (from May to October). After ovulation, internal fertilization is possible without a new mating because spermatozoa are released from the spermatheca (Trauth *et al.*, 1994; Joly *et al.*, 1994). Therefore, these structures may compensate for the asynchrony between the time of copulation and ovulation. This could be also a strategy when the probability of finding males is low. Once inseminated, females can wait until adequate environmental conditions for ovulation and development of embryos occur (Trauth, 1983; Trauth *et al.*, 1994). During breeding season, when a male deposits the spermatophore near a female, she takes up the spermatophores by her cloacal lips. Inside the female's cloaca, the cap matrix of the spermatophore degenerates and spermatozoa are released. The spermatozoa then migrate along the dorsal wall of the cloaca into the spermatheca, maintaining them stored and quiescent (Guex and Greven, 1994; Joly *et al.*, 1994; Sever *et al.*, 1995, 1996). Spermatozoa may be kept in the spermatheca during various periods of time, lasting from few hours, several days, or weeks, up to six months as in *Ambystoma opacum* (Sever *et al.*, 1995), or even exceed two years as in *S. salamandra* (Guex and Greven, 1994).

After storage, the sperm is released from the spermatheca onto the egg jelly-coats surface, where the selection of spermatozoa occurs (Watanabe and Onitake, 2003. The oocytes can be fertilized when they pass into the cloacal cavity before oviposition or inside the *pars convoluta* of the oviduct, when spermatozoa leave the spermatheca and migrate to the anterior regions of the oviducts (Joly *et al.*, 1994; Sever *et al.*, 1995). The secretory activity of the epithelial cells of spermatheca is evident by the presence of secretory granules during the sperm storage period (Duellman and Trueb, 1986; Sever, 1991; Sever *et al.*, 1995). The function of the secretions of the spermatheca is unknown. Sever and Klopfer (1993) proposed that it provides the chemical conditions necessary for sperm quiescence in the lumen. Kruczyński *et al.* (1986) detected the presence of zinc by histochemical methods in the epithelial cells of the spermatheca of *S. salamandra*. They proposed that zinc may be important for long sperm storage and for possible reactivation of spermatozoa.

Oogenesis

Early Oogenesis

Primary oogonia are scattered in the periphery of ovarian cortex as individual cells, whereas secondary oogonia form small groups (nests). The oogonia are the earliest stage of female germ cells. They attain 9-12 µm in diameter, with lightly stained cytoplasm. The nucleus is ovoidal in shape, containing a nucleolus and fine granular chromatin. The primary oogonia divide mitotically, increasing the number of germ cells (some of them will stay as primary oogonia, whereas the other give rise to the secondary oogonia). The oogonial proliferation increases after ovulation, giving rise to new cell nests for the next generations of oocytes, as was described for *Taricha torosa* (Miller and Robbins, 1954). Secondary oogonia transform into primary oocytes and enter meiosis (Figs. 2B,C,D, 11C). During leptotene-pachytene stages the homologous chromosomes are clearly visible. Leptotene-pachytene oocytes transformed from secondary oogonia, which formed nests of cells connected by intercellular bridges. At the end of pachytene, or beginning of diplotene, the intercellular bridges disconnect and each diplotene cell becomes surrounded by its own follicle cells.

Early diplotene oocytes are surrounded by epithelial prefollicular cells

(Tokarz, 1978; Chieffi and Pierantoni, 1987; Wallace and Selman, 1990; Uribe, 2003). The meiotic oocytes advance until diplotene stage, at which they are arrested. The oocytes remain arrested until they receive the hormonal signal to reinitiate meiosis (Tokarz, 1978; Bement and Capco, 1990; Wallace and Selman, 1990). Similarly to the observations provided by Grier (2000), who extensively examined and discussed the events that lead to the initiation of the folliculogenesis in the teleost *Centropomus undecimalis*, we observed early follicles in the species of ambystomatids *Ambystoma mexicanum* and *A. dumerilii* (Uribe, 2003). The folliculogenesis is completed when somatic prefollicular cells surround the oocyte and become follicle (or *granulosa*) cells. During this process, the basal lamina completely encloses the external surface of the follicle cells. Consequently, according to Grier (2000), the structure of an ovarian follicle consists of an oocyte surrounded by a layer of follicular cells and the basal lamina. Subsequently, a thin thecal layer, derived from the stroma, develops around the basal lamina of the follicle. The theca is composed of vascularized connective tissue and secretory cells.

The presence of oogonia in the cortex of the ovaries of adult female urodeles gives rise to successive generations of oocytes during the adult life, resulting in a great number of oocytes during each reproductive season. This aspect differentiates urodeles from mammals, birds and elasmobranch fish where oogonia initiate the meiotic process in young females, consequently the proliferation of oocytes is not possible in mature females (Dodd, 1977; Tokarz, 1978; Norris, 1997; Grier, 2000). Also in anuran amphibians a stock of early diplotene oocytes forms a pool sufficient for the whole life of a female; every year one portion of the oocytes is recruited for vitellogenesis, and this portion will give rise to the mature oocytes in the next season (see 'Oogenesis and Female Reproductive System in Anura' in this volume).

Diplotene Oocytes

During oogenesis, the oocytes of urodeles increase greatly in volume attaining approximately 1.50-3 mm in diameter in the majority of species studied so far. However, in some species the oocytes could attain 5 mm, as in *Necturus maculosus* and *Bolitoglossa subpalmata*, or 6 mm as in *Dicamptodon ensatus* (see review by Duellman and Trueb, 1986). Duellman and Trueb (1986) observed that the ova of tropical species are generally smaller than those of the temperate zone species. Wake (1993) reported that viviparous amphibians have small ova and small clutches. Progression from oogonia to mature oocytes may require three years or more for completion (Norris, 1997). The oocyte shows progressive growth divided into the two main stages of the oogenesis: previtellogenesis (Figs. 11, 12), and vitellogenesis

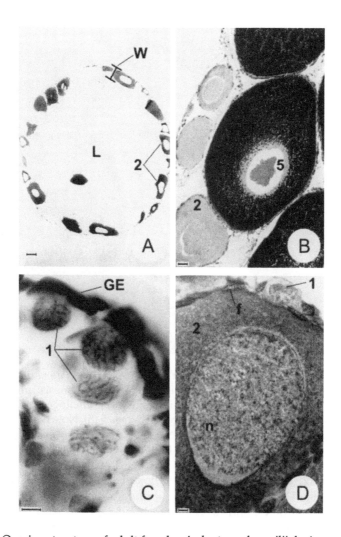

Fig. 11. Ovarian structure of adult females *Ambystoma dumerilii* during previtello-genesis (A) and various stages of vitellogenesis (B-D). Wide ovarian lumen (L) limited by the wall (W) containing previtellogenic stage 2 oocytes (2). B— Previtellogenic oocyte, stage 2 (2), and vitellogenic oocyte stage 5 (5). Note the abundant yolk platelets in the late vitellogenic oocytes. C—Nest of early oocytes showing spherical nuclei with pachytene chromosomes (Previtellogenesis, stage 1) (1), and somatic cells of the germinal epithelium (GE). D—Oocyte during previtellogenesis, stage 2 (2), containing abundant ooplasm and spherical nuclei (n) with lampbrush chromosomes. Squamous follicular (f) cells are seen around the oocyte. Compare the size of the oocyte in previtellogenesis, stage 1 (1), situated in the germinal epithelium (GE), with that in previtellogenesis, stage 2 (2). (A) H+E staining. Bar 0.1 mm. (B) Masson's trichrome. Bar 20 m. (C, D) H+E staining. Bar 10 μm.

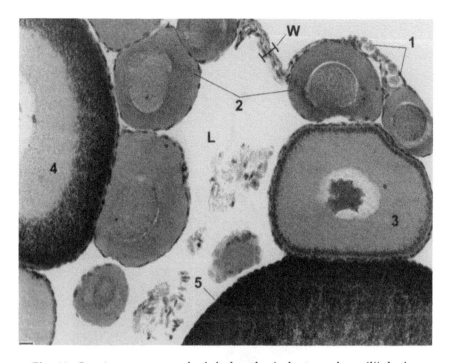

Fig. 12. Ovarian structure of adult female *Ambystoma dumerilii* during vitellogenesis. Oocytes during various stages of oogenesis: previtellogenesis, stage 1 (1), previtellogenesis, stage 2 (2), vitellogenesis, stage 3 (3), vitellogenesis, stage 4 (4), vitellogenesis, stage 5 (5). Note the irregularly folded contour of the nucleus in the oocyte in vitellogenesis, stage, Wall (W) of the ovary and lumen (L). H+E staining. Bar 20 μm.

(Figs. 11B, 12, 13, 14). During previtellogenesis, the cytoplasm volume increases (Fig. 11D, 12) accumulating abundant messengers RNAs (in the form of mRNA, tRNA and rRNA), mitochondria, endoplasmic reticulum, enzymes and diverse precursors of macromolecular synthesis.

Vitellogenesis is a process of accumulation of yolk precursors and formation of yolk platelets (Figs. 11B, 12, 13A,C, 14). The overall material stored in the ooplasm for the nutrition and the high metabolic activity during development of the embryo is composed of about 45% phosphoproteins, 25% lipids, and 8% glycogen. Energy stores are obtained during vitellogenesis *via* the uptake of yolk and the formation of lipid droplets. The precursors of the yolk platelets are synthesized in the liver and are transported by the circulatory system to the follicles in the ovary. An active pinocytotic transport of materials across the endothelial cells of the thecal capillaries of the growing follicles was observed in several species, such as *Necturus maculosus, Triton alpestris, Triturus viridescens, T. vulgaris,* and *T.*

pyrrhogaster (see review by Wallace, 1978). In he next step the follicular cells permit this material to go through large intercellular spaces to the oocyte surface where it is incorporated by endocytosis. The oocyte surface forms numerous projections, the microvilli, which increase the area available for the uptake of yolk precursors. The deposition of yolk begins at the periphery of the oocyte, advancing near to the periphery of the nucleus. During vitellogenesis, the animal-vegetal polarity is established. The yolk platelets are larger and more abundant in the vegetal pole; the animal pole contains the nucleus.

The follicular cells, which surrounds the oocyte, form a single layer of squamous (Figs. 11D, 13A) or cubic cells (Figs. 13C,D), closely apposed to the oocyte membrane. The follicular cells remain as a monolayer throughout the entire period of oocyte development. These cells are active in synthesis of lipids, carbohydrates and proteins, substances that pass to the oocyte by diffusion and endocytosis. The space occupied by microvilli form the *zona radiata*, and a thin layer of amorphous material begins to accumulate between the oocyte and the follicular cells, forming the homogeneous layer of oocyte envelope (vitelline membrane). The *zona radiata* together with the oocyte envelope are seen in the light microscope as a transparent zone and is often referred as *zona pellucida*. Externally to the follicular epithelium there is a thin theca formed by vascularized and fibrous connective tissue. Diplotene oocytes in *Ambystoma mexicanum* and *A. dumerilii* are divided into six stages according to Uribe (2003), who modified and adapted staging developed by Dodd (1977) and Sharon *et al.* (1997). Two stages that occur during previtellogenesis (Stage 1 and 2), and four stages that occur during vitellogenesis (Stages 3-6) are described below. Although it is convenient to divide oogenesis into stages, the process is certainly continuous (Hope *et al.*, 1963).

Stage 1: The early diplotene oocyte, 50-55 μm in diameter, contains the central nucleus, which contains one or two nucleoli. It is surrounded by a follicular layer, which consists of squamous cells covered by the basal lamina. Diplotene chromosomes have numerous loops extended laterally, and representing sections of transcriptional DNA. Because of the morphology, they are called lampbrush chromosomes. The ooplasm is lightly stained with the exception of a basophilic irregular structure adjacent to the nucleus. This structure is known as the Balbiani body, and sometimes it is also called the vitelline body or yolk nucleus. It is composed of mitochondria, ribosomes, fibrogranular material and Golgi elements. As was suggested by Guraya (1978), this region is the site of metabolic activities involved in the initial formation of cytoplasmic organelles. Subsequently, the Balbiani body migrates to the oocyte periphery where it disperses.

294

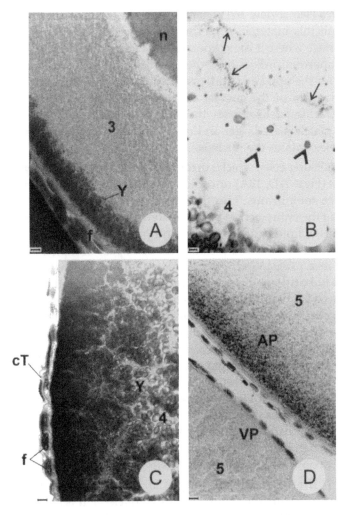

Fig. 13. Ovarian structure of adult female *Ambystoma dumerilii* during vitello-genesis. A—Oocyte at stage 3 (3). Early deposition of yolk platelets (Y) at the periphery of the ooplasm. Nucleus (n) of the oocyte. The follicular cells (f) show flattened nuclei. B—Nucleus of an oocyte in vitellogenesis, stage 4 (4). Lampbrush chromosomes (arrow) and nucleoli (arrowhead) are seen. C—Periphery of an oocyte at vitellogenesis stage 4 (4). The yolk platelets (Y) are abundant. Ovoidal nuclei of follicular cells (f). Squamous nuclei of fibroblasts of connective tissue (cT). D—Peripheral regions of two oocytes at vitellogenesis stage 5 (5). One of them, in the top-right side of the figure shows pigment granules and small yolk platelets, characteristic of the animal pole (AP). The other oocyte, at the bottom-left side of the figure does not contain pigment granules and the yolk platelets are bigger, characteristic of the vegetal pole (VP). (A, C) H+E staining. Bar 20 μm; (B) H+E staining. Bar 10 μm; (D) Alcian-blue. Bar 20 μm.

Stage 2: The oocyte diameter increases attaining 200-300 μm. The oocyte contains a larger number of cytoplasmic organelles, and abundant RNAs and proteins. The nucleus is large and round, containing numerous nucleoli situated along the periphery of the nuclear membrane, and chromosomes, which show the characteristic lampbrush appearance (Fig. 11D). The ooplasm is basophilic, homogeneous and fine granular. The Balbiani body is diffused. The follicular cells are squamous in shape, limited by the basal lamina. Follicular cells are intimately associated with the oocyte by numerous interdigitated microvilli, emanating from both the oocyte, and the follicular cells, that form the *zona pellucida*.

Stage 3: The oocyte diameter attains a dimater of 600-700 μm. The contour of the nucleus is irregularly folded (Fig. 12), the nucleoli increase in number and are located at the periphery of the nucleus. The ooplasm contains the earliest acidophilic yolk platelets at the periphery (Figs. 12, 13A). The *zona pellucida* is evident as a consequence of the development of the microvilli from the surfaces of the oocyte and the follicular cells projected into the amorphous material between both types of cells. Desmosomes can be seen on the projections of the oocyte and the follicular cells. The microvilli of the *zona pellucida* are responsible for the uptake of materials from follicle cells to oocytes. During this stage, micropinocytotic vesicles can be seen on the peripheral oolemma, near the base of the microvilli. The follicular epithelium is a layer of squamous or cubic cells. The connective tissue adjacent to the follicular layer forms a thin vascularized and fibrous theca.

Stage 4: The oocyte attains a diameter of 800-1000 μm. The nucleus shows lampbrush chromosomes, nucleoli situated at the periphery and a folded membrane (Fig. 13B). The ooplasm shows progressive accumulation of acidophilic ovoidal yolk platelets (Fig. 12). The animal hemisphere is distinguished by the initial deposition of the melanin pigment in the peripheral cytoplasm of this hemisphere (Fig. 13C). The follicular cells are slightly thicker (Fig. 13C) than these seen in the previous stage.

Stage 5: The oocyte continues growth and reaches 1800 μm in diameter. The oocyte becomes filled with yolk platelets deposited from the periphery and extending to the ooplasm adjacent to the nucleus (Fig. 11B, 12, 13D). The yolk platelets situated in the vegetal pole increase in size and in number. The animal pole contains smaller yolk platelets than those seen in the vegetal pole (Fig. 13D). Melanin deposition at the periphery of the animal pole is seen (Fig. 13D). The follicular epithelium is similar to that seen in previous stage. Besides late vitellogenic oocytes, the ovary contains numerous previtellogenic oocytes that will become mature in the next vitellogenic period (Fig. 11B).

Stage 6: The oocyte reaches a diameter of 1800-2000 μm. Finally, the

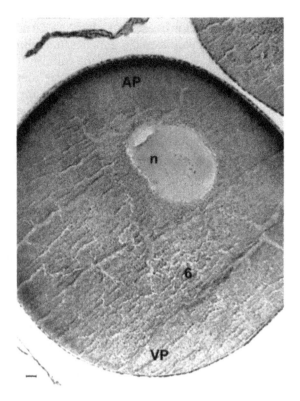

Fig. 14. Ovarian structure of adult female *Ambystoma mexicanum* during vitellogenesis. Oocyte during vitellogenesis, stage 6 (6). The oocyte nucleus, the germinal vesicle (n) moves to the periphery of the animal pole. The animal pole (AP) and the vegetal pole (VP) are well defined. Alcian-blue. Bar 0.3 mm.

fully developed oocytes (Fig. 14), arrested in diplotene of meiosis, are stimulated by progesterone to progress to metaphase, a process referred as meiotic maturation (Bement and Capco, 1990). Meiotic maturation includes a global biochemical reorganization of the oocyte (Bement and Capco, 1990). The animal pole and the vegetal pole are well defined (Fig. 14). The oocyte nucleus (germinal vesicle) moves to the periphery of the animal pole, the meiotic spindle is formed, the chromosomes that are completely condensed, attach to the meiotic spindle and move to the equator of the spindle. The meiosis continues to telophase I when the primary oocytes divide and form the first polar body. After ovulation, the meiosis continues until metaphase II, where the oocytes remains again arrested until fertilization, when the meiosis continues to anaphase II and telophase II. At this stage the second polar body is emitted. Consequently, the meiotic process in oocytes is not completed until fertilization occurs.

Postovulatory Ovary

At the end of ovulation the mature oocytes are gone from the ovary and the ovary contains small oocytes that will become mature during the next reproductive season (Saidapur, 1982; Xavier, 1987; Norris, 1997). After ovulation in oviparous species, the discharged follicles collapse and become a mass of lipid-containing cells. Follicular cells hypertrophy, accumulate cholesterol, and are surrounded by a fibrous capsule derived from the thecal layer, thus forming a postovulatory follicles called *corpora lutea*. The theca becomes multilayered and may show an inner zone and a well vascularized outer zone (Dodd, 1977). The thecal cells remain at the periphery and do not invade the hypertrophied follicular cells at any stage of the postovulatory *corpora lutea* (Saidapur, 1982). The postovulatory *corpora lutea* of *Necturus maculosus* (Saidapur, 1982) and *Triturus cristatus* (Saidapur, 1982; Norris, 1997) contain 3β-hydroxysteroid dehydrogenase (3β-HSD) activity and may produce steroids. The presence of an intense 3β-HSD activity has been demonstrated histochemically (Saidapur, 1982; Chieffi and Pierantoni, 1987; Norris, 1997). However, no functional endocrine role for postovulatory *corpora lutea* has been demonstrated in oviparous species.

The postovulatory *corpora lutea* soon become resorbed. The signs of resorbtion are the decrease in size of the cells, pyknosis of the nucleus, and an accumulation of lipid droplets in the cytoplasm of the luteal cells (Xavier, 1987). The process is different in viviparous species, such as *Salamandra salamandra* and *S. atra* (Saidapur, 1982; Lofts, 1984), where the postovulatory follicles form a well-organized secretory *corpora lutea*, which remain functional throughout two-thirds of gestation period (Xavier, 1987). All viviparous urodeles have *corpora lutea* that appear to be involved in maintenance of the pregnancy (Wake, 1993). The *corpora lutea* of the viviparous species *Salamandra salamanda terrestris* apparently functions during pregnancy. The progesterone content in the ovary is high at the beginning of gestation and decreases during gestation (Joly *et al.*, 1994). The postovulatory *corpora lutea* of viviparous species are capable of converting pregnenolone to progesterone, and similar to those described in oviparous species, also have 3β-HSD activity and may be steroidogenic (Lofts, 1984; Xavier, 1987). Development of new oocytes may be arrested by the presence of *corpora lutea*, in both oviparous and viviparous species, whereas the new oocytes begin development only after degeneration of the postovulatory *corpora lutea* (Norris, 1997). In *S. atra*, the decrease of the *corpora lutea* and the growth of oocytes were seen, indicating inhibition of folliculogenesis by *corpora lutea* during gestation (Xavier, 1987).

Atretic Follicles

Atresia is a widespread process occurring in the ovaries of all vertebrates. Atresia is the degeneration and resorbtion of follicles containing diplotene oocytes at various stages of development. It is not well known why some oocytes are able to develop until ovulation, whereas others undergo degeneration. The precise mechanisms that cause such selection of germ cells is not well understood (Saidapur, 1978). Atresia occurs during various stages of oocyte development. Follicles may become atretic at previtellogenic (Fig. 15A) and vitellogenic stages (Fig. 15B). Even though the causes and functions of atresia have still to be determined more precisely, it has been suggested that atretic follicles contribute to steroid secretions and to regulate the number of ovulated oocytes (Guraya, 1969; Byskov, 1978; Saidapur, 1978). Thereby, atresia plays an important role in fecundity, determining the number of developing oocytes.

During atresia the follicle undergoes degeneration. The previtellogenic oocyte become shrunken (Fig. 15A), suggesting a relative loss of water. With the advance of atresia, the oocyte shows vacuoles at its periphery. The ooplasm and the nucleus are gradually resorbed by phagocytosis of the follicular cells. The yolk platelets and pigment granules form abnormal aggregations as a result of fusion (Fig. 15B). *Zona pellucida* becomes broken and diffuses, and follicular cells proliferate and hypertrophy. The retraction of the microvilli connecting follicular cells and oocyte is seen (Byskov, 1978). At late stages of atresia, the oocyte with yolk is digested and removed by the follicular cells, leaving an irregular mass surrounded by the theca. Final stages of atresia consist of follicular cells filled with phagocytosed pigment and surrounded by hypertrophied theca cells. Adams (1940) observed atretic follicles in the ovaries of *Triturus viridescens*, describing phagocytosis of yolk of the atretic oocytes by follicular cells and leucocytes, deposition of pigment and fat, hypertrophy of follicular cells, and finally the gradual disappearance of the follicle. According to Guraya (1976), the atretic yolky oocytes are transformed into structures called preovulatory *corpora lutea* or *corpora atretica*. In *corpora atretica* the yolk is in process of resorption, where both the fat and protein yolk coalesce to form a mass of triglycerides, gradually broken up and digested by the follicular cells leaving aggregations of pigmented granules.

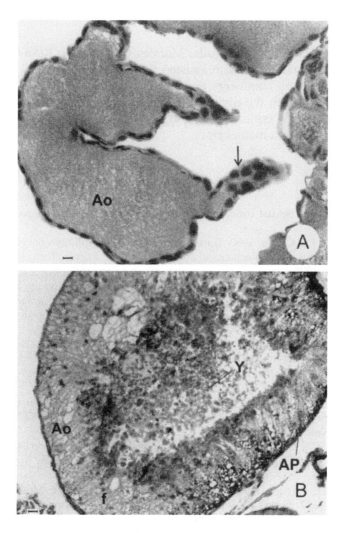

Fig. 15. Ovarian structure of adult female *Ambystoma mexicanum* during vitellogenesis. A—Atretic oocyte (Ao), during previtellogenesis, the oocyte is shrunken,
becoming very irregular in shape. Some nuclei (arrow) of follicular cells show
initial hypertrophy. B—Atretic oocyte (Ao), during vitellogenesis, the oocyte is
in advanced stage of disintegration, the remaining yolk platelets (Y) are dispersed,
or coalescence in small masses of yolk in the central portion. The animal pole
(AP) and the cortex of the animal hemisphere contains a layer of pigment
granules. The *zona pellucida* is not seen. The follicular cells (f) are abundant,
hypertrophied and irregular in shape. H+E staining. Bar 20 μm.

300

Acknowledgements

I thank Marcela Esperanza Aguilar Morales, Gabino De la Rosa-Cruz and Carlos López Arellano, whom assisted tissue processing for histological stains, and valuable comments, and José Antonio Hernández Gómez who kindly assisted with the digital preparation of figures. Many thanks to Maria Ogielska for critical reading and valuable comments that significantly improved this manuscript.

References

Adams AE. 1940. Sexual conditions in *Triturus viridescens*. Am. J. Anat. 66: 235-275.

Bement WM, DG Capco. 1990. Transformation of the amphibian oocyte into the egg: Structural and biochemical events. J. Electron Micros. Tech. 16: 202-234.

Brandon RA. 1970. Courtship, spermatophores and eggs of the Mexican achoque, *Ambystoma (Bathysiredon) dumerilii* (Duges). Zool. J. Linn. Soc. 49: 247-254.

Brizzi R, G Delfino, MG Selmi, DM. Sever. 1995. Spermathecae of *Salamandrina terdigitata* (Amphibia: Salamandridae): Patterns of sperm storage and degradation. J. Morphol. 223: 21-33.

Bruce RC. 1975. Reproductive biology of the mud salamander, *Pseudotriton montanus*, in western South Carolina. Copeia. 1975: 129-137.

Byskov AG. 1978. Follicular atresia. In: The Vertebrate Ovary (Ed.) RE Jones. Plenum Press, New York, pp. 533-555.

Chieffi G, R Pierantoni. 1987. Regulation of ovarian steroidogenesis, In: Hormones and Reproduction in Fishes, Amphibians and Reptiles, (Eds) DO Norris, RE Jones, New York, Plenum Press, pp. 117-144.

Dodd JM. 1977. The structure of the ovary of nonmammalian vertebrates. In: The Ovary, (Eds) S Zuckerman, BJ Weir. Academic Press New York, Vol. 1, pp. 219-263.

Duellman WE, L Trueb. 1986. Biology of Amphibians. New York, McGraw-Hill.

Greven H. 1977. Comparative ultrastructural investigations of the uterine epithelium in the viviparous *Salamandra atra* Laur. and the ovoviviparous *Salamandra salamandra* (L.) (Amphibia, Urodela). Cell Tiss. Res. 181: 215-237.

Greven H, GD Guex. 1994. Structural and physiological aspects of viviparity in *Salamandra salamandra*. In: *Mertensiella*, Biology of *Salamandra* and *Mertensiella*, (Eds) H Greven, B Thiesmeier. Bonn, pp. 139-160.

Greven H. 1998. Survey of the oviduct of Salamandrids with special reference to the viviparous species. J. Exp. Zool. 282: 507-525.

Greven H. 2002. The urodele oviduct and its secretions in and after G. von Wahlert's doctoral thesis "Eileiter, Laich und Kloake der Salamandriden". Bonner Zool. Monogr. 50: 25-61.

Greven H. 2003. Oviduct and egg-jelly. In: Reproductive Biology and Phylogeny of Urodela. Series: Reproductive Biology and Phylogeny, D Sever (volume ed.), BGM Jamieson (Series ed.). Vol. 1, Science Publishers, Inc. Enfield. NH, USA, pp. 151-181.

Grier HJ. 2000. Ovarian germinal epithelium and folliculogenesis in the common snook, *Centropomus undecimalis* (Teleostei: Centropomidae). J. Morphol. 243: 265-281.

Guex GD, PS Chen. 1986. Epitheliophagy: Intrauterine cell nourishment in the viviparous alpine salamandre, *Salamandra atra* (Laur.). Experientia. 42: 1205-1218.

Guex GD, H Greven. 1994. Structural and physiological aspects of viviparity in *Salamandra atra*. In: *Mertensiella*, Biology of *Salamandra* and *Mertensiella*, (Eds) H Greven, B Thiesmeier. Bonn, pp. 161-208.

Guillette LJ Jr, DO Norris, M Norman. 1985. Response of amphibian (*Ambystoma tigrinum*) oviduct to arginine vasotocin and acetylcholine *in vitro*: Influence of steroid hormone pretreatment *in vivo*. Comp. Biochem. Physiol. 80C: 151-154.

Guraya SS. 1969. Histochemical study of follicular atresia in the amphibian ovary. Acta Biol. Acad. Sci. Hung. 20(1): 43-56.

Guraya SS. 1976. Recent advances in the morphology, histochemistry, and biochemistry of steroid-synthesizing celllular sites in the non-mammalian vertebrate ovary. Rev. Cytol. 4: 365-409.

Guraya SS. 1978. Maturation of the follicular wall of non-mammalian vertebrates, In: The Vertebrate Ovary: Comparative Biology and Evolution, (Ed.) RE Jones. Plenum Press, pp. 261-329.

Heatwole H. 1961. Rates of desiccation and rehydration of eggs in a terrestrial salamander, *Plethodon cinereus*. Copeia 1961: 110-112.

Heller HE, E Ferreri, DHG Leathers. 1970. The effect of neurohypophysical hormones on the amphibian oviduct *in vitro*, with some remarks on the histology of this organ. J. Endocrinol. 47: 495-509.

Hope J, AA Humphries Jr, GH Bourne. 1963. Ultrastructural studies on developing oocytes of the salamander *Triturus viridescens*. I. The relationship between follicle cells and developing oocytes. J. Ultrastruct. Res. 9: 302-324.

Humphries AA Jr, WN Hughes. 1956. A study of the polysaccharide histochemistry of the oviduct of the newt, *Triturus viridescens*. Biol. Bull. 116: 446-451.

Jego P. 1974. Composition en glucides des differents segments de l'oviducte et des gangues ovulaires chez *Pleurodeles waltlii* Michah (Amphibien, Urodèle). Comp. Biochem. Physiol. 488: 435-446.

Joly J, B Picheral. 1972. Ultrastructure, histochimie et physiologie du follicle pre-ovulatoire et du corps jaune de l'urodele ovo-vivipare *Salamandra salamandra* L. Gen. Comp. Endocrinol. 18: 235-259.

Joly J, F Chesnel, D Boujard. 1994. Biological adaptations and reproductive strategies in the genus *Salamandra*, In: *Mertensiella*, Biology of *Salamandra* and *Mertensiella*, (Eds) H Greven, B Thiesmeier, Bonn, pp. 255-269.

Kambara S. 1956. Histochemical studies on the distribution of phosphatases in the oviduct of the newt, *Triturus pirrhogaster*. Annotationes Zoologicae Japonenses 29: 86-90.

Kruczyński D, H Greven, D Passia. 1986. Histochemical demonstration of zinc in the spermatheka of *Salamandra salamandra* (L.) (Amphibia, Urodela). Acta Histochem. 79: 181-186.

Licht P, DA Porter. 1987. Role of gonadotropin-realising hormone in regulation of gonadotropin secretion from amphibian and reptilian pituitaries, In: Hormones and Reproduction in Fishes, Amphibians and Reptiles, (Eds) DO Norris, RE Jones Plenum Press. New York, pp. 61-85.

Lofts B. 1974. Reproduction, In: Biology of the Amphibia, (Ed.) B Lofts,Vol. 2. Academic Press, New York and London, pp. 107-218.

Lofts B. 1984. Amphibians, In: Marshall's Physiology of Reproduction, Vol. 1. Reproductive Cycles of Vertebrates. (Ed.) GE Lamming, Churchill Livingstone, Edinburgh, pp. 127-205.

Lofts B. 1987. Testicular function, In: Hormones and Reproduction in Fishes, Amphibians and Reptiles, (Eds) DO Norris, RE Jones, Plenum Press, New York, pp. 288-298.

McCurdy HM. 1931. Development of the sex organs in *Triturus torosus*. Am. J. Anat. 47: 367-403.

Miller MR, ME Robbins. 1954. The reproductive cycle in the *Taricha torosa* (*Triturus torosus*) J. Exp. Zool. 125: 415-446.

Norris DO. 1997. Reproduction in Amphibians, Vertebrate Endocrinology, Academic Press, New York, pp. 428-446.

Potemkina DA. 1967. Formation of the Müllerian ducts in the urodele (*Ambystoma mexicanum*). Dokl. Biol. Sci. 174(1): 305-308.

Saidapur SK. 1978. Follicular atresia in the ovaries of non-mammalian vertebrates. Inter. Rev. Cytol. 54: 225-244.

Saidapur SK. 1982. Structure and function of postovulatory follicles (corpora lutea) in the ovaries of non-mammalian vertebrates. Int. Rev. Cytol. 75: 243-285.

Saidapur SK, VB Nadkarni. 1974. Steroid-synthesizing cellular sites in amphibian ovary. A Histochemical Study. Gen. Comp. Endocrinol. 22: 459-462.

Salthe SN. 1963. The egg capsules in the Amphibia. J. Morphol. 113: 161-171.

Sever DM. 1988. The ventral gland in female salamander *Eurycea bislineata* (Amphibia: Plethodontidae). Copeia. 1988: 572-579.

Sever DM. 1991. Comparative anatomy and phylogeny of the cloacae of salamanders (Amphibia: Urodeles). I. Evolution at the family level. Herpetologica. 47: 165-193.

Sever DM. 1994. Observations on regionalization of secretory activity in the spermathecae of salamanders and comments on phylogeny of sperm storage in female salamanders. Herpetologica. 50: 383-397.

Sever DM. 2002. Sperm storage in female amphibians. J. Exp. Zool. 292: 165-179.

Sever DM. 2003. Courtship and mating glands. In: Reproductive Biology and Phylogeny of Urodela. Volume 1 of Series: Reproductive Biology and Phylogeny, D Sever (volume ed.), BGM Jamieson (Series ed.). Science Publishers, Inc. Enfield. NH, USA, Chapter 9, pp. 323-381.

Sever DM, NM Kloepfer. 1993. Spermathecal cytology of *Ambystoma opacum* (Amphibia: Ambystomatidae) and the phylogeny of sperm storage in female salamanders. J. Morphol. 217: 115-127.

Sever DM, JD Krenz, KM Johnson, LC Rania. 1995. Morphology and evolutionary implications of the annual cycle of secretion and sperm storage in spermatheca of the salamander *Ambystoma opacum* (Amphibia: Ambystomatidae). J. Morphol. 223: 35-46.

Sever DM, LC Rania, JD Krenz. 1996. Reproduction of the salamander *Siren intermedia* Le Conte with especial reference to oviducal anatomy and mode of fertilization. J. Morphol. 227: 335-348.

Sharon, R, G Degani, MR Warburg. 1997. Oogenesis and the ovarian cycle in *Salamandra salamandra infraimmaculata* Mertens (Amphibia; Urodela; Salamandridae) in Fringe Areas of the Taxon's Distribution. J. Morphol. 231: 149-160.

Tokarz RR. 1978. Oogonial proliferation, oogenesis, and folliculogenesis in nonmammalian vertebrates, In: The Vertebrate Ovary. Comparative Biology and Evolution, (Ed.) RE Jones, Plenum Press, New York, pp. 145-179.

Trauth SE. 1983. Reproductive biology and spermathecal anatomy of the dwarf salamander (*Eurycea quadridigitata*) in Alabama. Herpetologica. 39(1): 9-15.

Trauth SE. 1984. Spermathecal anatomy and the onset of mating in the slimy salamander (*Plethodon glutinosus*) in Alabama. Herpetologica. 40(3): 314-321

Trauth SE, DM Sever, RD Semlitsch. 1994. Cloacal anatomy of paedomorphic female *Ambystoma talpoideum* (Urodeles: Ambystomatidae), with comments on intermorph mating and sperm storage. Can. J. Zool. 72: 2147-2157.

Uribe MCA. 2001. Reproductive systems of Caudata Amphibia. In: Vertebrate Functional Morphology, (Eds) HM Dutta, JSD Munshi, Science Publishers, Inc. Enfield (NH) USA, Plymouth, UK, Chapter 9, pp. 267-293.

Uribe MCA. 2003. The ovary and oogenesis. In: Reproductive Biology and Phylogeny of Urodela. Volume 1 of Series: Reproductive Biology and Phylogeny, D Sever (volume ed.), BGM Jamieson (Series ed.). Science Publishers, Inc. Enfield, NH, USA, Chapter 4, pp. 135-150.

Uribe MCA, G Gómez Ríos, C López Arriaga. 1991. Cambios morfológicos del testículo de *Ambystoma dumerilii* durante un ciclo anual. Bol. Soc. Herpetol. Mex. 3: 13-18.

Uribe MCA, LF Mena, MRC Carrubba, EF Estrada, BD Palmer, LJ Guillette, Jr. 1989. Oviductal histology of the urodele, *Ambystoma mexicanum*. J. Herpetol. 23(3): 230-237.

Vilter V. 1967. Histologie de l'oviducte chez *Salamandra atra* mature, Urodele totalment vivipare de haute montagne. C. R. Soc. Biol. Paris. 161: 260-264.

Vilter V, R Thorn. 1967. Histologie de l'oviducte et mode de reproduction d'un urodèle cavernicole d'Europe: *Hydromantes genei* (Temminck et Schlegel) C. R. Soc. Biol. Paris. 161: 1222-1227.

Wake M. 1985. Oviduct structure and function in non-mammalian vertebrates. In: Vertebrate Morphology, (Eds) H Duncker, G Fleisher,Fortschr. Zool. Gustav Fischer, Stuttgart, pp. 427-435.

Wake M. 1993. Evolution of oviductal gestation in Amphibians. J. Exp. Zool. 266: 394-413.

Wake M, R Dickie. 1998. Oviduct structure and function and reproductive modes in Amphibians. J. Exp. Zool. 282: 477-506.

Wallace RA. 1978. Oocyte growth in nonmammalian vertebrates, In: The Vertebrate Ovary. Comparative Biology and Evolution, (Ed.) RE Jones, Plenum Press, New York, pp. 469-502.

Wallace RA, K Selman. 1990. Ultrastructural aspects of oogenesis and oocyte Growth in fish and amphibians. J. Electron Microsc. Tech. 16: 175-201.

Watanabe A, K Onitake. 2003. Sperm activation. In: Reproductive Biology and Phylogeny of Urodela. Volume 1 of Series: Reproductive Biology and

Phylogeny, D Sever (ed.), BGM Jamieson (Series ed.). Science Publishers, Inc. Enfield. NH, USA.

Whittier JM, D Crews. 1987. Seasonal reproduction: patterns and control, In: Hormones and Reproduction in Fishes, Amphibians and Reptiles, (Eds) DO Norris, RE Jones, Plenum Press, New York, pp. 385-409.

Ward D, OJ Sexton. 1981. Anti-predator role of salamander egg membranes. Copeia. 1981: 724-726.

Xavier F. 1987. Functional morphology and regulation of the corpus luteum, In: Hormones and Reproduction in Fishes, Amphibians and Reptiles, (Eds) DO Norris, RE Jones, Plenum Press, New York, pp. 241-282.

Zalisko EJ, RA Brandon, J Martan. 1984. Microstructure and histochemistry of salamander spermatophores (Ambystomatidae, Salamandridae and Plethodontidae). Copeia. 1984: 739-747.

7

Oogenesis and Female Reproductive System in Amphibia—Gymnophiona

Jean-Marie Exbrayat

The female reproductive system in Gymnophiona is composed of a pair of ovaries accompanied by fat bodies, a pair of oviducts, and a cloaca. The first anatomical description of the reproductive system in female Gymnophiona was provided by Müller (1832) and Rathke (1852). Fertilization is always internal, and species are oviparous or viviparous. Scolecomorphidae and Typhlonectidae are two viviparous families (Wake, 1993), but only very few species have been studied thoroughly. About half of the gymnophionan species are viviparous (Wake, 1977b, 1989, 1992, 1993), but this number is discussed (Wilkinson and Nussbaum, 1998). Studies on reproduction in Gymnophiona give some evidence for the evolution from oviparity to viviparity. Several intermediate solutions, as direct development, were described in *Idiocranium russeli* (Sanderson, 1937), *Hypogeophis rostratus* (Nussbaum, 1984) and in *Boulengerula taitanus* (Malonza and Measey, 2005).

The structure and function of female reproductive system, as well as sexual cycles, were studied throughoutly in *Dermophis mexicanus* (Wake, 1980b) and *Typhlonectes compressicauda* (Exbrayat, 1983, 1984, 1986, 1988a,b, 1992, 1996; Exbrayat and Collenot, 1983; Exbrayat and Delsol, 1985; Exbrayat and Flatin, 1985), and thereby these species will serve as models in this chapter.

Structure of Ovary in Gymnophiona

Structure of Ovaries in Adults

Ovaries of *Caecilia gracilis*, *Ichthyophis glutinosus* and *Siphonops annulatus* were described by early researchers (Spengel, 1876; Sarasin and Sarasin,

1887-1890), and later in *Uraeotyphlus menoni* (Chatterjee, 1936) and *Uraeotyphlus oxyurus* (Garg and Prasad, 1962). Berois and De Sa (1988) described ovaries of *Chthonerpeton indistinctum*. Exbrayat (1983, 1986) and Exbrayat and Collenot (1983) described the ovaries of *Typhlonectes compressicauda*. Recently, Raquet *et al.* (2006, 2007) gave data about *Boulengerula taitanus*'s ovaries and Exbrayat (2006a) gave bibliographical synthesis about ovaries of several species. The relationships between ovaries and fat bodies were studied in *Hypogeophis rostratus* (Tonutti, 1931). Comparative studies of female reproductive systems in a number of species and genera were provided in several papers by Wake (1968, 1970a,b, 1972, 1977a,b) and Exbrayat (2006a). Beyo *et al.* (2007a,b) gave ultrastructural studies of previtellogenic and vitellogenic oocytes in *Ichthyophis tricolor* and *Gegenophis ramaswamii*.

Ovaries in Gymnophiona are elongated and look like a pair of strings that lie parallel to kidneys and gut (Wake, 1968) (Fig. 1). The internal structure of ovaries is metameric, although irregular, composed of "germinal nests" (which are equivalents of "germ patches" in Anura) and follicles containing diplotene oocytes that are scattered between the nests. In *Typhlonectes compressicauda* each ovary contains about 20 germinal nests. The ovary is attached to the corresponding kidney by the mesovarium, which is composed of a sheet of peritoneal epithelium and underlying connective tissue. Another sheet of connective tissue attaches each ovary to the corresponding fat body that lies parallel to, and is as long as the ovary. Blood vessels transverse the connective tissue and join the fat body with the ovary. Each ovary is enveloped by a simple or stratified epithelium covered with the peritoneal epithelium. There is no lumen in ovaries, as was originally described by Tonutti (1931). This author described an empty space within an ovary, situated between mesovarium and follicles, and named it the "ovarian duct" suggesting that oocytes used this "duct" to reach the cloaca; however, the ovarian lumen described by Tonutti was certainly an artifact and his hypothesis has been abandoned.

Ovaries of *Typhlonectes compressicauda* that are 50 to 90 mm in length contain 50 to 100 small cream-colored oocytes that are less than 1 mm in diameter. The oocytes grow in size and change the coloration. The oocytes that are 2 mm in diameter are yellow-colored and the largest are orange and attain 2.5 mm in diameter. The follicles are usually arranged in a row, one behind the other. The largest oocytes are generally grouped in the anterior region of the ovary. Big oocytes are often clustered in the anterior region of the gonad (Wake, 1968). In 300-350 mm long adult female *Typhlonectes compressicauda*, ovaries range from 60 to 100 mm in length, and average mass of an ovary is 50-78 mg (Exbrayat, 1986). The gonadosomatic ratio (GSR) varies, reaching its maximum just before breeding (October to February). The number of ovarian follicles with diplotene

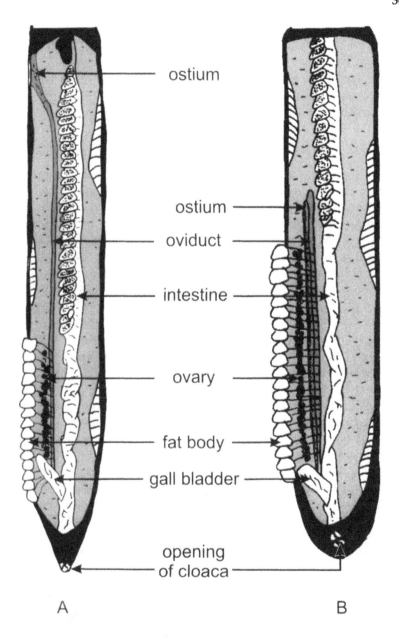

Fig. 1. Female urogenital tract in oviparous *Ichthyophis kohtaoensis* (A), and viviparous *Typhlonectes compressicauda* (B) (Modified after Exbrayat, 1989a).

oocytes varies individually (from 3 to 30), but is generally higher in pregnant females and is almost the same in the left and right ovary. At ovulation, the follicle breaks down and the oocytes are released to the body cavity. The broken follicles are seen as a discontinuity of the lateral wall of the ovary and the empty follicles transform into *corpora lutea*, both in oviparous and viviparous species.

Development and Differentiation of the Ovaries

The development of gonads in Caecilians has been known for some species for more than hundred years (Spengel, 1876; Sarasin and Sarasin, 1887-1890; Brauer, 1902; Tonutti, 1931; Marcus, 1939). More recently, general data on development of gonads in Gymnophiona have been published by Wake (1968) and studied in details in *Typhlonectes compressicauda* (Anjubault and Exbrayat, 2004, 2006; Exbrayat and Anjubault, 2005). Originally Marcus (1939) described two generation of gonads: the primary hermaphroditic that become fat bodies and the secondary that develop into the definite ovary. However, Anjubault and Exbrayat (2004) have shown recently that the gonad and corresponding fat body differentiate from the same gonadal anlage, although not anteriorly as in Anura, but parallel to the gonad (Fig. 2A,B). In that sense the "primary hermaphrodite gonad" never becomes a fat body; these results indicate that the hypothesis of Marcus (1939) cannot be considered as valid any longer.

Before hatching, primordial germ cells (PGCs) are observed on the dorsal region of endoderm, then in the dorsal part of gut inside the neural crest (Exbrayat *et al.*, 1981; Sammouri *et al.*, 1990). These cells are bigger than others, ovoid in shape, with numerous dense granules in the cytoplasm. Their chromatin is more or less condensed, probably due to the phase of cell cycle. PGCs leave endoderm by successive waves, their number increases due to mitoses observed in several of them. At successive stages before hatching they reach the dorsal mesentery close to primordial median genital ridge. The primordial median genital ridge is formed by the dorsal mesentery between the median region of endodermal roof and aorta. At these stages endoderm and mesoderm are in close contact, favoring migration of PGCs. Then PGCs leave the median ridge and migrate anteriorly into two small lateral ridges. The lateral ridges (gonadal anlages) are lobed and situated in coelomic cavity on both sides of aorta. At hatching, PGCs continue their migration into lateral ridges. Only few tiny connections are observed between the median and lateral ridges, and after emigration of PGCs the median ridge succesively degenerates.

In viviparous species *Typhonectes compressicauda* after hatching, free larvae undergo further development in uterus. Larval PGCs continue their migration to segmented gonadal anlages and, in result of growing number

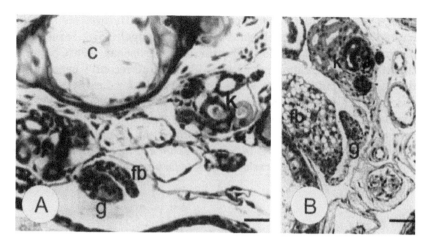

Fig. 2. Development of gonads in *Typhlonectes compressicauda*. A—Cross section at the level of mid-intestine, larval stage 30. Scale bar = 20 μm. B—Cross section at larval stage 32. Scale bar = 20 μm. c—centrum; fb—fat body; g—gonadal anlage; k—kidney.

of invading germ cells, each lobe increases in size. Growing gonads are situated close to mesonephros and Wolffian ducts. Fat bodies differentiate and increase in size. At that time undifferentiated gonads are composed of peripheral part containing large germ cells with big nuclei with granular chromatin and a central fibrous part with some smaller germ cells. These two parts will give rise to future cortex and medulla, respectively. At the beginning of metamorphosis, medulla and cortex are already well defined, although a gonad is still sexually undifferentiated. More germ cells are localized in lobes forming the middle part of a gonad. Germ cells still vary in size and are observed both in cortex and medulla. Somatic cells also vary in size and some big cells are situated in the periphery of a gonad. Some somatic cells, which began to surround small germ cells, are found in the cortical region. Later during metamorphosis germ cells start to proliferate, but it is still not possible to distinguish any signs of sexual differentiation, because gonia are localized both in the cortex and the medulla. At the same time blood vessels are observed in gonads. At this stage of development gonads resemble hermaphroditic structures of Marcus (1939).

Sexual differentiation of gonads in *Typhlonectes compressicauda* occurs just before or after the birth, i.e., about two months after metamorphosis (Anjubault and Exbrayat, 2004, 2006; Exbrayat and Anjubault, 2005). In *Dermophis mexicanus* gonads of newborns resemble a pair of small ribbons (Wake, 1980b). The development of ovaries in Gymnophiona is poorly known and was described in a general manner in some species (*Caecilia* gracilis, *Ichthyophis glutinosus, Siphonops annulatus, Hypogeophis rostratus, Uraeotyphlus menoni, Uraeotyphlus oxyurus* (Spengel, 1876; Sarasin and Sarasin, 1887-1890; Brauer, 1902; Tonutti, 1931; Seshachar, 1936; Chatterjee, 1936; Marcus, 1939; Garg and Prasad, 1962). More recently, general data have also been published by Wake (1968b, 1980b) for *Caecilia (Oscaecilia) ochrocephala, Gegenophis ramaswamii, Geotrypetes seraphinii, Icthyophis glutinosus, Rhinatrema bicolor, Schistometopum gregorii, Scolecomorphus kirkii, S. vittatus, S. uluguruensis, Siphonops annulatus, Uraeotyphlus menoni, Typhlonectes compressicauda, Idiocranium russeli, Boulengerula taitanus (Afrocaecilia taitana), B. uluguruensis (Afrocaecilia uluguruensis)* and *Dermophis mexicanus.*

Fat Body

Female fat bodies resemble these of the males (Fuhrmann, 1914, Sarasin and Sarasin, 1887-1890, Tonutti, 1931, Wake, 1968a). In *Typhlonectes compressicauda* the mass of fat bodies changes during sexual cycle. During quiescent period the size of fat bodies is relatively constant (Exbrayat, 1986, 1988b). During breeding period and at the beginning of pregnancy fat bodies are well developed and then decrease during gestation, becoming very small at the end of pregnancy, when embryos increase in length and weight. In this species, this suggests that the fat body reserves contribute to the development of embryos. No data are available from other oviparous or viviparous species.

Female Reproductive Tract in Gymnophiona

Oviduct and Uterus

The oviducts are paired elongated structures parallel to the ovaries and kidneys (Fig. 1). The most anterior part of the oviduct forms the funnel (infundibulum), which opens by the ostium. The structure of oviducts is adapted to the mode of reproduction: it is simple in oviparous species and specialized in viviparous species. In the latter, the posterior part of oviduct differentiates into uterus, inside which development of embryos occurs. During pregnancy the uterus undergoes some peculiar transformations related to feeding and development of embryos. The most evolved mode of development is that of *Typhlonectes compressicauda* (and perhaps of all Typhlonectidae) with formation of a placenta-like structure including both the uterine wall and the gills of the embryos.

    Oviduct of oviparous species. Oviducts were studied in the following species and genera: *Ichthyophis glutinosus* (Sarasin and Sarasin, 1887-1890), *Ichthyophis* and *Hypogeophis* (Marcus, 1939), *Idiocranium* and *Rhinatrema*

(Parker, 1934, 1936), *Uraeotyphlus oxyurus* (Garg and Prasad, 1956), *Ichthyophis beddomei* (Masood Parveez, 1987; Masood Parveez and Nadkarni, 1991), *Ichthyophis kohtaoensis* (Exbrayat, 1989a), *Caecilia (Oscaecilia) ochrocephala, Gegenophis ramaswamii, Geotrypetes seraphinii, Icthyophis glutinosus, Rhinatrema bicolor, Schistometopum gregorii, Scolecomorphus kirkii, S. vittatus, S. uluguruensis, Siphonops annulatus, Uraeotyphlus menoni, Typhlonectes compressicauda, Idiocranium russeli, Boulengerula taitanus (Afrocaecilia taitana), B. uluguruensis (Afrocaecilia uluguruensis), Dermophis mexicanus* (Wake, 1970b).

The oviducts of oviparous species are situated in some distance from the ovaries (Fig. 1A). In *Ichthyophis kohtaoensis* and *I. beddomei*, the funnel opens as an ostium that is situated at a large distance from the ovary and directed toward the head of the animal. During sexual rest, the funnel is bordered by a single layer of flattened cells, whereas before ovulation it is bordered by a single layer of cells with microvilli. During the period of sexual quiescence wall of the body cavity is lined by flat cells; these cells will produce a secret covering the body cavity in a female before ovulation (Exbrayat, 1989a). During breeding period the body cavity is covered by the secretions, which probably facilitates transport of oocytes to the ostium. Oocytes released from ovaries to the body cavity during ovulation are snatched by the ostium and enter the oviduct (Chatterjee, 1936; Oyama, 1952; Garg and Prasad, 1962; Wake, 1970a; Masood Parveez, 1987).

The oviducts are divided into three regions (*pars recta, pars convoluta,* and *pars utera*), which are equipped with glands producing mucous protein of egg envelopes and do not vary significantly from species to species. The *pars recta* is situated beneath the funnel and is lined by gland cells producing protein and acidic mucous secretions (Fig. 3A). The next portion of the oviduct, i.e., *pars convoluta*, is lined by some glands secreting acidic mucous and protein substances on only one side (Fig. 3B), whereas the other side is lined by cells that become ciliated during breeding period (Fig. 3C). The posterior portion of the oviduct, i.e., *pars utera*, is lined by the ciliated and glandular cells that produce neutral mucous substance during breeding period (Fig. 3D). The changes of oviducts in respect to the phase of sexual cycle were well studied in *Ichthyophis beddomei* (Masood Parveez, 1987; Masood Parveez and Nadkarni, 1991). In *Uraeotyphlus oxyurus*, the posterior two-thirds of the oviducts are superficially fused but each one possesses its own aperture in the cloaca (Garg and Prasad, 1962).

Oviduct of viviparous species. The first studies describing viviparity in Gymnophiona were provided by Peters (1874a,b, 1875) in *Caecilia (Typhlonectes) compressicauda*. Other viviparous species studied in regard to the anatomy of genital ducts belong to the following genera: *Chthonerpeton* (Parker and Dunn, 1964; Parker and Wettstein, 1929),

Geotrypetes (Parker, 1956), *Schistometopum* (Parker, 1956), *Dermophis* (Parker, 1956; Wake, 1980b), *Gymnopis* (Dunn, 1928; Taylor, 1955; Wake, 1967, 1977a,b). More recently, several studies were devoted to *Typhlonectes compressicauda* (Exbrayat, 1984, 1986, 1988a; Hraoui-Bloquet, 1995; Hraoui-Bloquet *et al.*,1994).

The funnel in *Typhlonectes compressicauda* is apposed to the ovary and oocytes are ovulated directly into the genital ducts (Fig. 1B). The ostium undergoes considerable changes according to the phase of sexual cycle (Exbrayat, 1984, 1986, 1988a). Each funnel is elongated, the ostium is

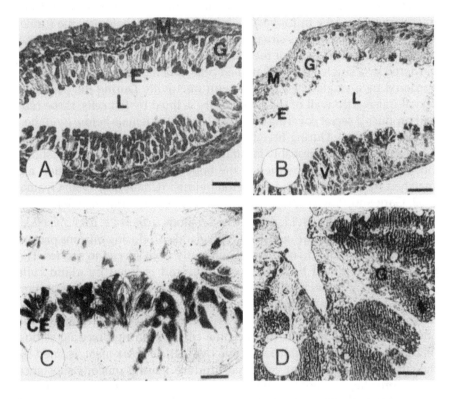

Fig. 3. Transverse sections on the oviduct in *Ichthyophis beddomei*. A—pars recta, August (pre-breeding phase of the reproductive cycle). B—pars convoluta, August (pre-breeding phase of the reproductive cycle). C—pars convoluta, January (breeding phase of the reproductive cycle). Ciliated cells of epithelium contain secretions stained with alcian blue. D—pars uterina, January (breeding phase of the reproductive cycle). Secretions are PAS positive. Scale bar = 10 µm. CE—ciliated epithelium; E—cuboidal epithelium; G—glandular cells; L—lumen; M—outer muscular layer; (From Masood-Parveez and Nadkarni, 1991).

situated parallel to the oviduct, to which it is attached by connective tissue. During sexual rest, the ostium is poorly developed and bordered with a single layer of flattened cells (Fig. 4A,B). During breeding period, the funnel develops and closely apposes the ovary. In this way oocytes are ovulated directly into the funnel without passing through the body cavity. Ciliated epithelium and some gland cells border the portion of funnel adjacent to the ostium (Exbrayat, 1986, 1988a, 1989a) (Fig. 4C,D).

The oviducts of viviparous species are divided into two regions: anterior and posterior. The anterior part of the oviduct corresponds to the *pars recta* of the oviparous species, i.e., the portion of oviduct, in which fertilization occurs and that contains glands involved in production of mucous envelopes. The envelopes produced by this part of oviduct are thin and will be resorbed during intrauterine development of embryos (Parker,

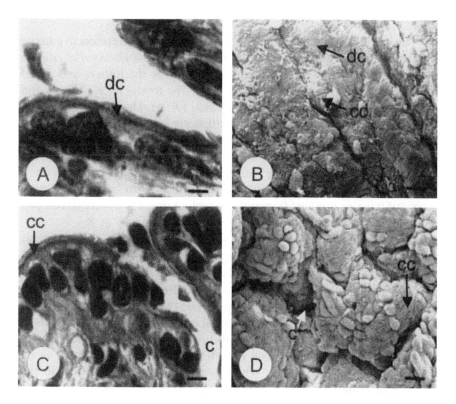

Fig. 4. Oviductal funnel in adult *Typhlonectes compressicauda*. A and B—In April (breeding season, after ovulation). A—Light micrograph, scale bar = 10 μm. B—SEM, scale bar = 15 μm. C and D—In February at ovulation. C—Light micrograph, scale bar = 110 μm. D—SEM, scale bar = 15 μm. c—crypt; cc—ciliated cells; dc—cell in desquamation.

1956). In adult *Typhlonectes compressicauda*, the different regions of oviduct are well differentiated and change during the biennial cycle. During breeding period, several glands are observed in the wall of the anterior part of oviducts. These glands are lined with cells containing PAS-positive secretion. The wall of the oviduct between the glands is bordered by ciliated columnar cells. The anterior part of oviduct contains several eggs and unfertilized oocytes. During pregnancy, when embryos are inside the uterus, the anterior part of oviduct regresses and epithelial cells lack cilia. The posterior region is modified into uterus and corresponds to both *pars convoluta* and *pars utera* of the oviparous species. The uterus wall undergoes modifications according to developmental stage of embryos (Fig. 6A). The wall permits the exchange of gases, water and other molecules between the embryo or foetus and its mother, as was shown in *Typhlonectes compressicauda* with use of ^{3}H-thymidine (Exbrayat, 1992; Hraoui-Bloquet, 1995).

The oviduct of *Scolecomorphus kirkii* is composed of a connective tissue wall surrounded by a thick layer of muscle cells. A single layer of ciliated epithelial cells with centrally situated nuclei lines the lumen, and gland cells are scattered among the ciliated cells. The oviductal wall thins and stretches when filled with embryos (Wake, 1970a). The oviductal wall in *Dermophis mexicanus* and *Gymnopis multiplicata* is poorly developed at the beginning of gestation, then connective tissue cells proliferate and some crests develop. Epithelial cells are filled with numerous inclusions, but not lipids, which appear in epithelial secretions at the end of pregnancy. The epithelium regenerates after parturition. The uterine wall of non-pregnant females remains thin and connective tissue is poorly developed (Wake, 1993; Wake and Dickie, 1998).

In other parts of the oviduct, branched connective tissue crests surrounded by unciliated cells and well-developed gland cells are observed. In *Typhlonectes compressicauda* the connective crests develop during the first year of biannual sexual cycle and are lined with epithelial secreting cells (Fig. 5A). The secretions and the epithelial cells serve as nutrition for embryos during pregnancy. The epithelial cells are degraded by fetal teeth (Fig. 8.) (Hraoui-Bloquet and Exbrayat, 1996). When embryos grow in size the wall of oviduct undergoes modification (Fig. 5B,C). There is no epithelium lining the lumen, which is now in direct contact with the connective tissue, the crests are no longer found, the gland zones are reduced, and some muscle tissue is degraded, perhaps as a result of the action of fetal teeth (Wake, 1980a). Fetal teeth found in embryos of viviparous species were well studied by Parker (1956), Parker and Dunn (1964), Wake (1970a, 1976, 1980a), Hraoui-Bloquet (1995), and Hraoui-Bloquet and Exbrayat (1996).

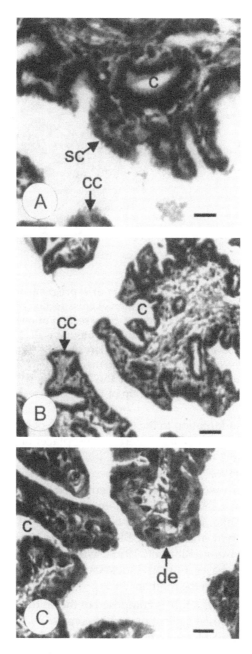

Fig. 5. Cross sections of uterus in adult *Typhlonectes compressicauda* showing changes in epithelium lining. A—Just before pregnancy (February). B—At the beginning of pregnancy (April). C—At the end of pregnancy. Scale bar = 100 μm, c—crypts; cc—ciliated cells; de—degraded epithelium; sc—secretory cells.

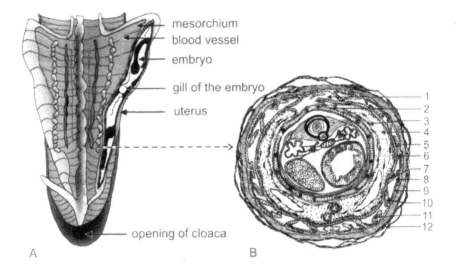

mesorchium
blood vessel
embryo
gill of the embryo
uterus

opening of cloaca

A   B

1
2
3
4
5
6
7
8
9
10
11
12

Fig. 6. A—Schematic of the posterior part of a pregnant female *Typhlonectes compressicauda*. B—Schematic representation of cross section of a pregnant female uterus The dotted line shows the level of cross section of one of the embryos in uterus. Gills are narrowly applied against the uterine wall. 1—lumen of uterus; 2—larval epidermis; 3—gill epithelium on the larval side; 4—gill connective tissue; 5—uterus muscle; 6—gill epithelium on the uterine side; 7—blood vessels; 8—uterine crypt; 9—uterine epithelium at the bottom of a crypt; 10—uterine connective tissue; 11—flattened tip of the uterine crest; 12—serosa.

An additional adaptation to the intrauterine life of embryos was found in *Typhlonectes compressicauda*, where a true placentation involves the embryonic gills and the uterine wall. Gills of *Gymnopis* embryos were studied by Wake (1967, 1969). In *Typhlonectes compressicauda*, the gills are transformed into a vascular structure, mentioned for the first time in the 19th century (Peters, 1874a,b, 1875). More recently, gills were studied by several authors (Wake, 1969; Delsol *et al.*, 1986; Exbrayat, 1986; Sammouri *et al.*, 1990; Exbrayat and Hraoui-Bloquet, 1991; Hraoui-Bloquet and Exbrayat, 1994; Hraoui-Bloquet, 1995). Gills may be involved in the exchanges of water, gases, nutritive substances, and some other molecules between the embryo and the mother.

The placenta-like structure is composed of the epithelium of embryonic gill that is tightly apposed against the naked connective tissue of the uterine wall (Fig. 6B) (Hraoui-Bloquet and Exbrayat, 1994; Hraoui-Bloquet, 1995; Exbrayat and Hraoui-Bloquet, 1992a,b). After parturition the uterus becomes undifferentiated. During the year of sexual quiescence, the uterus again undergoes seasonal changes, although it does not receive any eggs.

The oviducts in young *Typhlonectes compressicauda* are tiny elongated structures. During second year of life, when a female becomes sexually mature, some slight morphological changes of oviducts indicate the beginning of sexual cycle (Exbrayat, 1988a). Similar changes are also seen in the wall of cloaca (Exbrayat, 1996).

Case of Microcaecilia unicolor. An unclear situation was described in *Microcaecilia unicolor*, where some characters of the oviduct resemble that of oviparous species (e.g., the oviduct is divided into three parts), whereas other characters resemble that of viviparous species (e.g., the presence of developed crests in the posterior part). The wall of the anterior region provides some connective crests with undifferentiated cells. Gland cells scattered among the epithelial crests produce acidic mucus. The median part of the lumen is bordered on one side by a straight and smooth wall and by a wall with small developed crests on the other side (Exbrayat, 1989a). The funnel, resembling that of viviparous *Typhlonectes compressicauda*, is long, groove-shaped, lies parallel to the oviduct, to which it is attached by connective tissue, and is associated with the ovary. Its wall is lined with glandular cells that secrete PAS-positive mucus.

In *Microcaecilia unicolor*, some large follicles were observed in all females that were studied. Vitellogenesis starts in oocytes that are 300 μm in diameter. Some compact corpora lutea resembling those of *Typhlonectes compressicauda* were also observed (Exbrayat, 1989a). Study of oviducts of *Microcaecilia unicolor* suggests that this species is viviparous (Exbrayat, 1989a). However, the reproduction of *Microcaecilia unicolor* is unknown, because neither intra-uterine embryos nor egg clutches or free larvae have been observed so far, and only this would allow the proper classification of this species as ovi- or viviparous.

Cloaca

The cloaca is adapted for intromission of the male copulatory organ (phallodeum) during copulation and for internal fertilization. The cloaca in adult female Gymnophiona was studied in the following species: *Caecilia (Oscaecilia) ochrocephala, Gegenophis ramaswamii, Geotrypetes seraphinii, Icthyophis glutinosus, Rhinatrema bicolor, Schistometopum gregorii, Scolecomorphus kirkii, S. vittatus, S. uluguruensis, Siphonops annulatus, Uraeotyphlus menoni, Typhlonectes compressicauda, Idiocranium russeli, Boulengerula taitanus (Afrocaecilia taitana)* and *B. uluguruensis (Afrocaecilia uluguruensis* (Wake, 1972), and, in more detail in *Typhlonectes compressicauda* (Exbrayat, 1991, 1996). The anterior region of the cloaca is attached to the digestive tract, bladder, Wolffian ducts, and oviducts. The anterior region is homologous to both median and posterior (phallodeum) regions in males. Cloacal wall is lined by stratified epithelium. During the period of sexual rest this epithelium becomes keratinized with an apical squamous layer. During the breeding season some apical cells produce a carboxylic acid secretion. During pregnancy cells flatten and mucus is less abundant. Connective tissue is loose during the breeding period and pregnancy (Exbrayat, 1996). Like in males, a *musculus retractor cloacae* is associated with the female cloaca in several species (Wilkinson, 1990).

At birth the cloaca of a female *Typhlonectes compressicauda* is a small duct and its anterior region is lined by mucous epithelium with carboxylic secretions, like in males. The posterior part is lined by an undifferentiated epithelium with some mucous cells and cells with a large vacuole. During breeding period the posterior region is bordered by thick stratified epithelium that produces abundant amounts of mucus. At the end of breeding period this region is lined by keratinized epithelium resembling that of females during sexual rest (Exbrayat, 1991).

Oogenesis in Gymnophiona

Oogenesis in Gymnophiona is rather poorly studied. It is cyclic, with the renewal of oogonia after each breeding season (Fig. 10) and formation of new follicles, accompanied by degeneration of some of them when they achieve maturation stage. The sexual cycle is annual in oviparous species and biennial (every two years) in viviparous species. The sexual cycle is probably a protective adaptation of a female who carries several embryos during long pregnancy (6-7 or 12 months, depending on species); after parturition reserves of a female are exhausted and needed to be renewed.

Morphology of Germ Cells and Ovarian Follicles

Oogonia and early meiocytes. It is generally difficult to distinguish all stages of germ cells in an adult ovary. The onset of oogenesis in young females after metamorphosis was not studied in details and the data are scarce. Primary oogonia have light cytoplasm with dark nuclei, when stained with iron haematoxylin and eosin or Masson-Goldner's trichrome (Fig. 12A). The ovaries of newborn female *Typhlonectes compressicauda* contain some nests of oogonia and primary oocytes up to the pachytene stage (Exbrayat, 1986, 2006a). In more advanced ovaries oogonia form groups of germinal nests surrounded by flat somatic cells. Germinal nests are scattered among diplotene oocytes.

Diplotene oocytes. Diplotene oocytes are classified according to their size and stage of yolk accumulation, as previtellogenic and vitellogenic. In young females, early diplotene oocytes measuring 150 to 300 μm in diameter are surrounded individually by flat follicle cells. At this stage the connective tissue starts to form the theca that surrounds the follicle. In 20-month-old animals some 600 to 750 μm diplotene oocytes with thin vitelline membrane (i.e., oocyte envelope) are observed. In 25-month-old animals vitellogenesis starts and first vitellogenic oocytes and atretic follicles are observed. The follicle cells are flat and tightly joined one to the other, separated from the oocytes by a thin vitelline (oocyte) envelope. At the end of previtellogenesis, the follicles are about 1000 μm in diameter. The biggest vitellogenic oocytes are 1000 to 2000 μm in diameter in the viviparous *Typhlonectes compressicauda* (Fig. 13); 1000 to 1400 μm in direct developing *Boulengerula taitanus* (Raquet *et al.*, 2006), but it can reach 2500 μm in *Typhlonectes* or other species. The cytoplasm of the oocyte is filled with numerous yolk platelets. In *Ichthyophis beddomei*, the follicles reach 750 μm in diameter (Masood Parveez, 1987). Mature oocyte contains large amount of yolk, its nucleus (germinal vesicle) is localized just beneath the oolemma of the future animal pole and is surrounded by a layer of yolk-free cytoplasm.

Berois and de Sa (1988) classified the follicles of *Chthonerpeton indistinctum* as primary, secondary and tertiary, according to their position in the ovary, staining affinity of oocytes, and organization of the follicle wall. According to these authors, the primary follicles are situated in the periphery of an ovary and contain diplotene oocytes with central nuclei containing several nucleoli. Walls of these follicles are composed of flat epithelial (follicle) cells. Secondary follicles are localized deeper in the ovary and contain diplotene oocytes with cytoplasm filled with small PAS-positive granules. Nuclei of oocytes in secondary follicles are localized eccentrically and contain numerous nucleoli and granular chromatin. The PAS-positive vitelline membrane is situated between the oocyte surface and follicle cells. The largest are the tertiary follicles situated deep inside the ovary. The walls of the tertiary follicles are composed of follicle cells and connective tissue (theca). Two types of granules are found in the large oocytes: the cortical granules that are small and situated under the cell membrane, and granules that are intensely PAS-positive and fill the entire cytoplasm.

Beyo *et al.* (2007a) gave some ultrastructural observations of previtellogenic follicles in both *Ichthyophis tricolor* and *Gegenophis ramaswamii*. At transition between primordial and previtellogenic oocytes, the connective theca is differentiated into external and internal layers, and light and dark types of follicle cells are observed.

Eggs and Their Envelopes

Oocyte envelope. Vitellogenic ocytes in the least developed follicles are surrounded by a thin oocyte envelope (vitelline membrane). This envelope is secreted by the oocyte itself and therefore can be classified as the primary envelope. It is PAS-positive and thus contains carbohydrates (Berois and de Sa, 1988). In growing diplotene oocytes the oocyte envelope is separated from the follicle cells by a space, which is penetrated by microvilli of the oocytes, and thereby has an appearance of a striated zone. This zone together with the oocyte envelope is often referred as *zona pellucida*.

In *Ichthyophis tricolor* and *Gegenophis ramaswamii*, Beyo et al. (2007b) gave some ultrastructural information of vitellogenic follicles. Vitellogenesis begins with the appearance of precursors of vitelline envelope in the perivitelline space. Pinocytic vesicles, coated pits and coated vesicles are observed at oocyte surface, indicating contributions of the oocyte to the vitelline envelope and of the granulose cells to yolk material. The follicle cell/oocyte interfaces of these two Caecilians resemble the observations given for anura and urodela.

In *Typhlonectes compressicauda* the oocyte envelope becomes increasingly thick, then becomes more dispersed and the oocytes seem to float inside the follicles. At that time the follicle cells transform from flat to cubical. Some of the follicle cells are disposed on two layers, one against the theca, and the other against the oocyte.

Jelly or mucus envelopes. At ovulation oocytes that are surrounded by the vitelline membrane (oocyte envelope) leave the ovary and reach the anterior part of oviduct where they are surrounded by a thin mucous envelope. This envelope is secreted by the oviduct and thereby can be classified as secondary envelope. In oviparous species the eggs are laid on land after passing the oviduct, where ovulated oocytes are additionally surrounded by several jelly envelopes and fertilized. They are glued together and form a string. In viviparous species the envelope is thin and is called a mucous envelope. In *Typhlonectes compressicauda* 20 to 50 eggs are surrounded by a thin mucous PAS-positive and alcian blue-positive envelope that shows its acidic carbohydric nature. This envelope is secreted by the gland of the anterior region of the oviduct (Exbrayat, 1986, 1988a; Exbrayat and Delsol, 1988) and is the equivalent of the jelly envelope in oviparous species. The diameter of eggs together with envelops in oviparous species ranges from 4 to 10 mm, whereas in viviparous species this diameter does not exceed 2 to 3 mm (Wake, 1977b; Exbrayat and Delsol, 1988).

The gland system of oviduct in Gymnophiona is less complex than that in Anura and Urodela (Salthe and Mecham, 1974), where eggs are surrounded by more layers (up to five), according to the number of glandular zones (Salthe, 1963). The envelopes produced by the oviduct in Gymnophiona resemble jelly coats in Anura and Urodela, but are less numerous.

In oviparous species eggs after oviposition are attached to each other by a gelatinous substance (*Gegenophis carnosus*, Seshachar, 1942; *Ichthyophis malabarensis,* Seshachar *et al.,* 1982; Balakrishna *et al.,* 1983). The same situation has been observed in the direct developing *Boulengerula taitanus* (Malonza and Measey, 2005). In *Siphonops annulatus,* the elastic string is not broken during oviposition and the eggs are not separated (Goeldi, 1899). Breckenridge and Jayasinghe (1979) revealed that the string, which kept the eggs together in *Ichthyophis glutinosus,* was broken on each side of an egg, and in that way each egg acquired its own individual envelope. The envelope is rough and elastic and its external part is constituted of several concentric layers of fibers that positively react to the specific staining of the elastic connective tissue. An extension of this region constitutes the string linking each egg to the other. The internal part stains weaker and corresponds to an area, which lacks elastic fibers. This envelope surrounds a cavity filled with fluid, in which the embryo develops.

Postovulatory Follicles and Corpora Lutea Formation

In the viviparous and several oviparous species, the follicle emptied after ovulation becomes a *corpus luteum.* Several *corpora lutea* were observed in *Hypogeophis rostratus* (Tonutti, 1931), *Ichthyophis beddomei* (Masood Parveez, 1987), *Boulengerula taitanus* (Raquet *et al.,* 2006), and other species (Wake, 1968). *Corpora lutea* were observed also throughout the pregnancy in *Dermophis mexicanus* (Wake, 1977a,b). The presence of *corpora lutea* during pregnancy in viviparous species is related to the intrauterine development of embryos. It has been shown that *corpora lutea* of *Typhlonectes compressicauda* are able to synthesize progesterone.

The *corpora lutea* are 1200 to 2000 µm in diameter and persist throughout pregnancy and are modified according to the developmental stages of the embryos that develop inside the oviducts (Exbrayat, 1986, 2006b; Exbrayat and Collenot, 1983). At this stage the blood vessels are limited to the peripheral theca. After ovulation the follicle cells proliferate and invade the central cavity of the follicle (Fig. 11A). These cells are spherical (30 µm in diameter) or elongated (20 × 40 µm), and arranged in about 10 layers. The center of the follicle cavity is filled by an amorphous substance, which is not digested by the follicle cells. In the periphery of the follicle the cells originating from the theca penetrate between the follicle cells and some blood vessels are observed between the cells (Fig. 11B). Then the *corpora lutea* decrease and their central cavities are filled by the vacuolated cells. The *corpora lutea* degenerate at the end of pregnancy (Fig. 11C,D). The invading cells become high or elongated with degenerated nuclei and numerous blood cells are observed inside the follicle. Finally, the degenerative *corpora lutea* resemble small masses that are quickly integrated with the surrounding connective tissue. In *Typhlonectes compressicauda,* about 20 *corpora lutea* are observed in each ovary. At the end of pregnancy the number of *corpora lutea* decreases as a result of resorption. The *corpora lutea* seem to be functional during a great part of gestation, but at the end they can be no longer essential.

This was not observed in *Rhinatrema* (Wake, 1968), in which *corpora lutea* are in most cases filled with fat masses. In numerous species, *corpora lutea* contain a fibrous connective tissue. The *corpora lutea* evolve according to the stage of the embryos found in uterus of viviparous species, such as *Dermophis mexicanus* (Wake, 1980b) and *Typhlonectes compressicauda* (Exbrayat, 1983, 1986, 1992; Exbrayat and Collenot, 1983), and they are involved in hormonal regulation of reproduction.

Several *corpora lutea* have been observed in the direct developing *Boulengerula taitanus*. Just after ovulation these structures are invaded with cells issued from granulosa and blood vessels issued from connective theca. After egg-laying, these *corpora lutea* degenerate in compact structures with a single type of cells disposed in a disorganized manner (Raquet *et al.*, 2006).

Atretic Follicles

In both oviparous and viviparous species some diplotene oocytes degenerate and such follicles transform into atretic follicles (Fig. 14). Two types of atresia were observed in *Typhlonectes compressicauda* (Exbrayat, 1986) and *Ichthyophis beddomei* (Masood Parveez, 1987). In the first that concern previtellogenic oocytes, a gap is formed between follicle wall and oocyte, then oocytes degenerate inside the follicles and are resorbed. The second type concerns the atresia of follicles that contain vitellogenic oocytes. The degeneration begins when hypertrophied follicle cells invade and phagocytize the oocyte, thereby digesting the ooplasm. At the same time numerous blood vessels invade the theca. The phagocytic follicle cells fill the space left by the degenerating oocyte and become filled with brown pigment. The thecal cells proliferate and the follicle becomes a mass integrated in the connective tissue of the ovary, and finally degenerate. Atretic follicles have been also observed in *Boulengerula taitanus* (Raquet *et al.*, 2006) and other species (Exbrayat, 2006b). The degeneration described in Gymnophiona resembles that in Anura (for details see the part 'Oogenesis and Female Reproductive System in Anura' in this volume).

Number of Eggs and Embryos

The number of oocytes ovulated during breeding season is species specific. The number of ovulated oocytes is more important in oviparous than in viviparous species (Wake, 1977b). Some species lay a low number of eggs: *Idiocranium russelii* (Wake, 1977b), *Siphonops annulatus* (Goeldi, 1899), and

Siphonops paulensis (Gans, 1961) lay only 5 to 6 eggs, whereas *Ichthyophis malabarensis* lays about 100 eggs (Balakrishna *et al.*, 1983). The number of mature eggs in viviparous species does not exceed 30. In *Geotrypetes seraphinii*, from 4 to 30 mature oocytes per female were observed (Wake, 1968b, 1977b). In *Gymnopis multiplicata, Dermophis mexicanus* (Wake, 1977b), and *Typhlonectes compressicauda* (Exbrayat, 1986), from 20 to 30 mature oocytes were counted per individual.

Among 11-28 oocytes ovulated by each ovary in *Typhlonectes compressicauda* only 6 develop into embryos (Exbrayat 1986). Then, just before birth, only 1-3 embryos are found in one uterus and dead fetuses were found in oviducts containing embryos at larval stages. Similar situation was described in *Dermophis mexicanus* (Wake, 1980b) and *Chthonerpeton indistinctum* (Barrio, 1969; Prigioni and Langone, 1983a,b). Although very few data are available about the number of eggs and embryos in viviparous Gymnophiona, it seems that females of viviparous species can give birth to about ten young, and that the number of ovulated eggs is higher than the number of young.

Females become sexually mature usually as 24- to 30-month-old animals. However, it cannot be ruled out that some females become sexually mature one year earlier. The smallest pregnant females of *Typhlonectes compressicauda* were 300 to 330 mm in length, which corresponds to the size of two years old animals. In *Geotrypetes seraphinii*, sexual maturity was also recorded in two years old females (Wake, 1977b).

Hormonal Control of Reproduction

Fertilization and Development

Fertilization in both oviparous and viviparous Gymnophiona occurs in the anterior part of oviduct (*pars recta*) (Wake and Dickie, 1998; Exbrayat *et al.*, 1981; Delsol *et al.*, 1981; Hraoui-Bloquet, 1995; Hraoui-Bloquet and Exbrayat, 1993, Exbrayat and Hraoui-Bloquet, 2006; Exbrayat, 2006c). Fertilization was only described in *Typhlonectes compressicauda* by Hraoui-Bloquet and Exbrayat (1993) and Hraoui-Bloquet (1995). During fertilization, several spermatozoa are groupped at the animal pole of the oocyte (Fig. 7A). A perivitelline space appears between the oolemma and vitelline membrane (Fig. 7B). This space, as well as the space situated between the mucous and vitelline membrane, contains some unknown hyaline substances. Spermatozoa are attached to the mucous envelope, several of them penetrate the oocyte, but only one is used for fertilization, whereas the others degenerate. These observations suggest physiological polyspermic fertilization in this species of Gymnophiona.

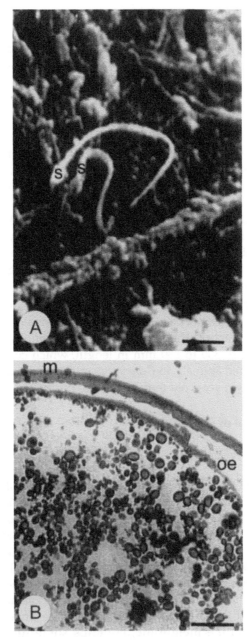

Fig. 7. A—Fertilization in *Typhlonectes compressicauda*. A—SEM view of the external part of a mucous envelope of the egg during fertilization. Scale bar = 2 μm. B—Transverse section of a fertilized ovum. Scale bar = 20 μm. m—mucous envelope; oe—oocyte envelope; s—spermatozoon.

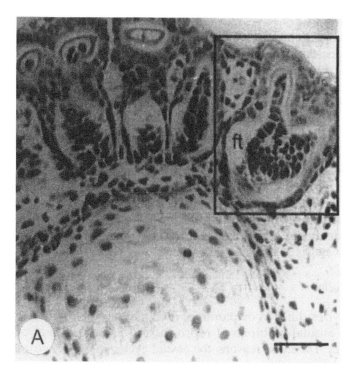

Fig. 8. Section of the lower jaw of the embryo *Typhlonectes compressicauda*
at stage 25-26 with developing fetal teeth (ft) shown in a frame.
Scale bar = 140 μm.

The period of pregnancy were more extensively studied in *Dermophis mexicanus* (Wake, 1980b) and *Typhlonectes compressicauda* (Exbrayat *et al.*, 1981; Exbrayat and Delsol, 1985; Exbrayat, 1986; Hraoui-Bloquet and Exbrayat, 1992, 1996; Hraoui-Bloquet *et al.*, 1994; Exbrayat and Hraoui-Boquet, 1992a,b; Hraoui-Bloquet, 1995). Although in the same pregnant female all embryos are of the same age, at the end of pregnancy the embryos can be at various stages of development (Delsol *et al.*, 1981). Some damaged ova and underdeveloped embryos are found among the properly developing ones. In this case, underdeveloped embryos are dead and more or less dehydrated. At the beginning of development, embryos are surrounded by a mucous envelope. The embryos are situated in the anterior part of the uterus, and are separated one from another by at least 10 mm distance. Then embryos after hatching, i.e., lacking mucous envelope, become fetuses. They are scattered along the uterus and occupy as much space as available. The yolk mass is resorbed and the embryos feed on the secretions produced by the uterus wall (Exbrayat, 1986). The large gills are spread out on each side of the embryo and applied tightly against the wall of the uterus. Finally, the uterine wall is entirely covered by gills of the larvae (Fig. 6B). Larvae are elongated and folded, becoming U or S shaped, inside the space limited by the gills. The gills form a cocoon-like structure with one gill disposed as a coat and the other as a hood. At the end of pregnancy, the uterus measuring 10 mm in length can contain three or four embryos measuring from 150 to 200 mm each.

The development of *Typhlonectes compressicauda* can be divided into several stages (Fig. 9 A-C). (Delsol *et al.*, 1981; Exbrayat *et al.*, 1981; Sammouri *et al.*, 1990). Stages 0 to 13 group the earliest stages of development. At stages 14 to 25, the embryos have more or less abundant yolk and are surrounded by a mucous envelope. Hatching is observed at stage 25 or 26, and at stages 26 to 31 the yolk mass is already resorbed and the fetuses are no longer surrounded by a mucous envelope. They are free inside uterus, where they can move. Metamorphosis occurs approximately at stages 30 to 32, and the birth occurs at stage 34.

Sexual Cycles

Sexual cycles differ in oviparous and viviparous species mainly because of their annual or biennial reproduction and hormonal control of pregnancy. For these reasons the sexual cycles in these two groups are described separately.

Oviparous species. The female sexual cycle in oviparous Gymnophiona is annual and discontinuous. The cycle of *Ichthyophis beddomei* was described by Masood Parveez (1987) and Masood Parveez and Nadkarni (1993a,b). Breeding lasts from December to February during the dry season when numerous ovulated oocytes are present in the oviducts. The oogonia undergo mitotic proliferation in January-February. From March until July the ovary regresses and a few fully grown follicles that were not ovulated become atretic and degenerate. At the same time a new wave of diplotene oocyte starts growing in size and first vitellogenic oocytes appear in August.

Breeding of *Ichthyophis* in Mysore was observed from December to March also during the dry season and hatching occurred in May-June at the first rains (Seshachar, 1936). On the other hand, in *I. glutinosus* from Sri Lanka the copulation and oviposition was observed during the wet season that lasts from May to September (Sarasin and Sarasin, 1887-1890; Breckenridge and Jayasinghe, 1979).

In *Boulengerula taitanus* (Raquet *et al.*, 2007), females are submitted to an annual reproductive cycle, rainfall or temperature being declenching factor. At dry season (January-March), during which the animals live deeply in the soil, germinal nests activity increased. This activity decreases

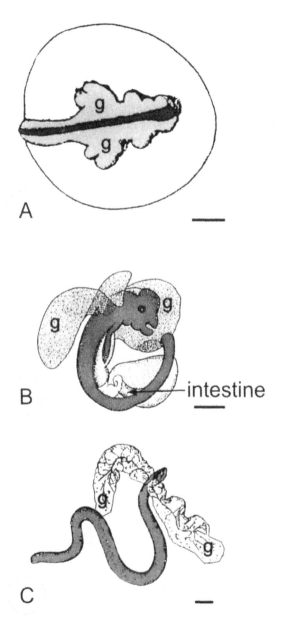

Fig. 9. Development of *Typhlonectes compressicauda*. A—Stage 20, two pairs of gill buds are situated on the posterior part of the head. Scale bar = 1 mm. Modified after Delsol *et al.* 1981. B—Stage 26, gills begins to develop as a pair of vascular? structure. Scale bar = 1 mm. C—Stage 34, a pair of vesiculous? gills on each side of the head. Scale bar = 5 mm (Modified after Toews and MacIntyre, 1971. g—gill).

at beginning of long rainfalls (March-May). In the long dry season (June-October), the vitellogenic follicle development is stopped. At the end of dry season and at the beginning of short rainy one (November-December), ovulation and fertilization occur.

Viviparous species. In *Chthonerpeton indistinctum* the sexual cycle is annual and pregnancy lasts 4 months (Barrio, 1969). Breeding occurs in August and September and births were observed in January-February. Other viviparous species such as in *Dermophis mexicanus* (Wake, 1980b) and *Gymnopis multiplicata* (Wake, personal communication) have biennial sexual cycles (Fig. 10A,B). This cycle is characterized by one season of

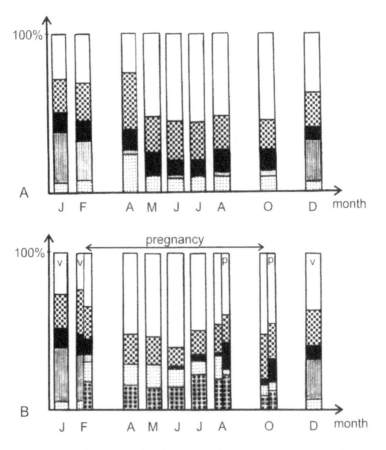

Fig. 10. Variation of relative quantities of follicles (based on size) during the reproductive cycle in female *Typhlonectes compressicauda*. A—Year of sexual quiescence; B—Year with pregnancy. p—post parturition; v—vitellogenesis. Follicle size: ☐ 150-600 μm; ▦ 600-750 μm; ■ 750-1200 μm; ⊞ more than 1200 μm; ▦ corpora lutea; ▦ atretic follicles.

reproduction followed by one season of sexual quiescent. The sexual cycle in *Dermophis mexicanus* is also biennial, and both pregnant and resting females were observed throughout the year (Wake, 1980a). In June (beginning of the cycle), the oviducts of some females contain very young embryos. The parturition occurs after one year of pregnancy in May-June. The next year of the cycle is a period of sexual quiescence

Typhlonectes compressicauda (Exbrayat, 1983, 1986, 1988a; Exbrayat *at al.*, 1981) may serve as another example of species with biennial cycle. Breeding and sexual activity occur during rainy season (January until May), then gestation starts during dry season (April until September). Females give birth to young in September after 6 or 7 months of pregnancy. During sexual rest the ovary mimics the activity characteristic of breeding period and a new wave of oocyte maturation occurs. However, the biggest oocytes degenerate and transform into atretic follicles, and during that time

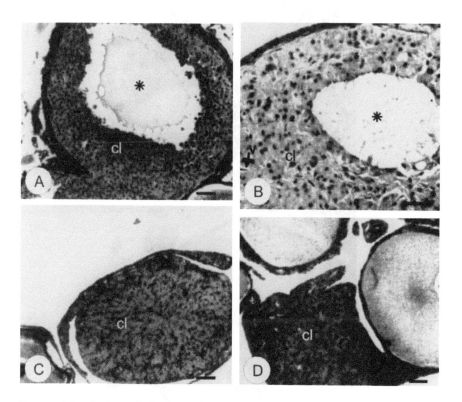

Fig. 11. Morphological changes of *corpus luteum* in *Typhlonectes compressicauda*. A—Beginning of pregnancy. Scale bar = 100 μm. B—Middle of pregnancy. Scale bar = 30 μm. C—End of pregnancy. Scale bar = 30 μm. D—Regressive corpus luteum after birth. Scale bar = 10 μm. *—cavity; cl—corpus luteum.

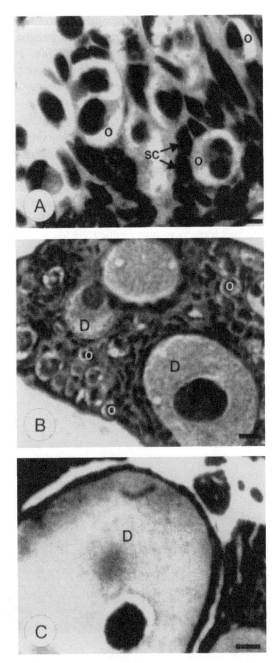

Fig. 12. Cross sections of ovaries of *Typhlonectes compressicauda* A—Germinal nest in a new-born. Scale bar = 10 µm. B—Diplotene oocyte. Scale bar for B and C = 30 µm. D—diplotene oocyte; o—oogonium; sc—somatic cells.

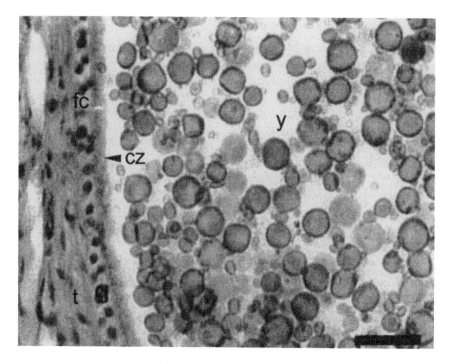

Fig. 13. Mature follicle in ovary of *Typhlonectes compressicauda*.
Scale bar = 100 μm. fc—follicular cells; cz—cortical zone;
t—connective tissue theca; y—yolk.

the ovary contains only small follicles with previtellogenic oocytes. At the time when breeding season occurs, germinal nests are conspicuous and some oogonia divide mitotically, although the female is resting. Most of oocytes are at pre-diplotene stages, but some big diplotene oocytes are also present and germinal nests increase in size. Germinal nests remain well developed and contain proliferating oogonia until the end of breeding season. Then, before the next reproductive season, the germinal nests become again mitotically active. During pregnancy the germinal nests in ovaries are reduced in size, but some dividing oogonia and early meiotic oocytes are still present, although they are less numerous.

A study of several individuals maintained in the laboratory gave evidence that female sexual cycle in *Typhlonectes compressicauda* was disturbed if the natural conditions were not respected. In particular the dry season seems to be a releasing factor for preparing the ovaries and genital ducts for breeding. On the contrary, the external conditions do not seem to be so important in regulation of a male sexual cycle, in which intrinsic or genetic factors seem to predominate (Exbrayat and Laurent, 1986).

Pituitary

The role of the pituitary in regulation of the sexual cycle in Gymnophiona is poorly known and most studies were devoted to its anatomy and histology (Stendell, 1914; Laubmann, 1927; Pillay 1957; Pasteel and Herlant, 1961; Kuhlenbeck, 1970; Gabe, 1972). Cytological changes of adenohypophysis (i.e., glandular part of the pituitary) were studied and correlated with sexual activity and cyclic changes of the target organs in *Chthonerpeton indistinctum* (Schubert *et al.*, 1977), *Ichthyophis beddomei* (Bhatta, 1987), and *Typhlonectes compressicauda* (Zuber-Vogeli and Doerr-Schott, 1981; Doerr-Schott and Zuber-Vogeli, 1984, 1986; Exbrayat, 1989b; Exbrayat and Morel, 1990-1991, 1995).

Four types of glandular cells were identified in the pituitary of Gymnophiona: the gonadotropic cells, the lactotropic cells, the corticotropic cells, and the somatotropic cells.

The gonadotropic cells in *Ichthyophis beddomei* are big with PAS-positive cytoplasm stained by alcian blue and aldehyde fuchsin. The size of gonadotropic cells reaches its maximum before, and decreases after breeding. The gonadotropic cells in *Typhlonectes compressicauda* secrete the gonadotropin LH, as was shown immunocytologically after binding the anti-LH serum. At the beginning of the biennial cycle the production of

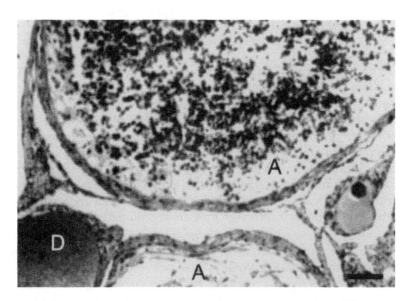

Fig. 14. Atretic follicles with degenerating oocytes in ovary of *Typhlonectes compressicauda*. Scale bar = 100 μm. Scale bar = 10 μm. A—atretic follicles; D—diplotene oocyte.

LH increases and the gonadotropic cells are progressively filled with pink or purple granules and some orange stained globules in the blue cytoplasm, when stained with trichrome. In females pregnant with young embryos, the size of these cells reaches their maximum. Later during pregnancy, the size of LH producing cells gradually decreases and reaches its minimum at parturition. At that time they contain no granules what suggests that secretion of LH decreases. The size of the cells is correlated with the size of their nuclei (Fig. 15A). At the beginning of the quiescent year of the sexual cycle, the gonadotropic cells develop again and look like those observed at the same time of the previous reproductive year.

The lactotropic cells produce prolactin PRL and their cytoplasm, filled with some big globules stains orange with Cleveland and Wolfe's trichroma, bind the anti-prolactine serum. Only some slight changes of the cell sizes are observed. The lactotropic cells reach the maximum size at parturition after a slight decrease when the ovulation occurs. Then the cells progressively decrease to reach the minimal size. At the beginning of the second (quiescent) year of the cycle, the lactotropic cells again increase and degenerate according to the same rhythm as during the reproductive year. During sexual quiescence, the size of lactotropic cells varies in the same pituitary, and before the start for a new cycle the size of these cells again reaches their maximum (Fig. 15B). The number and size of PRL secreting cells varies throughout the cycle. In females with vitellogenic oocytes, from the beginning to the middle of pregnancy, there are few big cells, whereas at the end of pregnancy there are many small cells. From the beginning to the end of pregnancy, an increase of mRNAs coding for PRL is observed. After parturition, during the vitellogenesis and sexual rest, the PRL mRNAs level remains constant and low (Exbrayat and Morel, 1995). It suggests that synthesis of PRL is low during vitellogenesis, as in the resting females. At the beginning of pregnancy the PRL level slightly increases. At parturition, the size of cells is reduced, and the cells do not synthesise the PRL mRNAs. During the second year of the cycle, the number of cells increases, but the cell size and the PRL mRNA synthesis is unchanged.

The corticotropic cells produce the adrenocorticotropic hormone and react with the anti-adrenocorticotropic hormone serum. The cycle of the corticotropic cells is correlated with the activity of adrenocortical cells in the interrenal glands. In females pregnant with embryos at the beginning or at the middle stages of development these cells are particularly numerous and big, thus suggesting high level of the hormone production. At the end of pregnancy and at parturition its level decreases and the cytoplasm looks very dense; it is pale purple after the Cleveland and Wolfe's staining. During the sexual quiescence the size of cells decreases.

Before the next season the volume of cell increases again and the cytoplasm is filled with granules.

The somatotropic cells produce the growth hormone GH and react with the anti-GH serum. Before the breeding season the somatotropic cells reach their maximal size. During pregnancy the cells sizes do not vary. During sexual quiescence some biometric variations are observed: the cells are large in May and August and small in April and June.

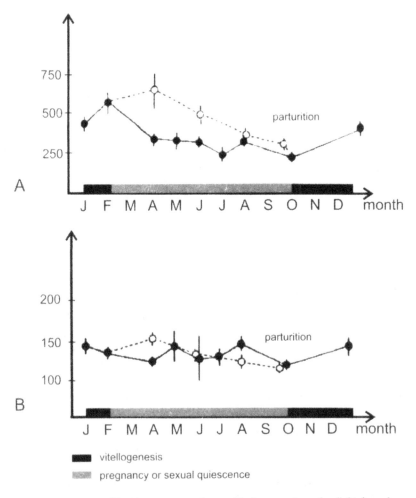

Fig. 15. Activity of hormone secreting cells in ovaries of adult female *Typhlonectes compressicauda* shown according to the size of their nuclei. Solid lines represent non pregnant females; dashed lines represent pregnant females. A—Gonadotropic cells. B—Lactotropic cells (Modified after Exbrayat and Morel, 1990-1991).

Ovary

The endocrine function of ovaries in Gymnophiona has not been well studied so far. However, steroid hormones were detected in follicle cells and sometimes in the connective tissue of follicular theca in two species, namely in the oviparous *Ichthyophis beddomei* (Masood-Parveez, 1987; Masood Parveez and Nadkarni, 1993a and b), and in the viviparous *Typhlonectes compressicauda* (Exbrayat, 1986, 1992; Exbrayat and Collenot, 1983).

A study of mRNAs coding for the PRL receptor by an *in situ* hybridization technique indicated an increase of PRL mRNAs in ovaries (Exbrayat and Morel, 2003). At the beginning of pregnancy, the level of these RNAs is low in early *corpora lutea* (with a cavity). The level of the RNAs is high in older *corpora lutea*, and the RNAs are absent in the degenerative ones (Exbrayat and Morel, 2003). On the other hand, mRNAs coding for PRL receptors were more abundant in the liver of females during the pregnancy than during the sexual rest. These observations show the importance of the role of PRL in regulation of the sexual cycle, gestation and also hydromineral balance like in Anura and Urodela (Exbrayat *et al.*, 1997), but the exact action of this hormone has not been still studied in Gymnophiona.

Steroidogenesis in follicles. In *Ichthyophis beddomei*, the histochemical detection of δ5 3β HSDH and 17β HSDH gave positive results in the follicle cells, cytoplasm of previtellogenic and vitellogenic oocytes, and in hypertrophied cells of the atretic follicles. The ovary of *Ichthyophis beddomei* contains the enzymes necessary to biosynthesis of the steroids (Masood-Parveez, 1987).

In *Typhlonectes compressicauda*, δ5 3β HSDH was detected in follicle cells surrounding the previtellogenic, vitellogenic and mature oocytes, and in cells of the internal theca. The use of an antibody directed against the estrogenic hormone (estriol and 17β estradiol) shows that the labeling is observed mainly in the follicles containing vitellogenic or mature oocytes. No label was observed in the atretic follicles (Exbrayat, 1992; Exbrayat and Collenot, 1983). The activity of δ5 3β HSDH activity was observed in cells of the *corpora lutea*, more specifically in the cells deriving from follicular cells. In the degenerative *corpora lutea* no δ5 3β HSDH was detected. The incubation of ovaries of pregnant females with radioactive pregnenolone (a precursor of the progesterone) has shown that the ovary is able to synthesize the progesterone, which is an important hormone responsible for maintenance of embryos in the uterus.

336

References

Anjubault E, JM Exbrayat. 2004. Contribution à la connaissance de l'appareil genital de *Typhlonectes compressicauda* (Dumeril and Bibron, 1841), Amphibien Gymnophione. I. Gonadogenèse. Bull. Mens. Soc. Linn. Lyon 73: 379-392.

Anjubault E, JM Exbrayat. 2006. Development of gonads. In: Reproductive Biology and Phylogeny of Gymnophiona (Caecilians). (Ed.) JM Exbrayat, Vol. 5 of Series Reproductive Biology and Phylogeny, (Ed.) BGM Jamieson Science Publishers, Enfield, USA, Jersey, Plymouth, UK, pp. 291-302.

Balakrishna TA, KR Gundappa, S Katre. 1983. Observations on the eggs and embryo of *Ichthyophis malabarensis* (Taylor)(Apoda: Amphibia). Curr. Sci. 52: 990-991.

Barrio A. 1969. Observaciones sobre *Chthonerpeton indistinctum* (Gymnophiona, Caecilidae) y su reproduccion. Physis. 28: 499-503.

Berois N, R De Sa. 1988. Histology of the ovaries and fat bodies of *Chthonerpeton indistinctum*. J. Herp. 22: 146-151.

Beyo RS, P Sreejith, L Divya, OV Oommen, MA Akbarsha. 2007a. Ultrastructrural observations of previtellogenic ovarian follicles of the caecilians *Ichthyophis tricolor* and *Gegenophis ramaswamii*. J. Morph. 268: 329-342.

Beyo RS, L Divya, OV Oommen, MA Akbarsha. 2007b. Stages in follicle cell/ oocyte interface during vitellogenesis in caecilians *Ichthyophis tricolor* and *Gegenophis ramaswamii*: A transmission electron-microscopic study. Cell Tiss. Res. 331: 519-528.

Bhatta GK. 1987. Some Aspects of Reproduction in the Apodan Amphibian *Ichthyophis*. Ph.D. Dissertation, Karnatak University, Dharwad, India.

Brauer A. 1902. Beitrage zur kenntniss der Entwicklung und Anatomie der Gymnophionen. III. Die Entwicklung der Excretionsorgane. Zool. Jb. Anat. 16: 1-176.

Breckenridge WR, S Jayasinghe. 1979. Observations on the eggs and larvae of *Ichthyophis glutinosus*. Ceylon J. Sci. (Biol. Sci.) 13: 187-202.

Chatterjee BK. 1936. The anatomy of *Uraeotyphlus menoni* Annandale. Part I: Digestive, circulatory, respiratory and urogenital systems. Anat. Anz. 81: 393-414.

Delsol M, JM Exbrayat, J Flatin, M Gueydan-Baconnier. 1986. Nutrition embryonnaire chez *Typhlonectes compressicaudus* (Duméril et Bibron, 1841) Amphibien Apode vivipare. Mem. Soc. Zool. Fr. 43: 39-54.

Delsol M, J Flatin, JM Exbrayat, J Bons. 1981. Développement de *Typhlonectes compressicaudus* Amphibien Apode vivipare. Hypothèse sur sa nutrition embryonnaire et larvaire par un ectotrophoblaste. C. R. Séanc. Acad. Sci. Paris, ser. 293: 281-285.

Doerr-Schott J, M Zuber-Vogeli. 1984. Immunohistochemical study of the adenohypophysis of *Typhlonectes compressicaudus* (Amphibia, Gymnophiona). Cell Tiss. Res. 235: 211-214.

Doerr-Schott J, M Zuber-Vogeli. 1986. Cytologie et immunocytologie de l'hypophyse de *Typhlonectes compressicaudus*. Mem. Soc. Zool. Fr. 43: 77-79.

Dunn ER. 1928. Notes on central American Caecilians. Proc. N. Engl. Zool. Cl. 10: 71-76.

Exbrayat JM. 1983. Premières observations sur le cycle annuel de l'ovaire de

Typhlonectes compressicaudus (Duméril et Bibron, 1841), Batracien Apode vivipare. C. R. Séanc, Acad. Sci. Paris 296: 493-498.

Exbrayat JM. 1984. Quelques observations sur l'évolution des voies génitales femelles de *Typhlonectes compressicaudus* (Duméril et Bibron, 1841), Amphibien Apode vivipare, au cours du cycle de reproduction, C. R. Séanc, Acad. Sci. Paris 298: 13-18.

Exbrayat JM. 1986. Quelques aspects de la biologie de la reproduction chez *Typhlonectes compressicaudus* (Duméril et Bibron, 1841), Amphibien Apode. Doctorat ès Sciences Naturelles, Université Paris VI.

Exbrayat JM. 1988a. Croissance et cycle des voies génitales femelles chez *Typhlonectes compressicaudus* (Duméril et Bibron, 1841), Amphibien Apode vivipare. Amphibia Reptilia 9: 117-137.

Exbrayat JM. 1988b. Variations pondérales des organes de réserve (corps adipeux et foie) chez *Typhlonectes compressicaudus*, Amphibien Apode vivipare au cours des alternances saisonnières et des cycles de reproduction. Ann. Sci. Nat., Zool. 13ᵉ ser. 9: 45-53.

Exbrayat JM. 1989a. Quelques observations sur les appareils génitaux de trois Gymnophiones; hypothèses sur le mode de reproduction de *Microcaecilia unicolor*. Bull. Soc. Herp. Fr. 52: 34-44.

Exbrayat JM. 1989b. The cytological modifications of the distal lobe of the hypophysis in *Typhlonectes compressicaudus* (Duméril and Bibron, 1841), Amphibia Gymnophiona, during the cycles of seasonal activity. I—In adult males. Biol. Struct. Morph. 2: 117-123.

Exbrayat JM. 1991 Anatomie du cloaque chez quelques Gymnophiones. Bull. Soc. Herp. Fr. 58: 30-42.

Exbrayat JM. 1992. Reproduction et organes endocrines chez les femelles d'un Amphibien Gymnophione vivipare, *Typhlonectes compressicaudus*, Bull. Soc. Herp. Fr. 64: 37-50.

Exbrayat JM. 1996. Croissance et cycle du cloaque chez *Typhlonectes compressicaudus* (Duméril et Bibron, 1841), Amphibien Gymnophione. Bull. Soc. Herp. Fr. 121: 99-104.

Exbrayat JM. 2006a. Oogenesis and folliculogenesis. In: Reproductive Biology and Phylogeny of Gymnophiona (Caecilians), (Ed.) JM Exbrayat Vol. 5 of Series Reproductive Biology and Phylogeny BGM Jamieson (Ed.) Science Publishers, Enfield, USA, Jersey, Plymouth, UK, pp. 275-290.

Exbrayat JM. 2006b. Endocrinology of reproduction in Gymnophiona. In: Reproductive Biology and Phylogeny of Gymnophiona (Caecilians), (Ed.) JM Exbrayat, Vol. 5 of Series Reproductive Biology and Phylogeny, BGM Jamieson (Ed.) Science Publishers, Enfield, USA, Jersey, Plymouth, UK, pp. 183-229.

Exbrayat JM. 2006c. Fertilization and embryonic development. In: Reproductive Biology and Phylogeny of Gymnophiona (Caecilians), (Ed.) JM Exbrayat, Vol. 5 of Series Reproductive Biology and Phylogeny, BGM Jamieson (Ed.) Science Publishers, Enfield, USA, Jersey, Plymouth, UK, pp. 357-386.

Exbrayat JM, E Anjubault. 2005. The development, differentiation and growth of gonads in *Typhlonectes compressicauda* (Amphibia, Gymnophiona). In: Herpetologia Petropolitana, (Eds) N Ananjeva, O Tsinenko. Russ. J. Herp. 12 (suppl.): 136-139.

Exbrayat JM, G Collenot. 1983. Quelques aspects de l'évolution de l'ovaire de *Typhlonectes compressicaudus* (Duméril et Bibron, 1841), Batracien Apode vivipare. Etude quantitative et histochimique des corps jaunes. Rep. Nutr. Dev. 23: 889-898.

Exbrayat JM, M Delsol. 1985. Reproduction and growth of *Typhlonectes compressicaudus*, a viviparous Gymnophione. Copeia: 1985: 950-955.

Exbrayat JM, M Delsol. 1988. Oviparité et développement intra-utérin chez les Gymnophiones. Bull. Soc. Herp. Fr. 45: 27-36.

Exbrayat JM, S Hraoui-Bloquet. 1991. Morphologie de l'épithélium branchial des embryons de *Typhlonectes compressicaudus* (Amphibien Gymnophione) étudié en microscopie électronique à balayage. Bull. Soc. Herp. Fr. 57: 45-52.

Exbrayat JM, J Flatin. 1985. Les cycles de reproduction chez les Amphibiens Apodes. Influence des variations saisonnières. Bull. Soc. Herp. Fr. 110: 301-305.

Exbrayat JM, S Hraoui-Bloquet. 1992a. Evolution de la surface branchiale des embryons de *Typhlonectes compressicaudus*, Amphibien Gymnophione vivipare, au cours du développement. Bull. Soc. Herp. Fr. 117: 340.

Exbrayat JM, S Hraoui-Bloquet. 1992b. La nutrition embryonnaire et les relations foeto-maternelles chez *Typhlonectes compressicaudus*, Amphibien Gymnophione vivipare. Bull. Soc. Herp. Fr. 61: 53-61.

Exbrayat JM, S Hraoui-Bloquet. 2006. Viviparity in *Typhlonectes compressicauda*. In: Reproductive Biology and Phylogeny of Gymnophiona (Caecilians), (Ed.) JM Exbrayat, Vol. 5 of Series Reproductive Biology and Phylogeny, BGM Jamieson (Ed.) Science Publishers, Enfield, USA, Jersey, Plymouth, UK, pp. 325-357.

Exbrayat JM, MT Laurent. 1986. Quelques observations sur la reproduction en élevage des *Typhlonectes compressicaudus*, Amphibien Apode vivipare. Possibilité de rythmes endogènes. Bull. Soc. Herp. Fr. 40: 52-62.

Exbrayat JM, G Morel. 1990-1991. The cytological modifications of the distal lobe of the hypophysis in *Typhlonectes compressicauda* (Dumeril and Bibron, 1841), Amphibia, Gymnophiona, during the cycles of seasonal activity. II—In adult females. Biol. Struct. Morph. 3: 129-138.

Exbrayat JM, G Morel. 1995. Prolactin (PRL)-coding mRNA in *Typhlonectes compressicaudus*, a viviparous gymnophionan Amphibian. An in situ hybridization study. Cell Tiss. Res. 280: 133-138.

Exbrayat JM, G Morel. 2003. Visualization of gene expression of prolactin-receptor (PRL-R) by in situ hybriodization in reproductive organs of *Typhlonectes compressicauda*, Gymnophionan Amphibian. Cell Tiss. Res. 312: 361-367.

Exbrayat JM, M Delsol, J Flatin. 1981. Premières remarques sur la gestation chez *Typhlonectes compressicaudus* (Duméril et Bibron, 1841) Amphibien Apode vivipare. C. R. Acad. Sci. Paris 292: 417-420.

Exbrayat JM, A Ouhtit, G Morel. 1997. Visualization of gene expression of prolactin receptors (PRL-R) by in situ hybridisation, in *Typhlonectes compressicaudus*, a Gymnophionan Amphibian. Life Science 61: 1915-1928.

Fuhrmann O. 1914. Le genre *Typhlonectes*. Mem. Soc. Neuch. Sci. Nat. 5: 112-138.

Gabe M. 1972. Contribution à l'histologie du complexe hypothalamo-hypophysaire d'*Ichthyophis glutinosus* L. (Batracien Apode). Acta Anat. 81: 253-269.

Gans C. 1961. The first record of egg laying in the Caecilian *Siphonops paulensis* Boettger. Copeia: 26-27.

Garg BL, J Prasad. 1962. Observations of the female urogenital organs of limbless amphibian *Uraeotyphlus oxyurus.* J. Anim. Morph. Physiol. 9: 154-156.

Goeldi E. 1899. Ueber die Entwicklung von *Siphonops annulatus.* Zool. Jb. 12: 170-173.

Hraoui-Bloquet S. 1995. Nutrition embryonnaire et relations materno-foetales chez *Typhlonectes compressicaudus* (Duméril et Bibron, 1841), Amphibien Gymnophione vivipare. Thèse de Doctorat E.P.H.E., Lyon.

Hraoui-Bloquet S, JM Exbrayat. 1992. Développement embryonnaire du tube digestif chez *Typhlonectes compressicaudus* (Duméril et Bibron, 1841), Amphibien Gymnophione vivipare. Ann. Sci. Nat. Zool. 13ᵉ ser. 13: 11-23.

Hraoui-Bloquet S, JM Exbrayat. 1993. La fécondation chez *Typhlonectes compressicaudus* (Duméril et Bibron, 1841), Amphibien Gymnophione. Bull. Soc. Zool. Fr. 118: 356-357.

Hraoui-Bloquet S, JM Exbrayat. 1994. Développement des branchies chez les embryons de *Typhlonectes compressicaudus*, Amphibien Gymnophione vivipare. Ann. Sci. Nat. Zool. 13ᵉ ser. 15: 33-46.

Hraoui-Bloquet S, JM Exbrayat. 1996. Les dents de *Typhlonectes compressicaudus*, Amphibia, Gymnophiona) au cours du développement. Ann. Sci. Nat. Zool. 13ᵉ ser. 17: 11-23.

Hraoui-Bloquet S, G Escudié, JM Exbrayat. 1994. Aspects ultrastructuraux de l'évolution de la muqueuse utérine au cours de la phase de nutrition orale des embryons chez *Typhlonectes compressicaudus*, Amphibien Gymnophione vivipare. Bull. Soc. Herp. Fr. 119: 237-242.

Kuhlenbeck H. 1970. A note on the morphology of the hypophysis in the Gymnophione *Schistometopum thomense*. Okajimas. Fol. Anat. Jap. 46: 307-319.

Laubmann W. 1927. Uber die Morphogenese von Gehern und Gereuchsorgan der Gymnophionen. Beitrag zur Kenntniss der Gymnophionen. Z. Anat. Entwick. 84: 597.

Malonza PK, J Measey. 2005. Life history of an African Caecilian: *Boulengerula taitanus* Loveridge 1935 (Amphibian Gymnophiona Caeciliilidae. Trop. Zool. 18: 49-66.

Marcus H. 1939. Beitrag zur kenntnis der Gymnophionen. Ueber keimbahn, keimdruusen, Fettkörper und Urogenitalverbindung bei Hypogeophis. Biomorphosis 1: 360-384.

Masood-Parveez U. 1987. Some aspects of reproduction in the female Apodan Amphibian Ichthyophis. Ph.D., Karnataka University, Dharwad, India.

Masood-Parveez U, B Nadkarni. 1991. Morphological, histological histochemical and annual cycle of the oviduct in *Ichthyophis beddomei* (Amphibia: Gymnophiona). J. Herpetol. 25: 234-237.

Masood-Parveez U, B Nadkarni. 1993a. The ovarian cycle in an oviparous gymnophione amphibian, *Ichthyophis beddomei* (Peters). J. Herpetol. 27: 59-63.

Masood-Parveez U, b Nadkarni. 1993b. Morphological, histological and histochemical studies of the ovary of an oviparous Caecilian, *Ichthyophis beddomei* (Peters). J. Herpetol. 27: 63-69.

340

Müller J. 1832. Beitrag zur Anatomie und Natursgeschichte der Amphibien. Z. Jb. Physiol. 4: 195-275.

Nussbaum RA. 1984. Amphibians of the Seychelles. In: *Biogeography and Ecology of the Seychelles Island*, (Ed.) DR Stoddard, Dr W. Junk Publishers. The Hague, Boston, Lancaster, pp. 379-415.

Oyama J. 1952. Microscopical study of the visceral organs of Gymnophiona, *Hypogeophis rostratus*. Kumamoto J. Sci. 1B: 117-125.

Parker HW. 1934. Reptiles and Amphibians from Southern Ecuador. Ann. Mag. Nat. Hist. 10: 264-273.

Parker HW. 1936. The Caecilians of the Mamfe Division, Cameroons. Proc. Zool. Soc. London 1936: 135-163.

Parker HW. 1956. Viviparous caecilians and amphibian phylogeny. Nature (London) 178: 250-252.

Parker HW. 1958. Caecilians of the Seychelles, a description of a new species. Copeia 12: 71-76.

Parker HW, ER Dunn. 1964. Dentitional metamorphosis in the Amphibia. Copeia 1964: 75-86.

Parker HW, O Wettstein. 1929. A new caecilian from southern Brazil. Ann. Mag. Nat. Hist. 10: 594-596.

Pasteels JL, M Herlant. 1962. Les différentes catégories de cellules chromophiles dans l'hypophyse des Amphibiens. Anat. Anz. 109: 764-767.

Peters W. 1874a. Observations sur le développement du *Caecilia compressicauda*. Ann. Sci. Nat. Zool. ser 5. 13: 2.

Peters W. 1874b. Derselbe las ferner über die Entwicklung der Caecilien und besonders der *Caecilia compressicauda*. Monatsb. Deutsch. Akad. der wiss. zu Berlin. 1874: 45-49.

Peters W. 1875. Uber die Entwicklung der Caecilien. Monatsb. Deutsch. Akad. der wiss. zu Berlin. 1875: 483-486.

Pillay KV. 1957. The hypothalamo-hypophyseal neurosecretory system of *Gegenophis carnosus* Beddome. Z. Zellforsch. Mikr. Anat. 46: 577-582.

Prigioni MY, JA Langone. 1983a. Notas sobre *Chthonerpeton indistinctum* (Amphibia, Typhlonectidae). IV Notas complementarias. Jornadas de Ciencias naturales, Montevideo 19-24 set. 1983, 3: 81-83.

Prigioni MY, JA Langone. 1983b. Notas sobre *Chthonerpeton indistinctum* (Amphibia, Typhlonectidae). V Notas complementarias. Jornadas de Ciencias naturales, Montevideo 19-24 set. 1983, 3: 97-99.

Raquet M, GJ Measey, JM Exbrayat. 2006. Premières observations histologiques de l'ovaire de *Boulengerula taitanus*Loveridge, 1935, Amphibien Gymnophione. Rev. Fr. Histotechnol. 19:

Raquet M, GJ Measey, JM Exbrayat. 2007. Structure of ovaries and ovarian cycle in *Boulengerula taitanus* Loveridge, 1935 (Amphibia, Gymnophiona). 14th Eur. Congress Herp., Porto (Portugal), Abstracts: 285.

Rathke H. 1852. Bemerkungen uber mehrere Korpertheile der *Coecilia annnulata*. Mullers Arkiv 1852: 334-350.

Salthe NS. 1963. The egg capsules in the Amphibia. J. Morph. 113: 161-171.

Salthe NS, JS Mecham. 1974. Reproductive and courtship patterns. In: Physiology of Amphibia, (Ed.) B Lofts, Vol. 2. Academic Press, NewYork, pp. 309-521.

Sammouri R, S Renous, JM Exbrayat, J Lescure. 1990. Développement embryonnaire de *Typhlonectes compressicaudus* (Amphibia Gymnophiona). Ann. Sci. Nat. Zool. 13ᵉ ser. 11: 135-163.

Sanderson IT. 1937. Animal Treasure, Viking Press, New-York, pp. 221-224.

Sarasin P, F Sarasin. 1887-1890. Ergebnisse Naturwissenschaftlicher Forschungen auf Ceylon. Zur Entwicklungsgeschichte und Anatomie der Ceylonischen Blindwuhle *Ichthyophis glutinosus*. C.W. Kreidel's Verlag, Wiesbaden.

Schubert C, U Welsch, H Goos. 1977. Histological, immuno and enzyme-histochemical investigations on the adenohypophysis of the Urodeles, *Mertensiella caucasica* and *Triturus cristatus* and the Caecilian, *Chthonerpeton indistinctum*. Cell Tiss. Res. 185: 339-349.

Seshachar BR. 1936. The spermatogenesis of *Ichthyophis glutinosus* (Linn.) I. The spermatogonia and their division. Z. Zellforsch. Mikr. Anat. 24: 662-706.

Seshachar BR. 1942. Origin of intralocular oocytes in male Apode. Proc. Ind. Acad. Sci. 15: 278-279.

Seshachar BR, TA Balakrishna, KR Katre Shakuntala and Gundappa. 1982. Some unique features of egg laying and reproduction in *Ichthyophis malabarensis* (Taylor) (Apoda: Amphibia. Curr. Sci. 51: 32-34.

Spengel JW. 1876. Das Urogenitalsystem der Amphibien. I. Theil. Der Anatomische Bau des Urogenitalsystem. Arbeit. Zool. Inst. Wurzburg 3: 51-114.

Stendell LW. 1914. Die Hypophysis Cerebri. In: Oppel's Lehrbuch Vergleich Mikroskopie Anatomie, (Ed.) Fisher, Jena, 8: 1-168.

Taylor EH. 1955. Additions to the known herpetological fauna of Costa Rica with comments on other species. Univ. Kansas Sci. Bull. 37: 499-575.

Tonutti E. 1931. Beitrag zur Kenntnis der Gymnophionen. XV. Das Genital-system. Morph. Jb. 68: 151-292.

Wake MH. 1967. Gill structure in the Caecilian genus *Gymnopis*. Bull. So. Calif. Acad. Sci. 66: 109-116.

Wake MH. 1968. Evolutionary morphology of the Caecilian urogenital system. Part I: The gonads and fat bodies. J. Morph. 126: 291-332.

Wake MH. 1969. Gill ontogeny in embryos of *Gymnopis* (Amphibia: Gymnophiona). Copeia: 183-184.

Wake MH. 1970a. Evolutionary morphology of the caecilian urogenital system. Part II: The kidneys and urogenital ducts. Acta Anat. 75: 321-358.

Wake MH. 1970b. Evolutionary morphology of the caecilian urogenital system. part III: The bladder. Herpetologica, 26: 120-128.

Wake MH. 1972. Evolutionary morphology of the caecilian urogenital system. Part IV: The cloaca. J. Morph. 136: 353-366.

Wake MH. 1976. The development and replacement of teeth in viviparous Caecilians. J. Morph. 148: 33-63.

Wake MH. 1977a. Fetal maintenance and its evolutionary significance in the Amphibia: Gymnophiona. J. Herpetol. 11: 379-386.

Wake MH. 1977b. The reproductive biology of Caecilians. An evolutionary perspective. In: The reproductive biology of Amphibians, (Eds) DH Taylor, S I Guttman, Miami University, Oxford, Ohio, pp. 73-100.

Wake MH. 1980a. Fetal tooth development and adult replacement in *Dermophis mexicanus* (Amphibia: Gymnophiona): Fields versus clones. J. Morph. 166: 203-216.

Wake MH. 1980b. Reproduction, growth and population structure of the central american Caecilian *Dermophis mexicanus*. Herpetologica 36: 244-256.

Wake MH. 1980c. The reproductive biology of *Nectophrynoides malcolmi* (Amphibia: Bufonidae) with comments on the evaluation of reproductive modes in the genus *Nectophrynoides*. Copeia: 194-209.

Wake MH. 1989. Metamorphosis of the hyobranchial apparatus in *Epicrionops* (Amphibia, Gymnophiona, Rhinatrematidae): Replacement of bone by cartilage. Ann. Sci. Nat. Zool. 13^e ser. Paris 10: 171-182.

Wake MH. 1993. Evolution of oviductal gestation in Amphibians. J.Exp. Zool. 266: 394-413.

Wake MH. 1992. Patterns of peripheral inervation of the tongue and hyobranchial apparatus in Caeclians (Amphibia: Gymnophiona). J. Morph. 212: 3753.

Wake MH, R Dickie. 1998. Oviduct structure and function and reproductive modes in Amphibians. J. Exp. Zool. 282: 477-506.

Wilkinson M. 1990. The presence of a Musculus retractor cloacae in female caecilians. Amphibia Reptilia 11: 300-304.

Wilkinson M, R Nussbaum. 1998. Caecilian viviparity and amniote origins. J. Nat. Hist. 32: 1403-1409.

Zuber-Vogeli M, J Doerr-Schott. 1981. Description morphologique et cytologique de l'hypophyse de *Typhlonectes compressicaudus* (Duméril et Bibron) (Amphibien Gymnophione de Guyane franţaise). C. R. Seanc. Acad. Sci. Paris ser. D. 292: 503-506.

8

Development and Reproduction of Amphibian Species, Hybrids, and Polyploids

Maria Ogielska

Amphibian development and reproduction is distinguished by a great variety of breeding, fertilization, and developmental modes, as well as differences in the degree of parental care and the use of aquatic or terrestrial habitats for reproduction. Because the ecology of amphibian reproduction is described in detail in several books and articles, among which that of Duellman and Trueb (1986) summarizes most of examples, I will focus mainly on early and postembryonic development. Our knowledge of amphibian development is based on detailed studies on only few species belonging to three genera: *Rana*, *Bufo*, and *Xenopus* in particular. Most of our knowledge about molecular and genetic regulation of developmental events comes from extensive studies of only one species, *Xenopus laevis*. For this reason, and because complicated processes of developmental control are best understood in anurans, the description of development provided in this chapter is based mainly on this amphibian species.

The entire development, from an egg to a metamorphosed individual, is usually divided into a number of stages, which are delineated in so-called developmental tables. There are diverse developmental tables prepared for particular species or genera, especially for anurans (reviewed and compared by McDiarmid and Altig, 1999). Currently two main tables for anuran amphibians are in use: Nieuwkoop and Faber's (1956) for *Xenopus laevis*, a species widely used in laboratories focused on developmental biology, and Gosner's (1960), which is a generalized table for anurans in general based on clear morphological and morphometric features. In this chapter I refer to the Gosner's staging, and convert these of Nieuwkoop and Faber to Gosner's staging according to McDiarmid and Altig (1999).

A fertilized egg is equipped with several programs of genetic control, namely one responsible for early embryonic development, and the other regulating the postembryonic development (Hall and Wake, 2000). Postembryonic development includes genetic and developmental processes that regulate the formation of larval structures, and their further degradation and remodeling according to an adult body plan.

Amphibians are bisexual animals, a characteristic of all vertebrates. Females and males produce gametes: ova during oogenesis and spermatozoa during spermatogenesis. Both ova and spermatozoa contain a haploid number of chromosomes (the 1n genome), i.e., a half of the diploid number (2n) characteristic of the somatic cells of the body. A diploid cell contains two versions of each chromosome (homologues). The reduction in chromosome number is a result of a unique cell cycle, meiosis, in which two successive cell divisions follows one phase S (DNA replication). During the first division the homologues (after replication each of them is composed of two sister chromatids join by non-divided centromere region) move to the opposite poles, whereas during the second division the centromeres of sister chromatids break and allow them to separate, giving rise to four haploid cells. The prophase of the first meiotic division is long, and homologue chromosomes become juxtaposed and physically aligned by a specific protein structure, the synaptonemal complex. Moreover, there are also physical contacts of DNA strands between chromatids of a homologous pair. This contact allows recombination to occur. Recombination is initiated by meiosis-specific DNA double strand breaks (DSBs), which are converted to chemically stable connections between homologues (Holliday junctions); cytologically they can be observed as chiasmata. After pachytene the synaptonemal complexes decompose, but chiasmata are still present. During diakinesis chiasmata disappear and each homologue with recombined chromatids enters metaphase I, and then complete meiosis. For details concerning several events during normal meiosis see reviews by Kleckner (1996) and Cnudde and Gerats (2005). The above description of meiosis is valid in cases when both chromosome sets: one from the mother and the other from the father belong to the same species and thereby are homologous. Homology enables proper exchange of chromatids during crossing-over and ensures the recombination between parental genomes; this in turn results in genetic variability of progeny. The majority of amphibians are diploid species with normal meiosis. The exceptions to this rule are hybrids and polyploids with an uneven number of genomes (3n; 5n), which display alternation of meiosis. Although such animals are rare in the nature, they are a focus of attention because of unusual ways of reproduction and modifications of meiosis (Archetti, 2004a,b).

The alternative modes of reproduction are known in three amphibian complexes: *Ambystoma laterale-jeffersonianum-texanum-tigrinum* (Urodela), *Rana esculenta, R. grafi,* and *R. hispanica,* and *Bufo viridis* complex (Anura). In *Ambystoma* gynogenesis (sperm-dependent parthenogenesis) and in *Rana* hybridogenesis are normal ways of reproduction. The unique mode of reproduction, being a mixture of normal bisexual reproduction and a modification of hybridogenesis, was also reported in the *Bufo viridis* complex. The modification of reproduction and gametogenesis in the above-mentioned three complexes are described in more detail later in this chapter.

Natural hybridization and polyploidy are of great value for evolutionary biologists because these phenomena serve as examples of some mechanisms leading to genetic isolation and speciation. On the other hand, all known gynogenetic or hybridogenetic taxons include at least one polyploid form (triploid, tetraploid or pentaploid). For this reason hybridization and polyploidy are often described together, although they are two separate phenomena, and there are many polyploid taxons without gyno- or hybridogenesis. Detailed reviews concerning these problems were provided by Hewitt (1988), Avise *et al.* (1992), Bullini (1994), Dowling and Secor (1997), Beukeboom and Vrijenhoek (1998), and Otto and Whitton (2000). Polyploidization is also discussed in respect to genome evolution and gene regulation, because polyploidization provides extra-copies of genes, which may lead to independent alterations and achieving new biological functions (reviewed by Beçak and Kobashi, 2004). In this chapter, which is devoted to reproduction, I will focus on the modifications of gamete formation, gonad structure, and development of progeny in natural hybrid and polyploid amphibians.

Early Development

Early development can be divided into two periods with respect to genetic control: cleavage, when development is controlled by products of maternal genes deposited in the ooplasm during oogenesis followed by gastrulation, neurulation and organogenesis that are controlled by the products of zygotic genes. Zygotic genes are gradually activated during gastrulation, between cleavage and organogenesis but embryonic development still relies in large part on transcripts and proteins from oocyte genes. The entire development is a hierarchical process, in which each stage must be preceded by very precise steps to be normal (schematically represented in Figures 1 and 2). In most species the sperm enter the egg at any point of the animal hemisphere, and this part will become the ventral side. Sperm entry causes rotation of the egg cortex, which displaces maternal mRNAs localized to the vegetal pole toward the presumptive dorsal side (Gerhart

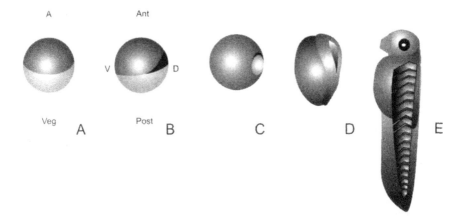

Fig. 1. The effect of egg polarity to establishment of the dorso-ventral and anterio-posterior axis of the embryo. A—Unfertilized oocyte is a polarized cell with the animal-vegetal axis. B—After fertilization and rotation, the former animal-vegetal axis (dotted line) is replaced by the anterio-posterior axis of an embryo. The dorsal side is indicated by the grey crescent (black), and the ventral side is indicated by V (Reprinted from Ogielska, 1999, with permission).

et al., 1984, 1986; Vincent *et al.*, 1986). Consequently the site of fertilization and the cortical rotation establish future dorso-ventral and anterior-posterior body axes (Riuz i Altaba and Melton, 1990; Tickle and Davey, 2003).

During cleavage (Gosner stages 1-9), 12 modified cell cycles give rise to about 4,000 cells (blastomeres) without growth or rearrangement of daughter blastomeres between divisions. Each cell diminishes in size after each cell cycle and at the end of cleavage an embryo resembles a ball with asymmetric walls (thick at the vegetal hemisphere and thin at the animal hemisphere) with a space (blastocoel) inside. At this moment the mitotic division rate slows down and the embryo achieves the mid-blastula-transition (MBT) stage (Gerhart *et al.*, 1986). The cell cycles before MBT are short and composed of S an M phases without G1 and G2 (Yasuda and Schubinger, 1992). Short cell cycles are possible due to the accumulation of regulatory molecules, such as cyclins and cyclin-dependent kinases, during oogenesis (Hartley *et al.*, 1997). After MBT the cell cycles become asynchronic and G1 and G2 phases appear under the control of zygotic genes. During gastrulation (Gosner stages 10-12) the blastomeres start to move actively, individually or as cell assemblages, resulting in the spatial reorganization of an embryo and formation of three basic germ layers, i.e., endoderm, ectoderm, and mesoderm (Gerhart *et al.*, 1986; Elinson and Kao, 1989; Keller, 1991; Winklbauer and Schürfeld, 1999).

The beginning of gastrulation is marked externally by the appearance of a blastopore, which is a slit formed by invagination of the dorsal equatorial cells along the blastocoel roof (schematically shown in Figures 1C and 2g,h). The thin sheet of invaginated cells constitutes the dorsal mesoderm, which will play the key role during early embryogenesis in dorsal structure patterning. The invagination of the dorsal part (lip) spreads into the lateral, and finally into the ventral blastopore lips, the blastopore becomes circular and the lateral and ventral parts of mesoderm are invaginated. Concomitantly with mesoderm invagination, the cells of the animal hemisphere form ectoderm, which will be the most external layer of an embryo. The more slowly proliferating vegetal cells, which are finally covered by fast proliferating ectoderm in a process called epiboly, form endoderm. Endodermal cells are not only passively covered by the expanding ectoderm, but move by active inrolling to the interior of the blastula in a process called vegetal rotation (Winklbauer and Schürfeld, 1999). Concomitantly to these processes, the diameter of the blastopore gradually decreases, and finally closes.

The proper position of the dorsal mesoderm is crucial for neural plate induction (neurulation, Gosner stages 13-16) and early organogenesis. The role of the dorsal mesoderm in early embryonic induction was originally described by Spemann and Mangold in 1924 (reviewed by Gerhart, 2001), and for this reason this region is called the Spemann-Mangold organizer. After neural induction the patterning of mesodermal axial structures (notochord and somites, which will give rise to axial skeleton and muscle) takes place. Next (tail-bud, Gosner stages 17-19), differentiation of eye and nasal anlagens, heart, gut and its derivatives occur.

Early development is controlled by regulatory molecules, which can be divided into nuclear transcription factors and signaling factors acting mainly at the level of cell membrane. Transcription factors have at least one DNA-binding domain (zinc finger, homeodomain, T-box, paired domain and others) and are able to activate or suppress several genes, thereby controlling transcription. Examples of such proteins involved in amphibian development are *Veg-T, goosecoid, Siamois, Xbrachury, Eomesodermin*. Signaling molecules, either secreted molecules or integral membrane proteins of neighboring cells, act through cell surface receptors and signal transduction pathways to activate intracellular second messengers and proteins which then regulate gene expression. The signaling molecules active during early development are members of the transforming growth factor- TGF-β (activin, BMP), fibroblast growth factor (FGF), and Wingless (Wnt) families of growth factor proteins. The following description is based primarily on observations and experiments with *Xenopus laevis*.

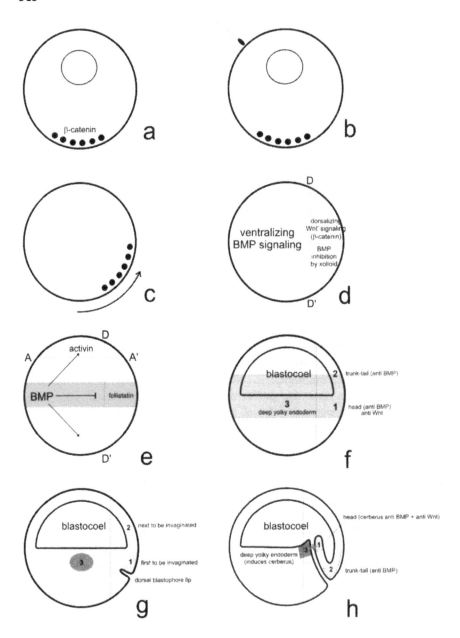

Fig. 2. See caption on the next page.

The fully grown oocyte is a polarized cell and some molecules important to early development (various RNAs and proteins) are localized to specific parts of its cytoplasm, or are unevenly distributed, thereby forming concentration gradients. During cleavage these molecules are localized in specific blastomeres and for this reason the animal and vegetal blastomeres differ in respect to their molecular content. Between the animal and vegetal hemispheres there is a ring of blastomeres (marginal zone, Fig. 2e,f) with specific properties. Moreover, blastomeres of the presumptive dorsal side of an embryo contain molecules, which were transferred there at the time of cortical rotation following fertilization (Fig. 2a-c). Additionally, dorsal blastomeres of the marginal zone contains unique composition of molecules, which will control the formation of the Spemann-Mangold organizer, and for this reason this region of a blastula was named the "organizer of the organizer" (or the Nieuwkoop's center; Fig. 2a-c; Christian and Moon, 1993; Gerhard, 2001). These pre-patterned parts of a blastula are precursors of three main germ layers: the animal cap will form the ectoderm, the marginal zone will form the mesoderm, and the vegetal hemisphere will form the endoderm. One of the key molecules during cleavage is the maternal transcription factor Veg-T (Zhang and King, 1996; Smith, 1997; Zhang et al., 1998), which forms a gradient, being present in the vegetal hemisphere, lower in concentration in the marginal zone, and absent in the animal cap (Fig. 3). After the MBT, Veg-T is activated, and cells with high concentration of this transcription factor start differentiating into endoderm (Kimelman and Griffin, 1998). Already differentiated endodermal cells start to produce FGF-β signaling molecules, which in turn act on the marginal zone and initiate the induction of mesoderm. The main genes, which are expressed in all mesodermal cells at this time code for the transcription factors *eomesodermin* and *Xbrachyury* (Smith et al., 1991;

Fig. 2. Generalized schematic representation of genetic control of early embryogenesis of an amphibian. Dorsalizing molecules (maternal β-catenins) are displaced during cortical rotation from the vegetal pole of an oocyte to the future dorsal side, where they are localized to cell nuclei (a-c). β-catenins are degraded at the future ventral side, which allows Bone Morphogenetic Protein (BMP) to act as a main ventralizing factor. The BMP molecules are inhibited at the dorsal side along the D-D' line (d). Neural tissue will be formed in regions where BMP action is blocked by zygotic protein follistatin expressed in the dorsal side (e), whereas epidermis is formed in regions of animal cap containing maternal growth factor activin, where BMP is not blocked, denoted by A-A' line. Three parts of the Spemann-Mangold organizer are shown in f-h: 1—head organizer; 2—trunk-tail organizer; and 3—deep yolky endoderm, which moves close to 1 during gastrulation and induces Cerberus, which is a multipotent BMP, nodal and Wnt inhibitor.

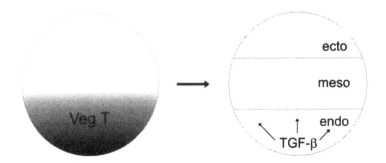

Fig. 3. Pre-patterning of a blastula. Maternal transcription factor Veg-T forms a gradient. Transcripts of Veg-T are present in the vegetal hemisphere blastomeres, absent in the animal cap cells, and weakly expressed in marginal zone cells. Such prepatterned regions will give rise to endoderm, mesoderm, and ectoderm, respectively. Already, differentiated endodermal cells start to produce FGF-β signaling molecules, which in turn acts on the marginal zone and initiates the induction of mesoderm.

Ryan *et al.*, 1996). During next steps the mesoderm will differentiate into small groups of cells, which vary in respect to spatio-temporal order of zygotic gene expression (Lane and Keller, 1997). The region of special interest is dorsal mesoderm (Spemann-Mangold organizer), which will control differentiation of axial structures (neural plate, notochord, and somites) (de Robertis, 1995; Gerhart, 2001; Smith and White, 2003).

The other molecules of great importance are those controlling the proper formation of dorsal and ventral structures. The genetic control of early embryogenesis is schematically shown in Figure 2. Dorsalizing molecules are displaced during the cortical rotation from the vegetal pole of an oocyte to the future dorsal side (Fig. 2a-c). They stabilize maternal β-catenins, which are localized to cell nuclei during early cleavage. This is possible due to the inhibition of a kinase GSK-3, which phosphorylates and causes degradation of β-catenins, which in the absence of this inhibition occurs on the ventral side (reviewed by Heasman, 1997). The β-catenins activate another maternal transcription factor, XTcf3, stored in the ooplasm during oogenesis, which activates the expression of zygotic genes *Siamois, goosecoid, noggin, chordin, follistatin* (for more detailed description of a variety of molecular events and signal transduction pathways see Gerhart, 2001; Smith and White, 2003; Key, 2003). The main ventralizing molecule is Bone Morphogenetic Protein (BMP-4), which appears at the future ventral side (Fig. 2d,e; Graff *et al.*, 1994; Graff, 1997). Dorsal structures can differentiate in regions, where BMP is not active. The inactivation of BMP signal molecules is caused by the products of the zygotic genes *noggin* and *chordin* (Harland, 1994; Zimmerman *et al.*, 1996). An enzyme protease (product of the *xolloid* gene) controls the cleavage of BMP-chordin complexes (Piccolo *et al.*, 1997).

The dorsal mesoderm can be subdivided into two regions: the head organizer and the trunk-tail organizer (Fig. 2f-h). The trunk is formed when BMP signaling is inhibited, whereas the head organizer requires additional inhibition of Nodal and Wnt signaling. The multivalent inhibitor of BMP, Nodal and Wnt is coded by the gene *cerberus* (Glinka *et al.*, 1997; Piccolo *et al.*, 1999; Yamamoto *et al.*, 2001). Proper formation of head structures are supported by induction of a specific region of endoderm called deep yolky endoderm, which moves close to the head organizer as gastrulation proceeds (Fig. 2f-h). The role, function, and genetic control of the three parts of the Spemann-Mangold organizer (namely head organizer, trunk-tail organizer, and deep yolky endoderm) was described in detail and discussed by Gerhart (2001).

The ectoderm is prepatterned to form neural structures, but in regions where molecules inducing neural formation are inhibited, the ectodermal cells differentiate into epidermis (Hemmati-Brivanlou and Melton, 1992; Hemmati-Brivanlou *et al.*, 1994; Chang *et al.*, 1997; McDowell and Gurdon, 1999). The mechanism underlying the epidermis formation resembles that of dorsalization. BMP-4 is a signaling molecule present in the animal cap cells containing maternal growth factor activin, a member of TGF-β superfamily that is necessary for epidermis formation. Neural tissue is formed in regions where BMP and activin are blocked by zygotic protein follistatin (expressed in the Spemann-Mangold organizer) (Fig. 2e).

After neurulation an embryo reaches the tail-bud stages (Gosner stages 17-20), after which it hatches from the egg jelly capsule, and becomes a larva. Hatching marks the end of embryonic, and beginning of post-embryonic development (Tata, 1993). However, not all anurans have the same or similar pattern and rate of development as was outlined above for *Xenopus*. Some species with big eggs that do not develop in aquatic environments display various modifications of early development, organogenesis, and postembryonic development. These species include marsupial frogs and direct developers. As was pointed out by del Pino (1989) and del Pino and Loor-Vela (1990), the cleavage pattern of *Xenopus* and the tight coupling of events at MBT are features of the accelerated development of small amphibian eggs with aquatic reproduction rather than generalized features of amphibian development. The large eggs of marsupial frog *Gastrotheca riobambae* display atypical and slow-rate cleavage, during gastrulation the blastopore lips close over the small archenteron, and the cells of the lips form the embryonic disc, which is an unusual feature of amphibian embryos. The embryonic disc will give rise to the embryo, whereas expanding epithelium covers the big vegetal cells. In result, an early embryo of marsupial frogs resembles that of birds rather than amphibians (del Pino, 1983; Elinson and del Pino, 1985; Gatherer and del Pino, 1992).

The early developmental stages (cleavage, gastrulation, and neurulation) of a direct developer *Eletherodactylus coqui* are irregular, with mesoderm formation localized more superficially toward the animal pole (Ninomyia et al., 2001). This is probably due to more restricted area of sperm entry and different distribution of dorsal determinants after fertilization (Fang et al., 2000). During further development the newly formed epidermis replaces ectoderm, and finally envelops large vegetal yolky mass. In that way an early embryo of a direct developer resembles that of marsupial frog.

Postembryonic Development

Postembryonic development in most amphibian species is biphasic, i.e., it includes larval stages of distinct morphology and physiology, followed by more or less dramatic metamorphosis, after which a species-specific adult body shape is achieved. The metamorphic changes concern only somatic tissues whereas gonads follow an independent rate of development, both in Urodela (Rosenkilde and Ussing, 1996) and in Anura (Ogielska and Kotusz, 2004).

Free Larvae and Metamorphosis in Anura

The free living aquatic larva is an ancient feature in amphibians, known also from paleontology research (for review see Hanken, 1999). During larval stages (Gosner stages 21-24) a sticky transient structure (cement gland) differentiates and disappears at Gosner stage 25. External gills are formed, and then disappear after a specialized skin fold (the operculum) has covered them (stages 23-25), and the next generation of internal gills develops. Larval stages end during relatively long stage 25 when posterior parts of the operculum fuse and only small opening (the spiraculum) remains open and enables water flow over the gills. Melanophores become visible at stage 21, and a species-specific chromatophore pattern of a tadpole is established. Concomitantly with operculum closure, the mouth opening and functional digestive tract with vent opening are formed. The mouth opening is equipped with several structures such as keratin teeth, beak, and papillae. The arrangement and size of these structures are species specific and serve as important taxonomic features (Altig and McDiarmid, 1999).

Beginning with stage 25 an anuran larva is called a tadpole. During stages 25-40 a tadpole mainly enlarges in size. Morphogenesis is restricted to only few organs, such as lungs and limbs. Various aspects of anuran tadpole biology are described in detail in a special issue edited by McDiarmid and Altig (1999). Generally, tadpoles can be divided into exotrophic and endotrophic. Exotrophes generate energy from ingested food when yolk is exhausted after embryogenesis. In endotrophs the energy continues to come from maternal sources, usually yolk. Typically, species with endotrophic larvae produce fewer, but larger yolky eggs in comparison to similar-sized frogs with exotrophic tadpoles. Endotrophic larvae occur in one or more species belonging to at least 10 families of anurans: Leiopelmatidae, Pipidae, Bufonidae, Leptodactylidae, Sooglossidae, Myobatrachidae, Hylidae, Microhylidae, Ranidae, and Rhacophoridae (for review see Duellman and Trueb, 1986, Thibaudeau and Altig, 1999, and Hanken, 1999).

Metamorphosis is the transition from larval to adult state, and is accompanied by change of body structures and functions. Metamorphosis in anuran amphibians is usually divided into 2 periods: prometamorphosis (Gosner stages 31-40) and metamorphic climax (stages 41-46) (Etkin, 1968; Dent, 1968), although some authors (Shi, 2000) include a premetamorphic period in this process (Gosner stages 26-30), thereby dividing metamorphosis into 3 periods. During prometamorphosis a tadpole reaches its maximal size, although the growth rate is not as fast as during premetamorphosis.

Among endotrophs only direct developers display highly modified postembryonic development, during which an embryo completes metamorphosis inside egg capsules, but outside the parental body (Thibaudeau and Altig, 1999). Typical anuran metamorphosis is exemplified by that of exotrophs. Most of changes are rapid, but those concerned with hind limb bud differentiation and elongation start early during larval period called premetamorphosis (stages 26-30). During that time the hind limb buds elongate until their length is twice as long as the width of the bud base (Gosner, 1960). Beginning with stage 31 limb patterning occurs and the toes begin to differentiate to attain their mature form (metatarsal and subarticular tubercles) at stage 40. The shape of 5 digits of hind limbs become visible at stage 35, and for this reason some authors (Dent, 1968; Shi, 2000) designate this stage as the beginning of prometamorphosis. During stages 40-45 the growth stops, and rapid morphological changes caused by a new wave of morphogenesis occurs. One of the most spectacular changes is an eruption of forelimbs through the opercular tissue. The larval mouth is lost and an adult-shaped mouth is formed along with remodeling of the head. The digestive tract is extensively rebuilt, the cloacal tailpiece is resorbed, and adult-shaped cloacal opening is formed. Some organs (eyes, nasal structures) remain unchanged from early tailbud stages until stage 46, at which metamorphosis is completed.

The rebuilding of body plan during metamorphic climax includes destruction of some fully functional larval organs (tail and gills), *de novo* formation of new organs (limbs), or remodeling of already existing organs (liver, intestine, brain and nervous system, and epithelium). The digestive tract must be remodeled because tadpoles are phytophagous and adults are carnivorous. This change is possible as a result of a precise program of cell proliferation, differentiation, and programmed cell death (apoptosis), all genetically controlled and triggered by the thyroid hormone action. Resorption of the tadpole tail is a well known phenomenon involving massive programmed cell death of muscles, connective tissue, epidermis, and notochord. The tissues of tadpole tail are genetically predetermined for destruction, and even when transplanted to the other region of the body, undergo destruction at the proper time. Along with body-plan remodeling, also biochemical changes occur (reviewed by Frieden, 1968, and Shi, 2000). For example, larval-type hemoglobin is replaced by adult-type and the change from ammonia to urea excretion is accompanied by the enzymatic changes required for urea excretion.

Free Larvae and Metamorphosis in Urodela

The urodelan larval body plan resembles that in adults and thereby metamorphosis in urodelans is not as dramatic as in anurans (Dent, 1968; Rosenkilde and Ussing, 1996; Rose, 1999). Urodelan larval gills are external and not covered by an operculum. During metamorphosis the gills are resorbed. The tail persists, and only tail fin is destroyed. Larval forelimbs, since they are not covered by an operculum, are visible from the beginning of their differentiation, similarly to hind limbs. The mouthparts are not dramatically changed because urodelan larvae are carnivorous like adults, but the head attains adult shape. Ossification of the cartilaginous skull starts after hatching and metamorphosis is marked by reorganization of the palate, hyobranchium, teeth, middle ear, as well as reorganization of capsules around the eyes and nasal sac (reviewed by Rose, 1999; Duelman and Trueb, 1986). Most of urodelan families display progressive morphological changes accompanied by progressive thyroid hormones (TH) concentration in the circulation. On the contrary, metamorphosis in plethodontids is very rapid and dependent on sharp rise of TH during short period.

Hormonal Control of Metamorphosis

The extensive changes in external body shape and internal rebuilding affecting almost all organs during metamorphosis must be precisely guided and regulated. Main molecules, which control metamorphosis, are those synthesized and released by the thyroid gland. The thyroid gland differentiates during late embryogenesis about the time of hatching as a median thickening of pharyngeal epithelium. The gland becomes functional when is composed of several follicles, it attains the greatest size during metamorphosis, and then slightly decreases in size. They key thyroid hormones are T_4 (thyroxine; 3,5,3′,5′-tetraiodothyronine) and T_3 (3,5,3′-triiodothyronine). T_4 may be regarded as prohormone, which is converted to more biologically potent T_3 by 5′-deiodinase. In the following section thyroid hormone (TH) is used to designate both the prohormone and T_4 and the more active T_3. T_4 and T_3 can be inactivated to reverse T_3 (rT_3) and T_2, respectively, by another enzyme, 5-deiodinase. In that way T_4 may be transformed into biologically potent T_3 or the inactive (i.e., with low affinity to thyroid hormone receptors) forms rT_3 and T_2, thereby regulating tissue-specific response to TH.

The synthesis of thyroid hormone starts during prometamorphosis, when its concentration rises and achieves the highest level at stage 40, which is the starting point for metamorphic climax. Recently Shi (2000) published a detailed and compendious study devoted to mechanisms regulating amphibian metamorphosis. The brief description, which is presented below, is based mainly on this publication. Many lines of evidence show that metamorphosis in anurans and urodeles is regulated and guided by basically the same mechanisms (reviewed by Rosenkilde and Ussing, 1996).

TH is released by thyroid gland to the circulation where is bind to specific TH-transport proteins, enters cells of the target tissues where is bound to cytosolic proteins. The exact mechanism, by which TH enters the cell, is still unknown, but the most plausible way is a passive transport across the cell membrane, because both T_4 and T_3 are hydrophobic molecules. Finally, TH molecules enter nucleus, where they bind to specific receptors (TRs), which are chromatin-associated nuclear proteins acting as transcription factors. Almost all metamorphic changes are controlled by repression or activation of several genes by TRs (described in detail by Shi, 2000). TRs belong to the superfamily of nuclear hormone receptors, including receptors of glucocorticoid, estrogen, retinoic acid, and vitamin D. As a transcription factors, TRs have several protein domains, among which one is the DNA-binding domain containing zinc finger motifs, and the other is the hormone-binding domain. TRs are coded by four genes: two TRα and two TRβ. Additionally, alternative splicing gives rise to several isoforms of TRβ. TRs acts as heterodimers of TRα, TRβ, and 9-*cis*-retinoic acid receptors (RXRs). A great variety of possible combinations of TRs-RXRs heterodimers makes TH a multipotent protein regulating gene activation and/or repression.

The TRs-RXRs complexes activated by TH may regulate several genes directly or indirectly. The genes, which are activated (or repressed) as a result of the direct action of TRs-RXRs are called immediate or early TH response genes. The activation (or repression) of these genes causes a cascade of indirect activation (or repression) of a variety of other genes, called late TH response genes. In other words, late response genes are those, which require syntheses of new proteins, which in turn will regulate these genes. As a rule, early response genes are not tissue- or organ-specific, whereas late response genes are tissue-specific. Under experimental conditions, the genes, which are regulated within 24 hours of treatment, are classified as early, and those regulated later are classified as late. Early TH response genes isolated from *Xenopus laevis* and *Rana catesbeiana* (reviewed by Shi, 2000) were detected during hind limb development (about 120 genes), the intestine and brain remodeling during metamorphosis, and tail resorption (about 35 genes). Late TH response genes involves those genes coding for the urea cycle enzymes in the liver and genes involved in maturation of adult skin.

Both early and late TH response genes are stage-dependent and can be activated only after the responding tissues or organs have achieved specific developmental stages. There must be appropriate level of a signal (TH) in the circulation, and its receptors (TRs and RXRs) in the target tissue. Before metamorphic changes, the levels of TH and its receptors must be low, during metamorphosis they must be high, and after metamorphosis they must decrease. It is thought that TRs-RXRs heterodimers have a dual function: they repress several genes before metamorphosis when TH is absent, and activate these genes during metamorphosis when TH is present. One of the other candidates for a regulating molecule is prolactin (PRL), which acts through its cell membrane receptors (JAK/stat pathway of signal induction) and can prevent autoinduction of TRβ genes, thereby inhibiting transcription of some other genes TRs.

The release of TH by the thyroid gland is controlled by other hormones acting on the hypothalamus-pituitary-thyroid and hypothalamus-pituitary-interrenal axes (reviewed by Shi, 2000). Neurons of the hypothalamus produce neurohormones, which are released by nerve ends in the median eminence. The neurohormones diffuse into blood vessels and reach the pituitary portal system (Rosenkilde and Ussing, 1996). The action of hypothalamus is regulated by environmental factors regulating pituitary action: thyrotropin-releasing hormone (TRH) and corticotropin releasing factor (CRF). TRH regulates the level of prolactin (PRL), whereas CRF regulates levels of thyrotropin stimulating hormone (TSH; thyrotropin) and adrenocorticotropin (ACTH). TSH regulates the level of TH released by the thyroid, and ACTH regulates the level of corticosteroids. The co-action of TH, corticosteroids, and PRL controls metamorphosis: TH enables it, corticoids can accelerate, and PRL can decelerate it. One of possible roles of PRL is to counteract high concentrations of TH, and thereby to coordinate the sequential transformations of various tissues and organs. Gonadal steroids (estradiol and progesterone) have no strong effects on metamorphosis, but seem rather to inhibit than accelerate it.

Environmental Influence on Metamorphosis

Environmental conditions may alter the rate of metamorphosis (reviewed by Shi, 2000). The most important factors are: temperature, diet, iodine content, and concentration of various chemical compounds in water. Temperature and food are well known and obvious factors, and iodine availability is essential for TH synthesis. Water composition may change due to the rising concentration of metabolites excreted by tadpoles in overcrowded conditions, or by evaporation from the pond or other body of water containing tadpoles. These changes can affect the rate of metamorphosis. Overcrowding usually prolongs the period of larval stages, whereas pond desiccation may accelerate it. It is still a matter of discussion whether pond desiccation causes a direct change in concentration of compounds, which in turn accelerates the synthesis of TH, or TH level rises as a result of an action of the stress neuropeptide (corticotropin-releasing hormone), which can directly stimulate the release of TSH by the pituitary, and finally the TH by the thyroid.

Direct Development

Direct development is found in amphibian species (both anuran and urodelan) that display short and modified postembryonic development and achieve the adult morphology without the free-living larval phase. Among anurans the best known direct developers are members of the genus *Eleutherodactylus* (Leptodactylidae), and *E. coqui* in particular (reviewed by Callery *et al.*, 2001). This species displays several unique developmental features in comparison to the more commonly studied anuran, *Xenopus laevis*. Several structural characteristic of anuran free-living larvae, such as the cement gland, gills, pharyngeal slits, operculum, mouthpart, coiled intestine, and lateral line organ are lacking or vestigial in direct developers. The tail is highly vascularized and most probably serves as respiratory organ. The cement gland is not formed because of the lack of competent ectoderm; the ectoderm is not pre-patterned by BMP-4, and thereby genes regulated by Otx2, which is a key gene in cement gland differentiation, are not expressed (reviewed by Callery *et al.*, 2001). No expression of NeuroD, a key gene in neural tissue differentiation, and the absence of lateral line organ were reported by Schlosser *et al.* (1999). Fore- and hind-limb buds appear early and are not dependent on TH synthesis, but later limb patterning is TH dependent. TRα and TRβ mRNAs are expressed in *E. coqui*, the thyroid gland develops, and some metamorphic events occur, such as cranial muscle and cartilage remodeling, reorganization of body muscles, and tail resorption. All these observations suggest that direct developers undergo a cryptic metamorphosis (Callery and Ellinson, 2000).

Direct development in Urodeles is known in the family Plethodontidae (Collazo and Marks, 1994). As is pointed out by several authors, direct development is the most common way of reproduction among urodelans owing to the fact that plethodontids constitute more than a half of extant species (Wake and Hanken, 1996; Hanken, 1999; Bruce, 2003). It is also believed that direct development contributed to the evolutionary success of this family of amphibians.

Reproduction of Larvae (Neoteny)

Urodelans are the best example of independence between gonad development and sexual maturity, and the rate of somatic development. Many urodelan species belonging to various families (Plethodontidae, Salamandridae, Ambystomatidae, Dicamptodonidae, Hynobiidae) do not fully metamorphose and can reproduce as sexually mature larvae, whereas all representatives of Cryptobranchidae, Sirenidae, Proteidae, and Amphiumidae reproduce as obligatory larvae (Dent, 1968; Rosenkilde and Ussing, 1996; Rose, 1999). This classical model of heterochrony is discussed by evolutionary biologists (Gould, 1977) as an example of paedomorphosis (retention of ancestral juvenile characters by ontogenetic stages of descendants), progenesis or paedogenesis (precocious sexual maturation of an organism at morphologically juvenile stage), or neoteny (peadomorphosis as an effect of retardation of somatic development). I will use the term neoteny defined as a phenomenon brought about by maturation of the reproductive system in a larval form that fails to undergo metamorphosis (after Rosenkilde and Ussing, 1996; Gilbert, 2000).

Neotenic reproducers can display none of remodeling processes characteristic of metamorphosis (Plethodontidae) or can undergo some metamorphic changes (Salamandridae and Ambystomatidae) (reviewed by Rose, 1999). Neotenic reproducers are usually classified into three categories (two facultative and one permanent) depending on the ability to respond to experimental hormone treatment: facultative forms with fully inducible remodeling that do not naturally metamorphose; facultative forms that undergo metamorphosis under favorable natural environmental conditions; and non-inducible permanent neotenic forms (Dent, 1968; Rose, 1999). Neoteny may be regarded as a lack of response to one of the steps associated with production of hormones of the hypothalamus-pituitary-thyroid axis. Gilbert (2000) noted examples of species that are blocked at specific steps in the production of releasing hormones, stimulating hormones or other hormones, although the processes involved in neoteny my be more complex (reviewed by Rosenkilde and Ussing, 1996, and Rose, 1999). *Ambystoma gracile* and *A. tigrinum* retain neotenic when hypothalamus does not produce enough TRH in low temperatures, and thereby the pituitary is not triggered to produce TSH. In cases when individuals of these species are placed in higher temperatures (experimentally or naturally within their geographical ranges), they undergo metamorphosis. The axolotl *Ambystoma mexicanum* does not metamorphose because its pituitary does not produce enough TSH, and thereby does not trigger the thyroid gland to produce TH, or the TH-induced activity of 5'-deionidase regulating conversion of T_4 to T_3 is not sufficient. When TH is experimentally provided, the axolotls can complete metamorphosis. Permanent neotenes (*Siren, Necturus, Eurycea neotenes*) do not respond to any of those hormones and are believed to have a widespread target tissues insensitivity, although precise experimental data are still not available (reviewed by Rose, 1999). It is accepted that various tissues have different threshold of TH sensitivity, and it is still not fully explained which tissues of particular species of permanent neotenes lack or have inadequate levels of thyroid hormone receptors.

In summary and with regard to hypothalamus-pituitary-thyroid axis, as well as the amount of receptors in target tissues, it can be concluded that neoteny is a highly complex phenomenon, which may be regulated at many levels. It is also worth noting that sexual maturation is independent of metamorphosis, which means that not all tissues need high levels of TH for maturation. For example cloacal glands in neotenic salamanders are as well developed as in metamorphosing ones (reviewed by Rosenkilde and Ussing, 1996). The somatic tissues of gonads are also well developed in neotenic animals.

Hybridization

Most species do not hybridize because of effective isolating mechanisms, which may be classified as prezygotic or postzygotic (Michałowski, 1964). The prezygotic isolation, such as geographical, ecological or behavioral, prevents mating between two, even closely related species. The postzygotic isolation acts when prezygotic mechanisms do not function and mating is possible.

Many species can hybridize when isolating barriers are overcome artificially and a variety of such crosses were obtained in many laboratories. The following list presents only some of examples of both anurans and urodeles: *Rana clamitans* × *R. sylvatica*; *R. clamitans* × *R. pipiens* (Moore,

1941; Denis, 1970); *Rana pipiens* × *R. sylvatica* (Moore, 1946, 1947, 1948); *R. dunni, megapoda* and *pipiens* (Moore, 1966); *Discoglossus pictus* × *Xenopus laevis* (Woodland and Gurdon, 1969); *R. pipiens* × *R. palustris* (Hennen, 1972, 1973; Liepins and Hennen, 1977); *R. pipiens* × *R. catesbeiana* (Reynhout and Kimmel, 1969); *R. esculenta* × *R. temporaria; R. pipiens* × *R. sylvatica; Bufo arenarum* × *R. temporaria;* (Denis, 1970); *R. catesbeiana* × *R. clamitans* (Elinson, 1975a,b; Elinson and Briedis, 1981); *Hyla arborea* × *Pelobates cultripes; H. arborea* × *Bufo calamita, Pleurodeles waltlii* × *Euproctus asper, P. waltlii* × *Triturus palmatus, P. waltlii* × *Ambystoma mexicanum* (reviewed by Subtelny, 1974); *P. waltlii* × *P. poireti* (Gallien and Aimar, 1971; Guillet and Aimar, 1971); *A. mexicanum* × *A. dumerilli* (Boucaut and Gallien, 1975); *P. waltlii* × *A. mexicanum* (Aimar *et al.*, 1976); *Triturus cristatus carnifex* × *T. vulgaris meridionalis* (Mancino *et al.*, 1978).

Closely related amphibian species that occur sympatrically and have overlapping breeding seasons sometimes form natural interspecific hybrids. Most of them die during developmental stages, but some survive, although the hybrids are often infertile. The examples of such natural hybrids are: *Bufo americanus* × *B. fowleri* (Volpe, 1952; Jones, 1972; Green, 1984); *B. boreas* × *B. punctatus* (Feder, 1979); *Xenopus gilli* × *X. laevis* (Kobel *et al.*, 1981); *Rana cascadae* × *R. pretiosa* (Green, 1985); *Phyllomedusa distincta* × *P. tetraploidea* (Castanho and de Luca, 2001); *Hyla arborea* × *H. meridionalis* (Internet ref. 1, 2005); *Bufo ictericus* × *B. paracnemis* (Azevedo *et al.*, 2003); *B. houstonensis (terrestris)* × *B. valliceps* (Internet ref. 2, 2005). Some species form stable hybrid zones in the areas of sympatry, such as *Bombina bombina* and *B. variegata* (Madej, 1964; Szymura, 1993; Szymura and Barton, 1986, 1991); *Bufo americanus* and *B. hemiophrys* (Green, 1983); *Triturus cristatus* and *T. marmoratus* (Valée, 1959; Arntzen and Wallis, 1991); *Triturus vulgaris* and *T. montandoni* (Kotlik and Zavadil, 1999; Litvinchuk *et al.*, 2003; Babik *et al.*, 2003; Babik and Rafiński, 2004), and plethodontis salamanders *Ensatina eschscholtzlii xanthoptica* and *E. e. platensis* (Alexandrino *et al.*, 2005). The evolutionary role of hybrid zones was discussed by Hewitt (1988) and Harrison (1993).

Among Anura two great complexes of green frogs form natural hybrids: one related to *Rana pipies* in the Nearctic (reviewed by Hillis, 1988), and the other related to *Rana ridibunda* in the Palearctic (reviewed by Graf and Polls Pelaz, 1989). *R. pipiens* complex is composed of at least 27 living or recently extinct species divided into the Alpha and Beta complexes, each subdivided into two groups, with stable hybrid zones between them (Hillis, 1988). The Palearctic green frog group (subgenus *Pelophylax*) is composed of 26 recognized species, including 3 hybridogenetic taxons: *Rana esculenta, R. grafi,* and *R. hispanica* (reviewed by Dubois and Ohler, 1994a,b). The Nearctic and Palearctic green frogs most probably were separated when land connections between North America and Eurasia disappeared during the Eocene. From this time on the two groups of frogs evolved separately. Hybridization occurs within both groups, and diploid ova are produced spontaneously by females of the *R.pipiens* (Richards and Nace, 1977), and *R. esculenta* (Berger *et al.*, 1986) complexes. However, triploid hybrids of *R. pipiens* are sterile, whereas triploid male and female *R. esculenta* can produce gametes owing to the special modification of meiosis (hybridogenesis). Among Urodela a system involving polyploids and hybrids is known in *Ambystoma laterale-jeffersonianum-texanum-tigrinum* complex (reviewed by Bogart, 2003). The hybrids among Gymnophiona have not been reported. The reproduction of the Palearctic water frogs and the *Ambystoma* group is described below in more detail.

Polyploidy

Polyploid individuals, both plant and animal, are divided into autopoly-ploids (all genomes of the same species), and allopolyploids (genomes of two or more species). In that way allopolyploids are also hybrids. Kawamura (1984) reviewed 19 polyploid species, and currently 24 natural polyploid amphibians are recognized (Oto and Whiton, 2000).

Autoployploidization is often regarded as one of the mechanisms of evolution. It has been proposed that the evolutionary role of polyploidi-zation is to provide redundant copies of genes, which may mutate independently, leading to species diversification (Ohno, 1970). Polyploidy in clonally reproducing animals provides probably a buffer against the accumulation of deleterious mutations (Vrijenhoek, 1990).

Several anuran species underwent autoployploidization, thereby giving rise to new species (reviewed by Beçak and Kobashi, 2004). The autoployploidization may be regarded as further diploidization (2n to 4n), as is believed in *Hyla chrysoscelis* (2n = 24) and *H. versicolor* (4n = 48), or even from 4n to 8n as is the case of *Ceratophrys dorsata* and *C. ornata* (both 8n = 104). In these cases meiosis is regular, but multivalents, instead of bivalents, are formed. Meiosis must be modified both in auto- and allotriploids owing to the uneven number of chromosomes. The naturally occurring autotriploid bisexual *Bufo pseudoraddei baturae* (*Bufo viridis* complex) was recently described by Stöck *et al.* (2002). Natural allotriploids were described in the *Rana esculenta* complex (Berger *et al.*, 1978; Ogielska *et al.*, 2005). Naturally occurring autotripolids have also been reported in urodelans of the genus *Ambystoma* (*A. gracile*, *A. jeffersonianum*, *A. laterale*, *A. texanum*), but they constitute very low percentage of the studied populations (Lawcock and Licht, 1990). Occasional triploids in the genus *Triturus* were also reported by Book (1940) and Fankhauser *et al.* (1942).

Taxonomic Problems with Hybrid and Polyploid Amphibians

The complexity of naturally occurring hybrid and polyploid taxons causes confusion among taxonomists. In case of Palearctic water frogs, Dubois and Günther (1982) and Dubois (1990) proposed the term "klepton" for special species-rank taxons dependent on other species in their reproduction and persistence in a population. The authors' goal was to provide nomenclature for gynogenetic and hybridogenetic vertebrate forms, which cannot be considered as species *sensu stricto*. These forms are of hybrid origin, reproduce clonally or hemiclonally, and their reproduction depends on gametes provided by one or more "good" species. Because Dubois and Günther (1982) and Dubois (1990) did not consider recombination or the possibility that gametes can be provided by another hybrid, Polls Pelaz (1994) proposed a modified definition of the klepton, which he named the Biological Klepton Concept: "a klepton is a community of populations of hybrid genome from the same parental species, in reproductive dependence (sexual parasitism either via hybridogenesis or gynogenesis) with respect to a good species, which plays the role of sexual host". In most cases hybridogenetic hybrids form mixed populations with one of the parental species. Such mixed populations are stable and form genetic systems in cases, when hybrids eliminate from their germ lines the chromosomes of the species present in the population, and transmit to the gametes a non-recombined chromosomes of the other species (absent in the population) (Fig. 4). To restore the hybrid constitution of the progeny, the hybrid must mate with a species, with which it coexists. This situation may be regarded as a theft (klepton in Greek means a thief) because the hybrid steals a missing genome from one of the parental species. This dependence is also regarded as a kind of sexual parasitism, in which a hybrid is parasitic, and a species plays a role of a sexual host (reviewed by Graf and Polls Pelaz, 1989; Polls Pelaz, 1994; Joly, 2001).

Although Dubois and Günther (1982), Dubois (1990), and Polls Pelaz (1994) proposed the term "klepton" also to be applied in gynogenetic hybrid systems, this nomenclature is not used for the gynogenetic mole salamanders of the genus *Ambystoma*. This highly complex system of Nearctic salamanders includes at least 20 different combinations (from diploid to pentaploid) of chromosome sets (genomes) of four species: *Ambystoma laterale* (LL), *A. jeffersonianum* (JJ), *A. texanum* (TT), and *A. tigrinum* (TiTi) (for references see Bogart, 2003). Lowcock *et al.* (1987) proposed that the best way to describe the hybrid composition of the salamanders is to use hyphenated Latin names of contributing species, preceded by a number in cases when more than one genome is involved; however, the simplest method is to designate the genomes by letters, for example *A. laterale-(2)jeffersonianum* as LJJ (known also under the name *A. platineum*).

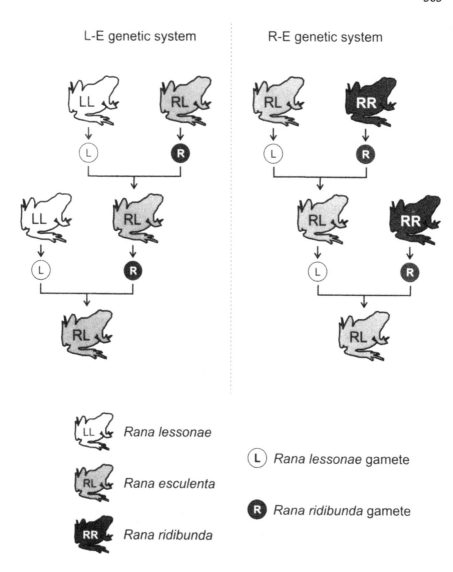

L-E genetic system

R-E genetic system

Rana lessonae

L *Rana lessonae* gamete

Rana esculenta

R *Rana ridibunda* gamete

Rana ridibunda

Fig. 4. Hybridogenesis as a mode of reproduction in the two genetic systems: L-E (composed of male and female *R. lessonae* LL and *R. esculenta* RL), and R-E (composed of male and female of *R. ridibunda* RR and *R. esculenta* RL). As a rule the hybrids produce clonal gametes containing non-recombined haploid chromosome set of one of the parental species (usually of the species missing from that particular population): R in the L-E system and L in the R-E system.

This nomenclature is very useful, because not all hybrid polyploid lineages form stable populations. However, separate Latin names are used in case of two triploid taxons, which form stable genetic systems with bisexual species: *A. platineum* (LJJ) accompanied by bisexual *A. jeffersonianum* and *A. tremblayi* (LLJ) accompanied by bisexual *A. laterale* (for references see Lowcock *et al.*, 1987; Spolsky *et al.*, 1992; Phillips *et al.*, 1997). Because all hybrid combinations always contain at least one *laterale* genome, this name was recommended to be introduced first (Lowcock *et al.*, 1987).

Reproduction of Hybrids and Allopolyploids

Both hybridization and polyploidization can in some cases lead to the better fitness of the progeny. Viable interspecies hybrids, especially those of the first generation (F1), are heterozygous in more alleles than progeny of the parental species and thereby may display the effect of heterosis. In that case, heterozygosity leads to expression of a phenotype that better tolerates environmental stresses and successfully competes with the related species. If the hybrid is fertile, the effect of heterosis will gradually disappear in next generations. However, in most cases the hybrids are sterile or have low fertility caused by disturbances during meiosis. The only escape from such disturbances and the maintenance of heterosis is the stabilization of the F1 generation. This situation is possible only in cases when reproduction is asexual or when some unusual alternative mechanisms appear before or during gametogenesis, such as parthenogenesis, gynogenesis or hybridogenesis. Asexual (vegetative) reproduction, in which gametes are not involved and the next generations originate from specialized somatic tissues of the parental body, does not occur in vertebrates. The sexual reproduction always involves gametes, at least an ovum. True (sperm-independent) parthenogenesis does not need the sexual act to ensure the clonal reproduction of all-female populations. However, this kind of reproduction does not occur in amphibians. The only alternative ways of sexual reproduction in polyploid and/or hybrid amphibians are gynogenesis (sperm-dependent parthenogenesis) and hybridogenesis.

Gynogenesis in Salamanders Belonging to Ambystoma laterale-jeffersonianum-texanum-tigrinum Complex

True parthenogenesis, i.e., the development of an embryo from an egg without a contribution from spermatozoa is not possible in amphibians because the egg does not have the functional centrosome able to form the mitotic spindle and asters. Centrosomes consist of two cylindrical centrioles composed of an array of 9 triplets of microtubules surrounded by a fibrillogranular perinuclear protein material. In *Xenopus laevis* the maternal centrosome becomes inactivated and lost during oogenesis in early stage I diplotene oocytes, but the ooplasm contains an extensive store of centrosomal proteins, such as γ-tubulin, necessary to the very fast formation of about 4000 centrosomes during cleavage before the mid-blastula transition (MBT) (Gard *et al.*, 1995a). The sperm not only supplies the centriole, which is a basal body for its flagellum, but also some of centrosome proteins (centrin, pericentrin, and CTR2611) (Stearns and Kirchner, 1994). The zygotic centromere forms a hybrid organelle, in which some components originate from the egg, whereas the others are supplied by spermatozoa (Klotz *et al.*, 1990). A functional centrosome is essential to microtubule-dependent cortical rotation (Gard *et al.*, 1995b), although Elinson and Paleček (1993) demonstrated that the vegetal array of microtubules is independent from the aster formation, whereas the aster is necessary for normal cleavage. *Xenopus* eggs will develop also when supplied by centrioles of another species: invertebrate sea urchin (Maller *et al.*, 1976) and vertebrate (Tournier and Bornens, 1994; Elinson and Paleček, 1993), but not *Drosophila* or yeast, most probably due to the non-compatible shape and structure of centrioles (Tournier *et al.*, 1999).

Gynogenesis is a major mode of reproduction in the all-female mole salamanders complex, which includes four species: *Ambystoma laterale* (LL), *A. jeffersonianum* (JJ), *A. texanum* (TT), *A. tigrinum* (TiTi), and two stable triploid hybrids known as *A. platineum* (LJJ) and *A. tremblayi* (LLJ). However, some eggs are occasionally fertilized (for references see Bogart, 2003). As was evidenced by Bogart and Licht (1987), only those females that are inseminated and have spermatozoa in their cloacae give rise to living progeny, whereas those with no spermatozoa spawn non-cleaving eggs. Sperm derive from males of the related species, with which hybrid females share the populations. Hybrid males are extremely rare and infertile (Bogart, 2003). In some cases the maternal nucleus may even be replaced by the sperm nucleus; this unusual mode of reproduction was named 'Kleptogenesis' (Bogart *et al.*, 2007).

Female sex in *Ambystoma* is determined by the dominant W chromosome (ZZ/WZ sex determination system; female heterogametic). The presence of at least one W chromosome in a polyploid determinates the female sex. Sex chromosomes in *A. laterale* and *A. jeffersonianum* constitute the 14th pair (n = 14). Because *laterale* sex chromosomes differ from *jeffersonianum* ones when the C-band technique is used, it was possible to detect that in both kinds of triploids (*A. tremblayi* LLJ and *A. platineum* LJJ) the W chromosomes are inherited from *A. jeffersonianum*. The putative ancestral diploid hybrids were females resulted from matings between female *A. jeffersonianum* (JJ) and male *A. laterale* (LL), and when the hybrid females backcrossed with respective males, the LLJ and LJJ triploid female offspring carrying the *jeffersonianum* W chromosome was produced

(Sessions, 1982). These results suggested that maternally inherited mitochondrial DNA (mtDNA) should be of the *jeffersonianum* type. Nevertheless, the analyses of mtDNA revealed that the mitochondrial genotype is unrelated to either of the nuclear genotypes and resembles that of *A. texanum*, thereby indicating *A. texanum* (or its sibling species *A. barbouri*) as the maternal ancestor (Kraus and Miyamoto, 1990; Bogart, 2003).

Under natural temperature conditions polyploid female *Ambystoma* tend to produce ova with unreduced chromosome number, *i.e.*, of the same genome composition as somatic cells of the mothers. The unreduced triploid ova are produced by endomitotic reduplication of chromosomes in zygotene/pachytene primary oocytes (Cuellar, 1976; Sessions, 1982). This mode of production of unreduced ova is classified as automixis (Suomalainen *et al.*, 1987).

The cytological mechanism of gynogenesis in the polyploid hybrid salamanders LLJ and LJJ was described by Elinson *et al.* (1992). As in other Urodela, the fertilization in *Ambystoma* is polyspermic (Elinson, 1986). It means that only one sperm nucleus, called the principal sperm nucleus, will contribute to further development and form male pronucleus, whereas the supernumerary (accessory) nuclei will be discarded in the vegetal hemisphere. In gynogenetic salamanders the principal sperm nucleus does not contribute genetically to the embryo, but provides centrioles, which makes the cleavage possible. After fertilization all sperm nuclei decondense, but only one (principal) becomes associated with the egg pronucleus, whereas the accessory nuclei will move to the vegetal hemisphere, where they become condensed and soon degenerate. Chromatin of the principal sperm nucleus, although decondensed, does not form chromosomes at the time when the egg nucleus enters the first mitotic division. During metaphase and anaphase the paternal chromatin remains as a clump, which is left between the two resulting nuclei, and then stays close to the first cleavage furrow where soon disappears. Although it is impossible to distinguish directly, to which species the chromosomes belong, the clump of chromatin most probably represents the rejected paternal genetic material.

In most all-female populations of *A. platineum* (LJJ) spermatozoa are supplied by males belonging to *A. jeffersonianum* (JJ), and for this reason the researcher cannot be sure, whether the paternal J genome is excluded from the embryo, or is incorporated and the maternal J is rejected (genome swapping, in this case the exchange of J). This situation is usually compared to hybridogenesis, but is very hard to verify because the electrophoretic pattern of specific proteins of the progeny would be indistinguishable from those resulted from gynogenesis (for references see Lowcock *et al.*, 1987; Bogart, 2003). The paternal chromosome loss, and thereby the gynogenetic (not hybridogenetic) reproduction, was confirmed indirectly by analyzing the electrophoretic patterns of selected enzymes of the mothers and their progeny from a unique population of *A. platineum* (LJJ), in which the accompanying bisexual species was *A. texanum* (TT) (Spolsky *et al.*, 1992). All the progeny had the maternal genotype LJJ, and none displayed the existence of T. The authors concluded that female *A. platineum* (LJJ) mainly produce unreduced ova, which develop gynogenetically, and even if other genetic combinations occur, such progeny does not survive and are not of reproductive relevance.

Although the production of non-reduced ova activated by gynogenesis is the major mode of reproduction of mole salamanders, some females produce also reduced ova (Sessions, 1982); moreover, both kinds of ova can be occasionally fertilized (Bogart, 2003). This leads to alternations in ploidy level in the progeny and enables some recombination between non-sister homologues. Phillips *et al.* (1997) reported a high percentage of tetraploids LJJT in a stable population of *A. platineum* (LJJ) accompanied by *A. texanum* (TT). The tetraploid females also produced pentaploids LJJTT offspring, but because of their very low viability, they were not numerous in a population.

To summarize this highly complex reproduction of hybrid all-female mole salamanders: the major way of reproduction is gynogenetical activation of unreduced automictic ova, although reduced ova are also produced; both of them can be occasionally fertilized. An exchange of a genome (genome swapping) sometimes occurs, but the mechanism underlying this process in unknown.

Hybridogenesis

The term "hybridogenesis" has been used for a long time to designate the origin of hybrids resulting from well-known hybridization between two taxons (reviewed by Dubois, 1990). However, Schultz (1969) used and propagated this term to designate a particular type of reproduction of hybrids of Mexican fishes of the genus *Poeciliopsis*, in which each generation renews the F1 hybrids. This unusual situation is possible owing to hybrid females (*P. monacha* × *P. lucida*) that produce haploid ova, which transmit the unrecombined genome of one of their parental species (*monacha*), whereas the other (*lucida*) is eliminated from the germ line prior to meiosis. To restore the F1 hybrid progeny, hybrid females must mate with *P. lucida* males. Hybridogenesis is only one aspect of the highly complicated reproduction in *Poeciliopsis* fishes, which includes polyploidization and gynogenesis (reviewed by Schultz, 1977; Vrijenhoek, 1984). The other fishes with hybridogenetic reproduction belong to the *Leuciscus alburnoides* complex (Alves *et al.*, 1998) and *Phoxinus eos-neogaeus* complex (Dawley *et al.*, 1987). Hybridogenesis was also reported in invertebrate stick insects of the *Bacillus rosius-grandii-benazzii* complex (Mantovani and Scali, 1992; Tinti and Scali, 1992; Mantovani *et al.*, 1999).

The definition of hybridogenesis still causes confusion, especially when defined by evolutionary biologists. Some of them classify hybridogenesis as a kind of parthenogenesis (Schmidt, 1993; Simon *et al.*, 2003), whereas others treat hybridogenetic animals as asexuals, which is hard to accept by developmental biologists. In a broad sense the term "parthenogenesis" is often used to designate an all-female reproduction lacking recombination (or with very little recombination), and without the contribution of paternal chromosomes to progeny (Dawley, 1989). Hybridogenesis in fishes and stick insects occurs only in females, and for this reason they are regarded as "all-female" taxons. On the other hand, male water frogs are common and hybridogenesis occurs in both sexes, but production of gametes (both sperm and ova) is clonal and recombination is very low (reviewed by Schmeller, 2004). For these reasons hybridogenetic frogs are often treated as "unisexuals" or "parthenogens", which is highly misleading. They should be regarded as bisexual hybrids with a special way of gametogenesis, during which one of the parental genome is eliminated before meiosis, and thereby gametes produced by a hybrid are clonal; when fertilized by gametes of a related species, the progeny is semiclonal.

Hybridogenesis in Water Frogs (Subgenus Pelophylax)

There are two concepts explaining the origin of hybridogenesis in water frogs. The first was proposed by Uzzell (1979, 1982), who suggested that *R. ridibunda* and *R. lessonae* were already separated species 9-15 billions years ago. During the Riss glaciations the interspecies hybridization occurred, and the resulting F1 hybrid by some mutation achieved the ability for hybridogenetic reproduction. This concept was supported by mtDNA analysis in the parental and hybrid species (Spolsky and Uzzell, 1986). The other concept was presented by Polls Pelaz (1994) as the Mutational Hypothesis of Hybridogenesis (MHH). The MHH considers the possibility that the cytological phenomena of hybridogenesis are genetically coded and could be present in some species before hybridization might have occurred. This mutation would have taken place after the Eocene, when the Nearctic was separated from the Palearctic. This is supported by the fact, that only Palearctic water frogs display the hybridogenetic phenomenon, whereas those from Nearctics do not.

In my opinion, the MHH better explains the already existing results of studies on hybridogenesis in water frogs. The possible ways of genome rejection described below clearly show that the cytological mechanism enabling hybridogenesis really exists. The presence of nucleus-like bodies (NLBs) in the parental species is of special interest, because it suggests that the genome rejection occasionally occurs in both parental species, namely *R. ridibunda* and *R. lessonae* (Ogielska, 1994a) (for NLBs description see the section *'Genome elimination in water frogs'*). However, other experiments concerning the examination of ovaries of 45 *R. ridibunda* from 9 populations revealed that the oocytes were not clonal, but recombined, and represented the same electrophoretic phenotype as the somatic tissues (Schmeller *et al.*, 2001). These results show that genome elimination is not a common phenomenon in *R. ridibunda*, but does not exclude the possibility that such elimination can occur at least in some individuals.

The other piece of evidence comes from somehow neglected results of the artificial hybridization between the western Palearctic *R. lessonae* (LL) males and eastern Palearctic *R. brevipoda* (BB) (Nishioka and Ohtani, 1984) and *R. nigromaculata* (NN) females (Ohtani, 1993). The former study contains careful characteristics of mitotic and lampbrush chromosomes, 14 electrophoretic loci, and histology of gonads of artificial BL and BBL one year old progeny. BL hybrids had underdeveloped gonads and were not analyzed, whereas BBL had normal ovaries and testes. The examination of lampbrush chromosomes and backcrossing experiments revealed that both males and females BBL were hybridogenetic. The two *brevipoda* genomes formed 13 bivalents and after normal meiosis produced haploid *brevipoda* eggs and spermatozoa, whereas the *lessonae* chromosome set was eliminated from germlines of both sexes. *Lessonae* chromosomes were also eliminated during spermatogenesis of NNL triploids. Ohtani (1993) described the existence of some aberrant mitoses in young, but sexually mature allotripoild hybrid males obtained by fertilizing diploid NN ova (deriving from experimental 4n *R. nigormaculata* female) with the *lessonae* sperm. In some metaphase plates he found one set of degenerating chromosomes (possibly *lessonae*), whereas in other cells two chromosome sets were eliminated; aneuploid spermatogonia were also observed. These results may be interpreted as the ability of *brevipoda* and *nigromaculata* to eliminate other chromosomes, the feature normally unrevealed because these species do not form natural hybrids. These results also suggest that the hybridogenesis may be an ancient feature of some Palearctic frogs of the subgenus *Pelophylax*.

Other western Palearctic water frog species (*R. shqiperica*, *R. kurtmulleri*, *R. epeirotica*, *R. bedriagae*, *R. cretensis*, *R. cerigensis*, *R. saharica*, and *R. terentievi*) are not susceptible and form non-hybridogenetic hybrids when crossed with *R. ridibunda* (Berger *et al.*, 1994; Plötner and Ohst, 2001). Moreover, not all *R. ridibunda* strains (or subspecies) are able to induce the hybridogenesis; those inhabiting the southern area of its wide geographical range (the Balkans) are not (Hotz *et al.*, 1985).

The Expansion of Ridibunda Genome

All hybridogenetic hybrids of western Paleartctic water frogs contain one or two *ridibunda* genomes along with *lessonae, perezi,* or *bergeri*. The *ridibunda* genomes are present also beyond the geographical range of the species *R. ridibunda*, specifically in southern and western Europe, where it constitutes genomes of hybridogenetic hybrids *R. hispanica* (*ridibunda* × *bergeri*) and *R. grafi* (*ridibunda* × *perezi*). The most probable explanation for expansion of the *ridibunda* genome is migration of hybrids (Arano *et al.*, 1994; Günther and Plötner, 1994). The unusual ability of the hybrids to migrate and establish reproductive systems with related species is well known in case of *R. esculenta* (Berger, 1970). The *ridibunda* genome present in European frogs (with exception of those from the Balkans) contains unknown "exortive" factors able to induce the elimination of any other genomes by hybridogenesis (Uzzell *et al.*, 1980; Hotz *et al.*, 1985). More recently Joly (2001) considered the *ridibunda* genome as a selfish one, containing segregation-distorter genes, which prevent the replication of other genes, and result in wide spreading of themselves. As a consequence, hybrids that contain the selfish *ridibunda* genome colonize a wide range of habitats, which are not available for *R. ridibunda* parents. In the Nishioka and Othani experiments presented in the previous paragraph, the chromosomes *brevipoda* and *nigromaculata* unexpectedly manifested their "extortive" behavior, characteristic of *ridibunda*, whereas the *lessonae* chromosomes retained their "susceptible" character.

Reproduction Patterns and Genetic Systems in Water Frogs

The group of Palearctic water frogs (subgenus *Pelophylax*) is composed of several species *sensu stricto* and 3 hybridogenetic species (kleptons) (Dubois and Ohler, 1994a,b). In this review I will focus on 3 species (*R. lessonae, R. perezi,* and *R. bergeri*), which form 3 kleptons (*R. esculenta, R. grafi,* and *R. hispanica,* respectively), all in an obligatory relation to the fourth species, *Rana ridibunda*.

The best-studied hybridogenetic water frog is *R. esculenta*. This common and widely distributed frog has long been regarded as a species with several subspecies or closely related species, until Berger (1966, 1968) reported that they have an unusual method of non-Mendelian inheritance. He crossed several individuals of *R. esculenta, R. ridibunda,* and *R. lessonae,* and stated that, at least in some regions of Poland, *R. esculenta* must be a hybrid between the two other species. These experiments were repeated in other regions of Europe with the same results (reviewed by Berger, 1983; Graf and Polls Pelaz, 1989). After a series of electrophoresis experiments with selected izozymes and blood proteins , Tunner (1974) suggested that the unusual reproduction in water frogs resembled that of *Poleciliopsis* (Schultz, 1969) and is probably hybridogenetic. According to this model, *R. lessonae* (genotype LL), *R. ridibunda* (genotype RR) are species, whereas *R. esculenta* (genotype RL) is a hybrid with *lessonae* (L) and *ridibunda* (R) genomes. Direct hybridization between *R. ridibunda* and *R. lessonae* in the nature is extremely rare because ecological preferences of the two parental species are different (Tunner and Nopp, 1979; Berger, 1982; Rybacki and Berger, 1994; Semlitsch *et al.*, 1996; Plénet *et al.*, 2000) and for this reason they very rarely coexist in the same population. Hybridization is possible in three basic kinds of mixed populations, which in most cases form genetic systems with predictable pattern of reproduction: *lessonae-esculenta* (L-E), *ridibunda-esculenta* (R-E), and homotypic *esculenta-esculenta* (E-E) (Uzzell and Berger, 1975; Berger, 1988; Plötner and Grunwald, 1991; Rybacki and Berger, 2001). The modes of reproduction in the L-E and R-E systems are schematically shown in Figure 4. In all systems, but specifically in E-E and R-E populations, diploid hybrid individuals (genotype RL) coexist with allotriploid hybrids (genotype RRL and/or LLR) (reviewed by Rybacki and Berger, 2001).

The second natural hybridogenetic hybrid *Rana grafi* (formerly known as *Rana* klepton RP), with chromosome constitution *ridibunda-perezi* (genotype RP), occurs in the northern range of *R. perezi* (genotype PP) in the Iberian Peninsula and in southern France, and constitutes the genetic system P-G (formerly designated P-RP). The third natural hybridogenetic hybrid *Rana hispanica* (formerly known as Italian hybrid) related to *R. bergeri* (formerly known as Italian taxon) with chromosome constitution *ridibunda-bergeri* (genotype RB) occurs in Sicily and constitutes the genetic system B-H (formerly designated I-RI). It should be emphasized that the geographical ranges of *R. perezi* and *R. bergeri* do not overlap with *R. ridibunda*, and that the *ridibunda* chromosomes persist and are transmitted only by the hybrids (*R. grafi* and *R. hispanica*, respectively) by backcrossing with the related species. This was confirmed by enzyme electrophoresis and crossing experiments (Graf *et al.*, 1977; Uzzell and Tunner, 1983; Graf and Polls Pelaz, 1989; Berger, 1988; Arano *et al.*, 1994; Hotz *et al.*, 1994; Günther and Plötner, 1994).

Gametogenesis in Water Frogs

The L-E system is the most common, most stable, and best studied, and for these reasons I will use a diploid *Rana esculenta* from this system as a model to describe gametogenesis and reproduction in water frogs. Both females and males *R. esculenta* from a L-E system transmit the *ridibunda* genome to the progeny (see Fig. 5). Fully grown diplotene oocytes of adult *R. esculenta* females have several enzymes characteristic of *R. ridibunda* (Vogel and

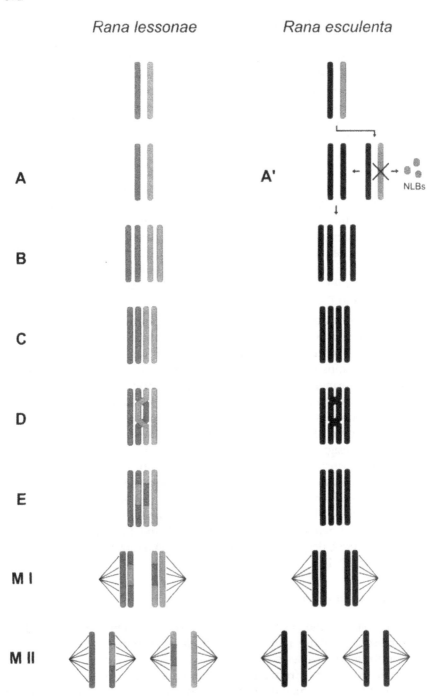

Fig. 5. See caption on the next page.

Chen, 1976; Graf and Müller, 1979; Chen and Stum-Zollinger, 1986), and nuclei of diplotene oocytes contain a haploid number (n = 13) of *ridibunda* bivalents with chiasmata (Graf and Müller, 1979; Bucci *et al.*, 1990). The production of sperm with the *ridibunda* genome was reported by Günther (1975a), Vinogradov *et al.* (1990a,b), and Bucci *et al.* (1990). The above studies confirmed the results of many crossing experiments (Berger, 1968, 1973, 1976) analyzed mainly in respect to morphological features. However, these authors dealt with sexually mature individuals and thereby could provide information only about the final result of meiosis, but not about mechanisms of genome elimination characteristic of hybridogenesis. The conclusion from all these studies was that elimination (or suppression) of the *ridibunda* genome must have taken place before meiosis (Uzzell *et al.*, 1980). Soon thereafter Tunner and Heppich (1981), and Tunner and Heppich-Tunner (1991) analyzed the chromosome composition in metaphase plates of oogonia and spermatogonia (Heppich *et al.*, 1982) of metamorphic frogs. They found that the onset of meiosis in *R. esculenta* females is prolonged, and during that time the gonial cells (both oogonia and spermatogonia) display variable numbers of chromosomes. They reported basically 3 types of cells with regard to the number and composition of chromosomes: haploid 1n *ridibunda*, 1n-2n aneuploid *lessonae-ridibunda* with the predominance of *ridibunda*, and diploid 2n *ridibunda*. On the basis of these observations they stated that the *lessonae* chromosomes are gradually lost, whereas the *ridibunda* chromosomes are

---

Fig. 5. A schematic representation of meiosis shown for one pair of homologous chromosomes (A) in a diploid species represented herein by *Rana lessonae* (left column), which is one of the parental species of a hybridogenetic hybrid *R. esculenta* (right column). During prophase I in *R. lessonae* each of the homologues after replication is composed of two sister chromatids (B). The prophase of the first meiotic division is long, and homologue chromosomes become juxtaposed and physically aligned during leptotene-zygotene (C). In pachytene this contact allows recombination between chromatids of paternal and maternal homologues, represented by different grey shades (D); the exchange sites form chiasmata. In diakinesis chiasmata become decomposed, and each homologue with recombined chromatids (E) enter metaphase I (M-I), during which recombined pairs of homologues move to the opposite poles. During the second division (M-II) the sister chromatids separate, giving rise to four cells (gametes), each with recombined haploid set of chromosomes. In the hybrid *R. esculenta* each homologue pair is composed of one *lessonae* (grey) and one *ridibunda* (black) chromosome. During a prolonged prophase (A') the *lessonae* chromosomes are rejected in form of nucleus-like bodies (NLBs) and the remaining *ridibunda* chromosome is reduplicated. In that way all chromatids are genetically identical. Next steps of hybrid meiosis resemble those in the species, but the resulting gametes form a non-recombined clone.

374

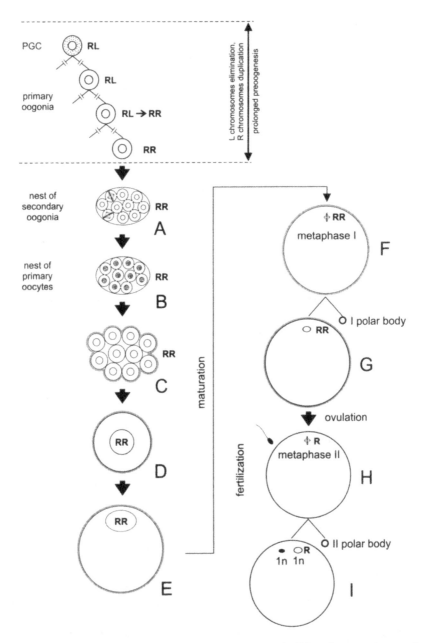

Fig. 6. A scheme representing hybridogenetic oogenesis (A) and spermatogenesis (B) in *Rana esculenta* from a L-E system. During prolonged preoogenesis and prespermatogenesis, oogonia and spermatogonia containing the *ridibunda-*

Contd. on next page

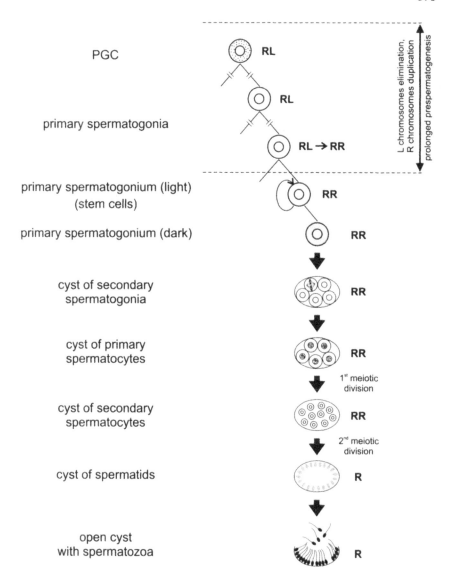

PGC

primary spermatogonia

primary spermatogonium (light)
(stem cells)

primary spermatogonium (dark)

cyst of secondary
spermatogonia

cyst of primary
spermatocytes

cyst of secondary
spermatocytes

cyst of spermatids

open cyst
with spermatozoa

Fig. 6 contd.

lessonae (RL) chromosome set proliferate and eliminate the *lessonae* chromosomes, which are replaced by reduplicated *ridibunda* (R) chromosomes. Oogonia and spermatogonia, which have properly rejected and reduplicated chromosomes enter meiosis and give rise to haploid gametes with *ridibunda* chromosomes. A, B—secondary oogonia and leptotene-pachytene primary oocytes in nests; C—early diplotene oocytes emerging from the nest became individual cells; D-F—growing diplotene oocytes; G-I—first and second meiotic division.

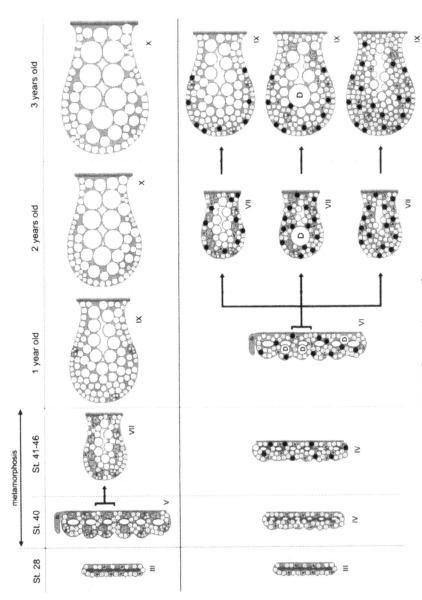

Fig. 7. See caption on the next page.

376

gradually reduplicated, and both these processes occur premeiotically.

It became clear that further analysis of dividing oogonia and spermatogonia was necessary. Günther (1975a) was among the first who focused on dividing primary spermatogonia, but in adult hybrids, and failed to find any mechanism that would explain the elimination. In my former studies I focused on the earliest stages of gonad development. My hypothesis is that elimination occurred "once for a lifespan" during the period of active gonial proliferation before the onset of meiosis (preoogenesis and prespermatogenesis). Meiosis starts in *R. ridibunda* and *R. lessonae* females very early in larval stages (around Gosner stage 30), and the pool of primary oogonia soon become exhausted and restricted to germ patches scattered in the peripheral layer of ovarian cortex (for details and ovarian staging see Ogielska and Kotusz, 2004, and Ogielska, 'Oogenesis and Female Reproductive System in Amphibia—Anura' in this volume).

The number of primordial germ cells (PGCs) in gonadal anlagen in *R. ridibunda*, *R. lessonae*, and *R. esculenta* is similar and ranges from 55 to 110, which is a standard for amphibians (reviewed by Hardisty, 1967). There is no difference in mitotic activity of gonial cells between the species and hybrids before Gosner stage 28, i.e., before the sexual differentiation of

---

Fig. 7. Schematic comparison of development and differentiation of ovaries in water frogs. The upper row represents stages of ovary differentiation in the parental species *Rana ridibunda* which attains sexual maturity when is 3 years old (*Rana lessonae* is mature when is 2 years old), whereas the lower row represents the rate of ovary differentiation in the hybridogenetic *Rana esculenta*. Staging of ovaries is based on Ogielska and Kotusz (2004). Sexual differentiation of gonads *Rana ridibunda* starts at Gosner stage 28 (ovarian stage IV). Nest of secondary oogonia and early meiocytes are abundant at ovarian stage V. The first diplotene oocytes appear in the parental species at Gosner stage 35 (ovarian stage VI). During metamorphosis (at Gosner stages 41-46) until early juvenile period (1-year-old female) diplotene oocytes gradually become predominant class, whereas less advanced meiotic stages are restricted to small areas at the periphery of ovarian cortex. During the second and third year of life primary oogonia form small "germ patches", and diplotene oocyte at various stages compose the ovarian cortex. In contrast, hybrid ovaries of *Rana esculenta* are highly underdeveloped, the cortex is composed of a mass of dividing primary oogonia (which corresponds to ovarian stage IV), many of them degenerating (indicated by black dots). Diplotene oocytes are lacking or their number is low. Ovaries of 1-year-old hybrids resemble those at metamorphosis in the parental species. In 2-year-old female hybrids, ovaries differ in respect to the number and classes of diplotene oocytes. Only these oocytes (marked with D) which appeared in 1-year-old females will be deposited when a female is 3 years old. Females achieve full fertility at the age of 4 or 5 years, depending on the degree of diplotene differentiation. The period of oogonial proliferation is prolonged and is observed also in 2- and 3-year-old females.

378

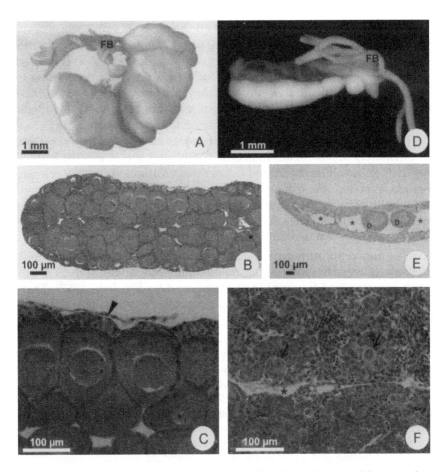

Fig. 8. Development and differentiation of ovaries of 1-year-old parental species *Rana ridibunda* (A-C) and hybridogenetic *Rana esculenta* (D-F). The total view of ovary of the parental species with well-developed ovarian lobes is shown in A, whereas an underdeveloped ovary is characteristic of the hybrid (shown in B). The degree of cortex differentiation can be compared in B and E. Arrows in C indicate thin layer of oogonia and early meiocyte nests (ovarian stage IX); the cortex in the hybrid ovary is composed of dividing oogonia (E and F) and few diplotene oocytes (arrows in F). Ovarian cavities are filled with diplotene oocytes in B, and well seen in poorly differentiated hybrid ovary (marked with asterisks). D—diplotene oocytes; FB—fat body.

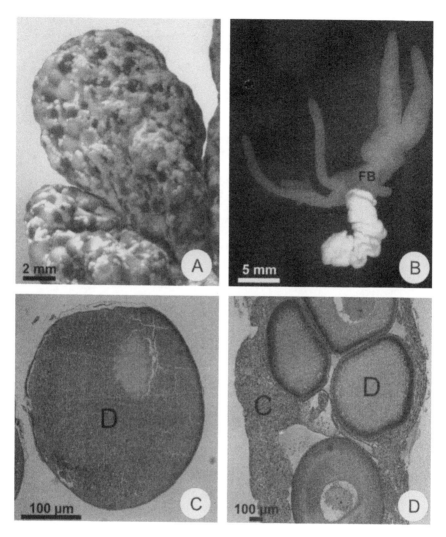

Fig. 9. Development and differentiation of ovaries of 3 years old parental species *Rana ridibunda* (A, C) and hybridogenetic *Rana esculenta* (B, D). Total view of the lobes of well-differentiated ovary of the parental species is shown in A, whereas ovary in the hybrid is poorly developed (B). The cortex of the parental species (shown in C) is filled with fully developed diplotene oocytes, whereas the cortex of the hybrid contains dividing oogonia and low number of small diplotene oocytes. C—cortex; D—diplotene oocytes; FB—fat body.

380

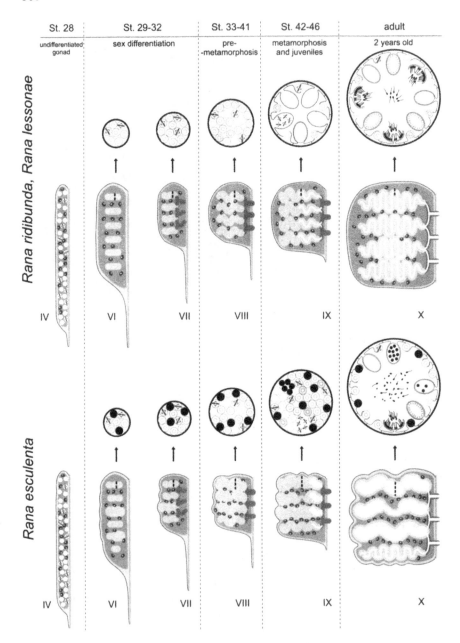

Fig. 10. See caption on the next page.

gonads. The number of gonial cells during that time increases about threefold, and ranges from 156 to 182 (Ogielska and Wagner, 1993). High mitotic activity is observed up to Gosner stage 33 in *R. lessonae* and 35 in *R. ridibunda*, when primary oogonia gradually transform into secondary oogonia and meiocytes. As a result, the number of primary oogonia gradually decreases, being replaced by the increasing number of diplotene oocytes at stage 46 (metamorphosis completed; ovarian stages VIII-IX, according to Ogielska and Kotusz, 2004). The detailed studies on the development and differentiation of *R. esculenta* ovaries (Ogielska and Wagner, 1993; Wagner and Ogielska, 1993; Ogielska, 1995; Rozenblut, 2006; Ogielska, unpublished data) confirmed that the period of oogonial proliferation before the onset of meiosis is prolonged up to one year or even longer in comparison to the parental species. This period may be considered as a prolonged preoogenesis in comparison to normal meiosis observed in the parental species and other anurans (Fig. 6A). The development of hybrid ovary is basically the same in all cross types (*lessonae* × *ridibunda*, *esculenta* × *ridibunda*, *esculenta* × *lessonae*, and reciprocals). The main difference between the species and hybrids is the degree of differentiation of the ovarian cortex. The comparison of development and differentiation of normal and hybrid ovary is schematically summarized in Figure 7, and shown in some detail in Figures 8 and 9. Sexual differentiation of gonads in water frogs starts at Gosner stage 28 (ovarian stage IV). First diplotene oocytes appear in the parental species at Gosner stage 35 (ovarian stage VI), and ovarian stage IX in which diplotene oocytes predominate (and less advanced meiotic stages are restricted to small areas at the periphery of ovarian cortex) is achieved at the time when metamorphosis is completed (stage 46). Hybrid ovaries are highly underdeveloped

---

Fig. 10. Schematic comparison of development and differentiation of testes in water frogs. The upper row represents stages of testis differentiation in the parental species *Rana ridibunda*, whereas the lower row represents the rate of testis differentiation in the hybridogenetic *Rana esculenta* (for staging see Ogielska, Chapter 2 of this book). The schematic drawings of cross section of seminiferous tubules are situated above the drawings of longitudinal sections of testes at successive stages. Seminiferous tubules (light grey) and tubules of rete testis (dark grey) are separated by mesenchymal cells (black). During larval and juvenile stages testes of hybrids are smaller and irregular. The seminiferous tubules are less numerous in comparisons to *R. ridibunda* and contain fewer germ cells. The mitotic activity of pale primary spermatogonia is high in hybrids, but most of germ cells degenerate (indicated by black dots). After metamorphosis testes of the parental species contains seminiferous tubules with many dark spermatogonia and cysts with leptotene-pachytene meiocytes, whereas hybrid testes contain only pale primary spermatogonia. Some spermatozoa appear in 1-year-old male, but sexual maturity is achieved in 2-year-old individuals.

and the thickness of their cortex at stage 46 is the same as in the species at stage 32 (90 and 91 µm, respectively) i.e. the cortex is about threefold as thick as at stage 28, whereas the cortex of non-hybrid species increases during that time about tenfold (from 48 to 465 µm). The thin cortex reflects its poor differentiation. The cortex of hybrid ovaries is composed of a mass of dividing primary oogonia (which corresponds to ovarian stage IV), it lacks or has a very low number of often degenerating nests with secondary oogonia and early oocytes (as is the case of ovarian stage V), and there are very few or no diplotene oocytes. The mitotic activity of primary oogonia is very high (see also Tunner and Heppich, 1981).

The differentiation of ovaries and dynamics of meiosis in hybrids differ from these processes in the parental species. Meiosis does not start or is restricted to very few nests of meiocytes, but oogonial mitoses are still active, and the number of primary oogonia increases. The period of oogonial proliferation is prolonged and is observed even in 2- and 3-years-old females. The onset of meiosis in majority of oogonia in hybrids is delayed about 1 year in comparison to species. During that period the oogonia multiply and 15-75% of them degenerate (Rozenblut, 2006). It seems very probable that meiosis can start only after one set of chromosomes is properly rejected, and the other is properly reduplicated. The delay seems to be an intrinsic characteristic of oogonia, and not the influence of the milieu of a gonad, because a small portion of oocytes (from one to several dozen, Fig. 9C) "escapes" from the arrest and these oocytes grow in a rate comparable to that of *R. ridibunda* oocytes (Ogielska and Wagner, 1990, 1993). This small portion (if present) will be ovulated after the third hibernation, i.e., when female water frogs reach sexual maturity (Socha, 2005; Rozenblut, 2006, Socha and Ogielska, unpublished). In that sense, hybrids mature at the same age as non-hybrid species, but their fecundity is very low or none during the first breeding season. Normal fecundity will be achieved within next one or two years, when new generations of diplotene oocytes will mature.

The development and differentiation of testes in *R. esculenta* are also delayed (Ogielska and Bartmańska, 1999). The comparison of development and differentiation of normal and hybrid testis is schematically summarized in Figure 10, and shown in some detail in Figure 11. During larval and juvenile stages, testes of hybrids are smaller and irregular. The seminiferous tubules are less numerous in comparisons to *R. lessonae* and *R. ridibunda* and contain fewer germ cells. The period of prespermatogenesis, when primary spermatogonia proliferate, is prolonged (Fig. 6B). The pool of primary spermatogonia seems to be lower in hybrids than in parental species but quantitative data are not yet available. The mitotic activity of pale primary spermatogonia is high in hybrids, but most of germ cells degenerate. After metamorphosis and before first hibernation most of hybrid testes contain only one class of primary spermatogonia (pale), whereas in the parental species of the same age many dark spermatogonia and cysts with leptotene-pachytene meiocytes are present inside seminiferous tubules. During summer before second hibernation spermatogenesis in parental species is very active, and spermatozoa appear in the lumen of tubules. Two-year old male *R. lessonae* and *R. ridibunda* are sexually mature. The first cysts containing secondary spermatogonia and primary

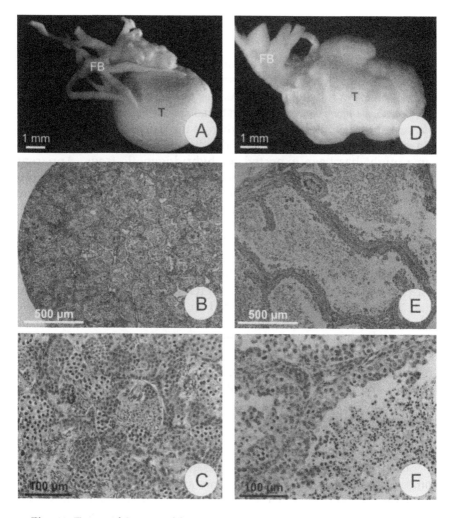

Fig. 11. Testes of 3-years-old *Rana ridibunda* (A-C) and *Rana esculenta* (D-F). Note the irregular surface of the hybrid testis (B) and the highly abnormal composition of seminiferous tubules (E) containing degenerating spermatozoa inside enlarged lumen (F).

spermatocytes in *R. esculenta* appear after first hibernation, and spermatozoa are present before and after second hibernation. However, hybrids after the second hibernation are often less fertile and have abnormal gonads. Testes of many fertile hybrid males become almost normal, but many remain abnormal also during adult ages. The seminiferous tubules of highly abnormal testes from the hybrid species are shown in Figure 11D-F. In such testes degeneration of all stages of spermatogenesis is very high, spermatozoa are scanty, and the tubules are abnormally organized.

Genome Elimination in Water Frogs

The question "when?" genome elimination occurs was roughly answered, but the question "how?" is still obscure, and the mechanism of genome elimination is not fully elucidated. As was reported for the hybridogenetic fish *Poecilopsis monacha-lucida* the maternal set of chromosomes (*monacha*) is attached to the mitotic spindle, whereas the paternal set (*lucida*) is not, and is discarded in the cytoplasm of dividing oogonia (Cimino, 1972). The same might has been expected for hybridogenetic frogs. Apparent differences in the constitution of pericentromeric heterochromatin between the *ridibunda* and *lessonae* chromosomes were reported, which might play a role in kinetochore structure, and thereby in the affinity to the mitotic spindle. These differences concern the amount of AT-rich heterochromatin visualized after Actinomycin D/DAPI staining, being characteristic of the *ridibunda*, and being absent from the *lessonae* chromosomes (Heppich *et al.*, 1982), and the presence of RrS1satellite family in the centromeres of 6 out of 13 haploid *ridibunda* chromosome set (Guerrini *et al.*, 1994; Ragghianti *et al.*, 1999). Ogielska (1994a) provided an analysis of shapes of mitotic spindles in water frogs in order to find whether they were bipolar or unipolar. The detailed analysis of the spindles of proliferating primary oogonia and spermatogonia were carried out during early oogenesis and spermatogenesis in tadpoles at stages during the period of most active oogonial and spermatogonial proliferation (Gosner stages 26-29 in the species, and 29-41 in hybrids) and later after metamorphosis; these studies revealed that mitoses were bipolar, and thereby the elimination of chromosomes was not a one-step event (Ogielska, 1994a) (Fig. 12C). Beginning with Gosner stage 29 and soon after the sexual differentiation of gonads, the increasing number of aberrant mitoses is observed, in which some chromosomes or their fragments are separated from the main anaphase or telophase chromosome mass (Fig. 12D). Aberrant mitoses are frequent in hybrids, but also sporadically observed in species (Ogielska, 1994a). The ultrastructural study of early gonads revealed unique structures in the cytoplasm of interphase oogonia and spermatogonia. They are small spherical structures 2 μm in mean diameter, morphologically resembling the nucleus (mean diameter 12 µm) and enveloped by a double membrane envelope (Fig. 13). They are localized at various distances from the "main" nucleus, and were named nucleus-like bodies (NLB) (Ogielska, 1994a). NLBs were never observed at the ultrastructural level in the parental species, and have never been reported in any amphibian species. They apparently are not nuclear lobes, because the shape of primary oogonia in Ranidae is mostly spherical or ovoidal, and only slightly irregular in outline, but not lobed (Ogielska and Kotusz, 2004). NLBs were most numerous between Gosner stages 28-41 (between 38 and 80 day after

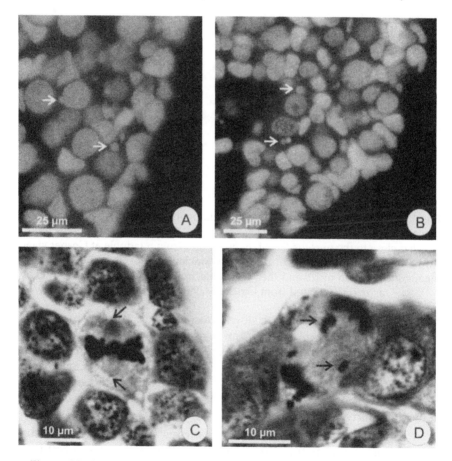

Fig. 12. Nucleus-like bodies (NLB) and mitoses in gonial cells of developing gonads of hybridogenetic *Rana esculenta* at Gosner stage 31. NLBs are pointed by arrows in testis (A) and ovary (B). Mitoses of gonial cells are bipolar (C) and often aberrant (D) with few chromosomes or chromosome fragments (arrows) left outside the future daughter nuclei (Reprinted from Ogielska 1994a, with permission).

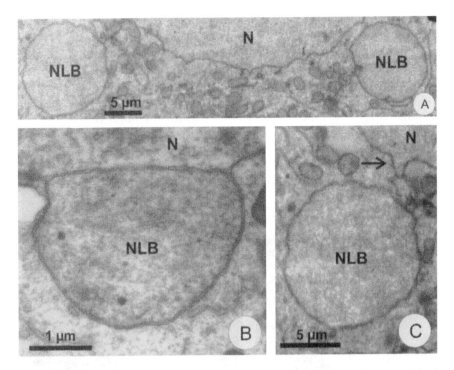

Fig. 13. Putative method of chromosome elimination from interphase nuclei of gonial cells of hybridogenetic *Rana esculenta*. A—Two nucleus-like bodies (NLB) situated close to the nucleus (N) of an oogonium from a tadpole at Gosner stage 31. NLBs most probably bud off from the nucleus by forming the separating membranes (shown in B), and then detaching from the nucleus (shown in C by arrow) (Reprinted from Ogielska, 1994a, with permission).

fertilization) and their number slowly decreased during metamorphosis (stages 42-46). Further cytochemical analysis at the level of light microscope after the Feulgen or DAPI staining revealed that NLBs contain DNA, and their number per cell ranges from 1 to 3 (Fig. 12A,B). However, one of the most striking results was the occasional presence of NLBs in oogonia and spermatogonia of the parental species *R. lessonae* and *R. ridibunda* observed at the light microscope level.

I considered two alternative mechanisms for NLB formation: by degradation of chromatin during interphase (Fig. 14A) or as a result of elimination of chromosomes or their fragments during sequential mitoses (Fig. 14B). The first possibility is based on possible sequence of budding off from the main nucleus during interphase after plausible genetically controlled degradation of chromatin in a process similar to apoptosis (see Fig. 13B,C). The second possibility is based on the presence of chromosomes (or their fragments) within anaphase and telophase spindles, but

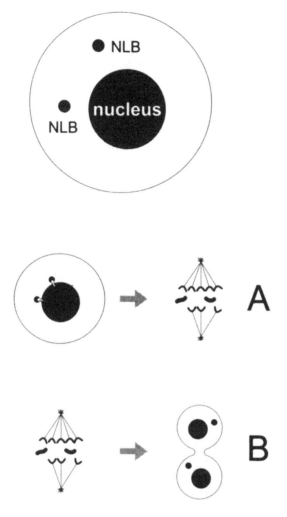

Fig. 14. Schematic interpretation of two possible mechanisms for the formation of nucleus-like bodies (NLB) and the origin of aberrant mitoses in gonial cells of hybridogenetic *Rana esculenta*. A—Elimination of chromosomes takes place during interphase and the chromosomes or their fragments are rejected from the nucleus as NLBs. During subsequent mitosis remaining fragments of rejected chromosomes do not attach to the mitotic spindle. B— Elimination of chromosomes takes place during mitosis, and rejected chromosomes are surrounded by nuclear envelopes in the daughter cells.

not attached to the mass of the dividing chromosomes; such single chromosomes could possibly generate their own nuclear envelope and form NLBs. In my opinion the first possibility is more probable for the following reasons. The dispersion of chromatin in NLBs and in the main nucleus was the same in all cases; if we consider the second possibility, that the rejected mitotic chromosomes would be the source of NLBs, their chromatin should be condensed at least for some time, but such situation was never observed. Elimination of chromosomes during mitosis may also be a secondary side effect of incomplete chromatin degradation during preceding interphase, which in turn results in its loss during subsequent mitosis. The formation of NLBs (which can be regarded as a kind of fragmentation of the nucleus) is not accompanied by the degradation of the cytoplasm. The degenerating oogonia and spermatogonia are easily distinguishable at the electron microscope level, and usually first the cytoplasm, and then the nucleus are destroyed. It is worth noting that a similar speculation (apoptosis-like degradation of chromatin) was also suggested by Vinogradov *et al.* (1990a) in their study devoted to DNA measurements in testes of adult diploid and triploid male *R. esculenta*. However, the model presented in Figure 14 is speculative and needs further verification.

Whatever the method of chromosome elimination, NLBs are detected in oogonia and spermatogonia at the same time when Tunner and Heppich (1981), Heppich *et al.* (1982), and Tunner and Heppich-Tunner (1991) reported the gradual loss of *lessonae* chromosomes from the germ line of *R. esculenta*. If we accept that NLBs are carriers of rejected chromosomes, their size and number also confirm gradual elimination of one of the parental genome. However, contradictory results were obtained by Vinogradov *et al.* (1990a) for some diploid and triploid males of *R. esculenta*. The analysis of DNA content in testes homogenate by flow cytometry revealed that the genome constitution of primary spermatogonia in adults resembled that in somatic tissue (or was a mixture of cells of various ploidy), whereas the spermatozoa were haploid. These results indicate that, at least in triploids, genome elimination might have taken place not only during juvenile period, but also in adults before the breeding season. These results do not agree with those of Ogielska (1994a), and should be studied further with more detailed methods.

Production of Haploid and Diploid Gametes by R. esculenta

It is generally accepted that the presence of the *ridibunda* genome is a prerequisite for hybridogenesis to occur. It is believed that the *ridibunda* genome contains a hypothetical gene (or genes), which induces obligatory elimination of any other genome from the germline of a hybrid (Uzzell and

Hotz, 1979; Uzzell *et al.*, 1980; Hotz *et al.*, 1985; Polls Pelaz, 1994; Joly, 2001). This explanation fits well to the L-E, G-P, and B-H systems, in which the *ridibunda* genome is transmitted, and *lessonae, perezi*, and *bergeri* are, respectively, rejected (for review, see Graf and Polls Pelaz, 1989). In other words, the *ridibunda* genome is "extortive", whereas the *lessonae, perezi*, and *bergeri* are "susceptible" to the elimination. However, there are exceptions to this rule. In some hybrid individuals the situation is reversed, which means that the *ridibunda* genome is not transmitted. Many years of crossing experiments (Berger, 1968, 1973, 1976; Berger and Czarniewska, 2002) unequivocally showed that the genomes transmitted by male and female hybrids into gametes vary. The best examples are *R. esculenta* of both sexes, which transmit the *lessonae*, and reject the *ridibunda* genome; their existence allows the R-E systems to occur (Uzzell and Berger, 1975; Berger, 1988; Berger *et al.*, 1986; Berger and Günther, 1992; Berger and Roguski, 2002; Socha, 2005). LLR triploids usually transmit the *lessonae* and reject the *ridibunda* genomes (Borkin *et al.*, 1989; Graf and Polls Pelaz, 1989; Brychta and Tunner, 1994; Mikuliček and Kotlik, 2001). Most curious, however, is the production of two types of gametes by the very same individuals, one type that contains the *lessonae*, and the other type that contains the *ridibunda* genomes (Günther, 1975a; Uzzell *et al.*, 1977; Kawamura and Nishioka, 1986; Vinogradov *et al.*, 1990b; Borkin *et al.*, 1989).

One of the well-documented phenomenons is the production of eggs of various size, ploidy, and genome composition by female *R. esculenta* (see Berger, 1994). Female *R. esculenta* usually produce 3 classes of eggs: small, medium, and large. As a rule, large eggs are diploid and give rise to triploid progeny. Production of diploid RL and RR ova was described by Berger *et al.* (1986), Berger and Roguski (2002), and Socha (2005). Much less is known about diploid sperm formation, although its production is plausible because some of triploid progeny derives from haploid ova (Rybacki, 1994a).

Sex Ratio

The XX/XY system of sex determination is common in Ranidae, and sex of the offspring is determined by the dominant Y sex chromosome transmitted by the heterogametic male (for review see Green and Sessions, 1991). When the eggs of *R. esculenta* are fertilized by *R. lessonae* or *R. ridibunda* male, the sex ratio of the resulting offspring usually is 1:1. When the eggs are fertilized by *R. esculenta*, the sex ratio is strongly biased in most cases toward females (Berger, 1971a, 1994). The analysis of these results indicates, that the clonally inherited *ridibunda* genome as a rule contains only X chromosomes, both in males and females. Unfortunately, the X and Y (chromosomes 4) in water frogs do not differ morphologically, although they have been distinguished after replication bands technique (Schempp and Schmid, 1981). For this reason the crossing experiments are of great value in searching for sex determination systems water fro:;s. As was revealed by Berger (1994), not only male hybrids, but also males of the parental species of Palearctic water frogs in some cases transmit only one type of sex chromosomes (either X or Y instead of 1:1 ratio of X and Y), and for this reason the progeny fathered by such males have strongly biased sex ratio. Hybrids and parental species in natural populations can also be over-represented by one sex, whereas the other sex may be even absent. The examples are the population inhabiting Neusidlersee in Austria (Tunner and Dobrowsky, 1976), where in the L-E population hybrids were mostly female, or L-E population from Latvia (Caune and Borkin, 1989), where all hybrids were diploid males. A similar population, but with triploid LLR males was described in France (Graf and Polls Pelaz, 1989). In some R-E population on the Wolin and Bornholm Islands in the Baltic Sea, the hybrids were represented mostly by males, whereas *R. ridibunda* was represented mostly by females (Rybacki, 1994b,c).

Development and Viability of Hybrid Progeny

Hybrids, both artificial and natural, provide a special kind of model organisms, in which the interaction between cytoplasm of maternal species and paternal genome can be studied. Other experiments were devoted to study of the induction ability and competence of embryonic tissues grafted between hybrid and parental embryos. Below I will analyze the abnormalities observed in hybrid embryos in respect to plausible molecular events. Although speculative, such analysis may provide some new interpretation of old experiments.

Subtelny (1974) proposed the classification of amphibian hybrids into 4 groups according to the time when abnormalities occurred during development. The first group involved early embryos, in which fertilization and/or cleavage is abnormal; the second group derived from embryos, in which cleavage was normal but embryos stopped development during gastrulation; the third group was composed of embryos with abnormal larval stages; and the fourth group was represented by progeny, which completed metamorphosis or even survived until sexual maturity. This classification reflects the most critical stages of development: mid-blastula transition (MBT), gastrulation, embryonic organogenesis, and metamorphosis.

Abnormal embryos are a rule in almost all experimental and natural crosses, in which at least one of the parents is *Rana esculenta*. Berger (1967, 1971b, 1976) described arrested blastulae, exogastrulae, larvae with protruding yolk plugs, body and tail curvatures, tail degenerations, eye

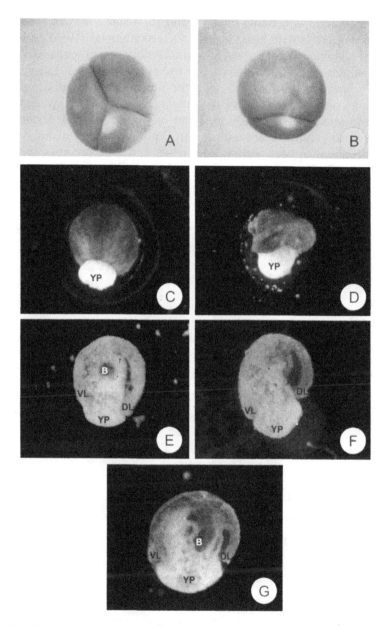

Fig. 15. *"Rana esculenta* developmental syndrome". Abnormal first cleavage giving rise to 3 (A) or 2 unequal blastomeres (B). C, D—Exogastrulae with non-resorbed endoderm forming yolk plug (YP). E-G—Manually bisected exogastrulae with small (E, G) or absent (F) blastocoel. B—blastocoel; DL—dorsal lip; VL—ventral lip (Reprinted from Ogielska, 1994b and Ogielska and Jurgowiak 1994, with permission).

malformations, and oedema (Figs. 15 and 16). Guex *et al.* (2001) reported some metamorphs with fused nostrils. Ogielska-Nowak (1985) provided a detailed histological analysis of abnormal embryos and revealed that the abnormalities can occur alone or in various combinations. Later experiments (Ogielska, 1991, 1994b), with monitoring of individual embryos selected according to the type of abnormality from the first cleavage furrow to their spontaneous death, revealed that very often an earlier abnormality can give rise to other abnormalities. An abnormality may disappear, and an embryo may appear normal for a while, but soon becomes abnormal again (Fig. 17). The occurrence of these abnormalities, together with very high mortality (up to 100% of embryos), was named *"esculenta* developmental syndrome" (Ogielska-Nowak, 1985; Ogielska, 1994b). Low viability

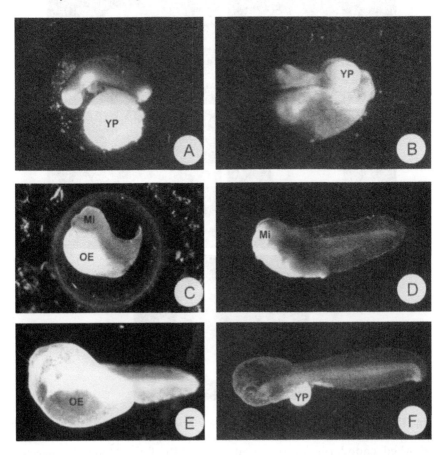

Fig. 16. *"Rana esculenta* developmental syndrome". A–F—Abnormalities at larval stages in *Rana esculenta*. Mi—microcephaly; OE—oedema; YP—protruding yolk plug (Reprinted from Ogielska, 1994b, with permission).

at stage 25 was also reported by Ohler (1988). In experimentally created green frog hybrids, Moore (1946, 1947, 1948) described arrested blastulae, abnormal invagination, failure of gastrulation, abnormal position of heart and sucker, formation of neuroid tissue, abnormal otic and optic vesicles and pronephric ducts in *R. pipiens* × *R. sylvatica* progeny. Exogastrulae and elongated exogastrulae were reported by Elinson (1977, 1981) in the progeny of crosses between *R. catesbeiana* and *R. clamitans*.

The earliest abnormality is the formation of irregular blastomeres during the first cleavage (Fig. 15A,B). Such embryos usually die, mostly as arrested blastulae or exogastrulae. In the case of the 3-blastomere embryo, the most probable cause is trispermy, i.e., the incorporation of 3, instead of 1 sperm. In anurans, which are physiologically monospermic, super-numerary sperm produce asters and induce additional cleavage furrows (reviewed by Elinson, 1986). In some instances they may give rise to mosaic embryos. Such mosaics 1n/3n *R. esculenta* were described by Berger and Ogielska (1999).

In cases when cleavage is disturbed, the most probable reason might be the quality of an egg cytoplasm, and possible aberrant localization of maternal molecules, which were deposited in egg cytoplasm during oogenesis. When cleavage is normal, but blastula is arrested, the most probable reason for the arrest is lack of appropriate zygotic genes expression during or soon after the mid-blastula transition (MBT). Such a situation can be induced experimentally in *Xenopus laevis*, when the appropriate zygotic genes are not expressed during MBT, blastomeres start to follow the maternally coded endogenous program of cell death, the blastula becomes arrested, and soon dies by cytolysis (Sible *et al.*, 1997).

Abnormal gastrulation may be caused by disturbances in both maternal and/or zygotic control. The abnormal maternal Veg-T activity causes too late or abnormal dorsal mesoderm specification, which results in lack of endoderm specification, abnormal blastopore formation, abnormal development of anterior-posterior axis, and underdevelopment or absence of head structures (acephaly or microcephaly) (Zhang *et al.*, 1998). Abnormal embryos may be ventralized when the process of dorsalization is disturbed. In the absence of β-catenins *Siamois* is not expressed, dorsal side of an embryo is not normally formed, but the ventral side is not affected (Heasman *et al.*, 1994). When these various signaling pathways are altered, the resulting embryo will display abnormalities during gastrulation or early axial patterning.

One of the most common developmental abnormalities of the *esculenta* developmental syndrome is exogastrulation (Fig. 15C-G). In all exogas-trulae (arbitrarily classified as I, II, and III depending on the degree of yolk plug protrusion and size of the blastopore), blastopore formation and

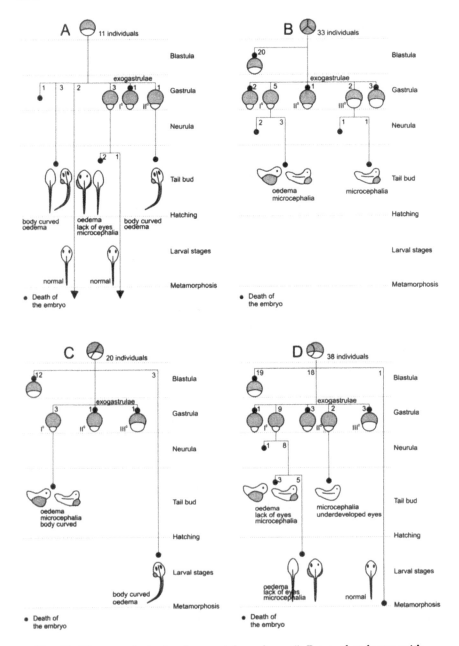

Fig. 17. "*Rana esculenta* developmental syndrome". Fates of embryos with normal (A) and abnormal (B-D) first cleavage furrow formation (Reprinted from Ogielska, 1994b, with permission).

invagination is delayed or inhibited. Exogastrulae are formed from both normally and abnormally cleaving embryos (Fig. 17), in which blastocoels are normal or reduced in size or even absent (Ogielska, 1994b; Ogielska and Jurgowiak, 1994). The histological analysis of the cells of the blastocoel roof and endoderm indicates that in some arrested blastulae the cells do not adhere to each other or do not form cytoplasmic processes (lamellipodia and filipodia); in opposite, in exogastrulae cells usually display high pseudopodial activity (Ogielska-Nowak, 1985). Apparently cells in the exogastrulae have the ability to move, but they cannot turn around the abnormally formed blastopore, and thereby they cannot move by involution to the interior, as is the case during normal gastrulation (Winklbauer and Schürfeld, 1999). In case when an embryo survives exogastrulation, the protruding yolk plug can be resorbed, can protrude further (Fig. 16B,F) or can be disintegrated and lost. The endodermal oedema (Fig. 16C,E) and underdevelopment of gut is most probably the result of the delay of vegetal rotation of endoderm. In all these cases, however, a part of endoderm is not at a proper place at a given time.

Many abnormal *R. esculenta* larvae and tadpoles display microcephalia or reduction of eyes (Ogielska, 1991) and nasal structures (Guex *et al.*, 2001). This kind of abnormality may be due to spatio-temporal disturbances in head organizer localization or abnormal position of deep yolky endoderm in exogastrulae (for comparison see Fig. 2e,f). Head underdevelopment is often accompanied by trunk/tail curvatures or shortening, most probably due to trunk organizer disturbances.

It is impossible to determine, which of the steps in a variety of complicated developmental pathways are disturbed in abnormal blastulae, gastrulae, larvae and tadpoles of water frog hybrids. But it is obvious that the genetic (zygotic) control must be somehow disturbed. The most common explanation of zygotic genes dysfunction in hybrid embryos is the incompatibility between the paternal genes of one species and maternal cytoplasm of the other species. In the case of *R. esculenta* this explanation seems not to be so simple, because the existence of viable hybrids obviously shows the high compatibility between the *lessonae* and *ridibunda* genomes. In these semiclonal water frogs, the lack of recombination during gametogenesis may lead to accumulation of deleterious recessive mutations in the *ridibunda* genome transmitted by hybrids through many generations of backcrossings with the parental species from the same population (reviewed by Graf and Polls Pelaz, 1989)

Reproduction Within the *Bufo viridis* Complex

Bufo viridis complex is composed of several diploid (*B.v. viridis, B.v. asiomantus, B.v. turanensis, B.v. latastii*), and tetraploid (*B. viridis, B.v. latastii*, and *B. danatensis*) species of green toads (for references see Stöck, 1997). In Middle Asia, *B. viridis* is represented by several diploid and tetraploid (Stöck, 1997; Stöck and Grosse, 1997), and mixed diploid-triploid-tetraploid (Castellano *et al.*, 1998) populations. Also European diploid *B. viridis* displays high genetic and morphological differentiation (Karakousis and Kyriakopoulou-Sklavounou, 1995). Tetraploid individuals are most probably autopolyploids, which produce diploid sperm and ova, and reproduce bisexually by fertilization.

More recently Stöck *et al.* (2002) described an all-triploid new species *Bufo pseudoraddei baturae*, another member of the *Bufo viridis* complex. This triploid species displays a combination of bisexual and hybridogenetic reproduction by production of diploid ova and haploid sperm, which means that the genetic contribution of male and females into the progeny is unequal. Triploid males reject premeiotically one of the three chromosome sets, and the resulting diploid primary spermatogonium will give rise to haploid spermatozoa. These results were possible to detect because triploid males have three chromosome sets, which differ in the presence of the nucleolus-organizing region (NOR) on the chromosome 6. Triploids have two NOR-positive and one NOR-negative chromosomes; during spermatogenesis they reject the NOR-negative sets, and in consequence each sperm receives one NOR-positive set. Diploid ova contain one NOR-positive and one NOR-negative chromosome sets. Most probably NOR-negative chromosome set in oocytes is transmitted clonally, whereas NOR-positive is recombined. Some recombination between selected genes loci suggests that two kinds of ova are produced: these containing one of the NOR-positive, and those with another NOR-positive chromosome set.

References

Alexandrino J, SJE Baird, L Lawson, JR Macey, C Mortiz, DB Wake. 2005. Strong selection against hybrids at a hybrid zone in the *Ensatina* ring species complex and its evolutionary implications. Evolution 59: 1334-1347.

Altig R, RW McDiarmid. 1999. Familial and generic characterization. In: Tadpole. The Biology of Anuran Larvae, The University of Chicago Press, pp. 295-337.

Altig R. 2003. Development. In: Reproductive Biology and Phylogeny of Anura, (Ed.) BGM Jamieson, Vol. 2, pp. 387-410.

Alves MJ, MM Coelho, MJ Collares-Pereira. 1998. Diversity in the reproductive modes of females of the *Rutilus alburnoides* complex (Teleostei, Cyprinidae): A way to avoid the genetic constrains of uniparentalism. Mol. Biol. Evol. 15: 1233-1242.

Arano B, GA Llorenta, P Herrero, B Sanchiz. 1994. Current studies on Iberian water frogs. Zool. Poloniae 39: 365-375.

Archetti M. 2004b. Recombination and loss of complementation: A more than two-fold cost for parthenogenesis. J. Evol. Biol. 17: 1084-1097.

Archetti M. 2004a. Loss of complementation and the logic of two-step meiosis. J. Evol. Biol. 17: 1098-1105.

Arntzen JW, GP Wallis. 1991. Restricted gene flow in a moving hybrid zone of the newts *Triturus cristatus* and *T. marmoratus* in western France. Evolution 45: 805-826.

Avise JC, JM Quattro, RC Vrijenhoek. 1992. Molecular clones within organismal clones. In: Evolutionary Biology, (Ed.) MK Hecht Wallace B, MacIntyre RJ, Vol. 26, Plenum Press, New York, pp. 225-246.

Azevedo MFC, F Foresti, PRR Ramos, J Jim. 2003. Comparative cytogenetic studies of *Bufo ictericus*, *B. paracnemis* (Amphibia, Anura) and an intermediate form in sympatry. Gene. Mol. Biol. 26: 289-294.

Babik W, J Rafiński. 2004. Relationship between morphometric and genetic variation in pure and hybrid populations of the smooth and Montandon's newt (*Triturus vulgaris* and *T. montandoni*). J. Zool. (London). 262: 135-143.

Babik W, JM Szymura, J Rafiński. 2003. Nuclear markers, mitochondrial DNA and male secondary sexual traits variation in a newt hybrid zone (*Triturus vulgaris* × *T. montandoni*). Mol. Ecol. 12: 1913-1930.

Beçak ML, LS Kobashi. 2004. Evolution by polyploidy and gene regulation in Anura. Gene. and Mol. Res. 3: 195-212.

Berger L, M Ogielska. 1999. Spontaneous haploid-triploid mosaicism in the progeny of *Rana* kl. *esculenta* female and *Rana lessonae* males. Amphibia-Reptilia 15: 143-152.

Berger L, E Czarniewska. 2002. Właściwości żab zielonych zachodniej Palearktyki. (Characteristics of western Palearctic water frogs). Przegląd Zool. 3-4: 189-202.

Berger L, R Günther. 1992. Inheritance patterns of water frog males from the environs of nature reserve Steckby, Germany. Zool. Polonia 37: 87-102.

Berger L, H Hotz, H Roguski. 1986. Diploid eggs of *Rana esculenta* with two *Rana ridibunda* genomes. Proc. Acad. Nat. Sci. of Philadelphia 138: 1-13.

Berger L, H Roguski, T Uzzell. 1978. Triploi F_2 progeny of water frogs (*Rana esculenta* complex). Folia Biol. (Kraków). 26: 135-152.

Berger L, H Roguski. 2002. Characteristics of diploid and triploid kin-tadpoles of Polish water frogs (*Rana esculenta* complex). Biol. Lett. 39: 29-42.

Berger L, T Uzzell, H Hotz. 1994. Postzygotic reproductive isolation between Mendelian species of European water frogs. Zool. Poloniae 39: 209-242.

Berger L. 1966. Biometrical studies on the population of green frogs from the environs of Poznań. Ann. Zool. 23: 303-324.

Berger L. 1967. Embryonal and larval development of F1 generation of green frogs different combinations. Acta Zool. Cracoviensia 12: 123-160.

Berger L. 1968. Morphology of F1 generation of various crosses within *Rana esculenta* complex. Acta Zool. Cracoviensia 13: 310-324.

Berger L. 1970. Some characteristics of the crosses within *Rana esculenta* complex in post-larval development. Ann. Zool. 27: 373-416.

Berger L. 1971a. Inheritance of sex and phenotype in F1 and F2 crosses within *Rana esculenta* complex. Genetica Pol. 12: 517-521.

Berger L. 1971b. Viability, sex and morphology of F2 generation within forms of *Rana esculenta* complex. Zool. Poloniae 21: 349-393.

Berger L. 1973. Some characteristics of backcrosses within forms of *Rana esculenta* complex. Genetica Pol. 14: 413-430.

Berger L. 1976. Hybrids of B2 generations of European water frogs (*Rana esculenta* complex) Ann. Zool. 33: 201-214.

Berger L. 1982. Hibernation of the European water frogs (*Rana esculenta* complex). Zool. Poloniae 29: 57-72.

Berger L. 1983. Western Palearctic water frogs (Amphibia, Ranidae); Systematics, genetics and population compositions. Experientia 39: 127-130.

Berger L. 1988. On the origin of genetic systems in European water frog hybrids. Zool. Poloniae 35: 5-32.

Berger L. 1994. Some peculiar phenomena in European water frogs. Zool. Poloniae 39: 267-280.

Bogart JP, LE Licht. 1987. Evidence for thee requirement of sperm in unisexual salamander hybrids (Genus *Ambystoma*). Can. Field Nat. 101: 434-436.

Bogart JP. 2003. Genetics and systematics of hybrid species. In: Reproductive Biology and Phylogeny of Urodela, (Ed) BGM Jamieson, Vol. 1, pp. 109-134.

Bogart JP, K Bi, J Fu, D Noble, J Niedzwiedzki. 2007. Unisexual salamanders (Genus *Ambystoma*) present a new reproductive mode for eukaryotes. Genome 50: 119-136.

Book JA. 1940. Triploidy in *Triton taeniatus*. Laur. Hereditas 26: 107-114.

Borkin LJ, R Gunther, YM Rozanov, AE Vinogradov. 1989. Inheritance in diploid and triploid hybridogenetic *Rana esculenta* males: Evidence from DNA flow cytometry. First World Congress of Herpetology, Kent, Canterbury, Abstracts S20. 11-19

Boucaut JC, LC Gallien. 1975. Transplantation intrablastocoelinne chez *Pleurodeles waltlii* Michah., de blastomeres isoles de blastula bloque entre *Ambystroma mexicanum* Shaw et *Ambystoma dumerilli* Duges (Amphibiens Urodeles). C. R. Acad. Sci. Paris 281: Series D: 571-574.

Bruce RC. Life histories. In: Reproductive Biology and Phylogeny of Urodela, (Ed.) BGM Jamieson, Vol. 1, pp: 477-525.

Brychta BH, HG Tunner. 1994. Flow cytometric analysis of spermatogenesis in triploid *Rana esculenta*. Zool. Poloniae 39: 507.

Bucci S, M Ragghianti, G Mancino, L Berger, HJ Hotz, T Uzzell. 1990. Lampbrusch and mitotic chromosomes of the hemiclonally reproducing hybrid *Rana esculenta* and its parental species. J. Exp. Zool. 255: 37-56.

Buekboom LW, RC Vrijenhoek. 1988. Evolutionary genetics and ecology of sperm-dependent parthenogenesis. J. Evol. Biol. 11: 755-782.

Bullini L. 1994. Origin and evolution of animal hybrid species. TREE 9: 422-426.

Callery EM, RP Ellinson. 2000. Thyroid hormone-dependent metamorphosis in a direct developing frog. Proc. Nat. Acad. Sci. USA 97: 2615-2620.

Callery EM, H Fang, RP Elinson. 2001. Frogs without polliwogs: Evolution of anuran direct development. BioEssays 23: 233-241.

Castanho LM, IMS De Luca. 2001. Moulting behavior in leaf-frogs of the genus *Phyllomedusa* (Anura: Hylidae). Zool. Anzeiger 240: 3-6.

Castellano S, C Giacoma, T Dujsebayeva, C Odierna, E Balletto. 1998. Morphometrical and acoustical comparison between diploid and tetraploid green toads. Biol. J. Linnean Soc. 63: 257-281.

Caune IA, LJ Borkin. 1989. An unusual population system (bisexual *Rana lessonae* with hybridogenetic male *R. esculenta*) in the European green frog complex. First World Congress of Herpetology, University of Kent at Canterbury, 11-19 September 1989. United Kingdom. Abstracts S20.

Chang CH, PA Wilson, LS Mathews, A Hemmati-Brivanlou. 1997. A *Xenopus* type I activin receptor mediates mesodermal but not neural specification during embryogenesis. Development 124: 827-837.

Chen PS, E Stumm-Zollinger. 1986. Patterns of protein synthesis in oocytes and early embryos of *Rana esculenta* complex. Roux's Arch. Dev. Biol. 195: 1-9.

Christian JL, RT Moon. 1993. When cells take fate into their own hands: Differential competence to respond to inducing signals generates diversity in the embryonic mesoderm. BioEssays 15: 135-140.

Cimino MC. 1972. Egg-production, polyploidization and evolution in a diploid all-female fish of the genus *Poeciliopsis*. Evolution 26: 294-306.

Cnudde F, T Gerats. 2005. Meiosis: Inducing variation by reduction. Plant Biol. 7: 321-341.

Cuellar O. 1976. Cytology of meiosis in the triploid gynogenetic salamander *Ambystoma tremblayi*. Chromosoma: 58: 355-364.

Dawley RM, RJ Schultz, KA Goddard. 1987. Clonal reproduction and polyploidy in unisexual hybrids of *Phoxinus eos* and *Phoxinus neogaeus* (Pisces; Cyprinidae). Copeia: 275-283.

Dawley RM. 1989. An introduction to unisexual vertebrates. In: Evolution and Ecology of Unisexual Vertebrates. (Ed.) RM Dawley, JP Bogart. Bulletin 466, New York State Museum, Albany, New York, USA, pp. 1-18.

De Robertis EM. 1995. Dismantling the organizer. Nature (London) 374: 407-408.

Del Pino EM, S Loor-Vela. 1990. The pattern of early cleavage of the marsupial frog *Gastrotheca riobambae*. Development 110: 781-789.

Del Pino EM. 1983. A novel development pattern for frogs: Gastrulation produces an embryonic disc. Nature (London) 306: 589-591.

Del Pino EM. 1989. Modifications of oogenesis and development in marsupial frogs. Development 107: 169-187.

Denis, H. 1970. Importance des interactions nucléo-cytoplasmatiques au cours du dévelopment embryonnaire. Arch. Int. Physiol. Biochim. 78: 367-380.

Dent JN. 1968. Survey of amphibian metamorphosis. (Eds) W Etkin, LI Gilbert, Appleton-Century-Crofts Division of Meredith Corporation. New York, pp. 271-311.

Dowling TE, CL Secor. 1997. The role of hybridization and introgression in the diversification of animals. Ann. Rev Ecol. Syst. 28: 593-619.

Dubois A, R Günther. 1982. Klepton and synklepton: Two new evolutionary systematics categories in zoology. Zool. Jb. Syst. 109: 290-305.

Dubois A, A Ohler A. 1994a. Frogs of the subgenus *Pelophylax* (Amphibia, Anura, genus *Rana*): A catalogue of available and valid scientific names with comments on name-bearing types, complete synonimes, proposed common names, and maps showing all type locations. Zool. Poloniae 39: 139-204.

Dubois A, A Ohler. 1994b. Catalogue of the names of frogs of the subgenus *Polyphylax* (Amphibia, Anura, genus *Rana*): A few additions and corrections. Zool. Poloniae 39: 205-208.

Dubois A. 1990. Nomenclature of parthenogenetic, gynogenetic and "hybrido-genetic" vertebrate taxons: new proposals. Alytes 8: 61-74.

Duellman WE, L Trueb. 1986. Biology of Amphibia, McGraw-Hill Book Company, USA.

Elinson PR. 1977. Amphibian hybrids: A genetic approach to the analysis of their developmental arrest. Differentiation 9: 3-9.

Elinson RP, JP Bogart, LE Licht, LA Lowcock. 1992. Gynogenetic mechanisms in polyploid hybrid salamanders. J. Exp. Zool. 264: 93-99.

Elinson RP, A Briedis. 1981. Triploidy permits survival of an inviable amphibian hybrid. Dev. Genet. 2: 357-367.

Elinson RP, EM del Pino. 1985. Cleavage and gastrulation in the egg-broodind marsupial frog, Gastrotheca riobambae. J. Embryol. Exp. Morph. 90: 223-232.

Elinson RP, H Fang. 1998. Secondary coverage of the yolk by the body wall in the direct developing frog, Eleutherodactylus coqui: An unusual process for amphibian embryos. Dev. Genes Evol. 208: 457-466.

Elinson RP, KR Kao. 1989. The location of dorsal information in frog early embryo. Develop. Growth Diff. 31: 423-492.

Elinson RP, J Paleček. 1993. Independence of two microtubule systems in fertilized frog eggs: the sperm aster and the vegetal parallel array. Roux's Arch. Dev. Biol. 202: 224-232.

Elinson RP. 1975a. Viable amphibian hybrids produced by circumventing a block to cross-fertilization (Rana clamitans ♀ × Rana catesbeiana ♂). J. Exp. Zool. 192: 323-330.

Elinson RP. 1975b. Fertilization of green frog (Rana clamitans) eggs in their native jelly by bullfrog (Rana catesbeiana) sperm. J. Exp. Zool. 193: 419-424.

Elinson RP. 1981. Genetic analysis of developmental arrest in an amphibian hybrid (Rana catesbeiana, Rana clamitans). Dev. Biol. 81: 167-176.

Elinson RP. 1986. Fertilization in amphibians: The ancestry to the block of polyspermy. Int. Rev. Cytol. 101: 59-100.

Etkin W. 1968. Hormonal control of amphibian metamorphosis. In: Meta-morphosis: A Problem in Developmental Biology, (Eds) W Etkin, LI Gilbert, Appleton-Century-Crofts Division of Meredith Corporation. New York, pp. 313-348.

Fang H, RP Elinson. 1999. Evolution alteration in anterior patterning: Otox2 expression in the direct developing frog Eleutherodactylus coqui. Dev. Biol. 205: 233-239.

Fankhauser G, R Crotta, M Perrot. 1942. Spontaneous and cold induced triploidy in the Japanese newt, Triturus pyrrhogaster. J. Exp. Zool. 89: 167-181.

Feder JH. 1979. Natural hybridization and genetic divergence between the toads Bufo borealis and Bufo punctatus. Evolution 33: 1089-1097.

Frieden E. 1968. Biochemistry of amphibian metamorphosis. In: Metamorphosis: A Problem in Developmental Biology, (Eds) W Etkin, LI Gilbert, Appleton-Century-Crofts Division of Meredith Corporation. New York, pp. 350-398.

Gallien CL, C Aimar. 1971. Sur un mode de gémellarite nouveau, réalisé par greffe nucléaire chez les Amphibiens Urodéles du gendre Pleurodeles. C. R. Acad. Sci. Paris Serie D, 272: 3348-3351.

Gard DL, D Affleck, BM Error. 1995a. Microtubule organization, acetylation, and nucleation in Xenopus laevis oocytes: II. A developmental transition in microtubule organization during early diplotene. Dev. Biol. 168: 189-201.

Gard DL, BJ Cha, MM Schroeder. 1995b. Confocal immunofluorescence microscopy of microtubules, microtubule-associated proteins, and microtubule-organizing centers during amphibian oogenesis and early development. In: Current Topics in Developmental Biology, (Ed.) DG Capco, Vol. 31, Academic Press, pp. 383-431.

Gatherer D, EM del Pino. 1992. Somitogenesis in the marsupial frog *Gastrotheca riobambae*. Int. Rev. Dev. Biol. 36: 283-291.

Gerhart J, S Black, S Scharf, R Gimlich, JP Vincent, M Danilchik, B Rowning, J Roberts. 1986. Amphibian early development. BioScience 36: 541-549.

Gerhart J, JP Vincent, S Scharf, S Black, R Gimlich, M Danilchik. 1984. Localization and induction in early development of *Xenopus*. Phil. Trans. R. Soc. Lond. B. 307: 319-330.

Gerhart J. 2001. Evolution of the organizer and the chordate body plan. Int. J. Dev. Biol. 45: 133-153.

Gilbert SF. 2000. Developmental Biology. Sinauer Associates, Inc., Publishers Sunderland, Massachusetts. Sixth Edition.

Glinka A, W Wu, D Onichtchouk, C Blumenstock, C Niehrs. 1997. Head induction by simultaneous repression of Bmp and Wnt signaling in Xenopus. Nature (London) 389: 517-519.

Gosner LK. 1960. A simplified table for staging anuran embryos and larvae with notes on identification. Herpetologica 16: 513-543.

Gould SJ. 1977. Ontogeny and Phylogeny. Belknap Press, Cambridge.

Graf JD, F Karch, MC Moreillon. 1977. Biochemical variation in the *Rana esculenta* complex: a new hybrid form related to *Rana perezi* and *Rana ridibunda*. Experientia 33: 1582-1584.

Graf JD, WP Müller. 1979. Experimental gynogenesis provides evidence of hybridogenetic reproduction in the *Rana esculenta* complex. Experientia 35: 1574-1576.

Graf JD, M Polls Pelaz. 1989. Cytogenetic analysis of spermatogenesis in unisexual allotriploid males from a *Rana lessonae-R. kl. esculenta* mixed populations. First World Congress of Herpetology, University of Kent at Canterbury, 11-19 September 1989. United Kingdom. Abstracts S20.

Graf JD, M Polls Pelaz. 1989. Evolutionary genetics of the *Rana esculenta* complex. In: Evolution and Ecology of Unisexual Vertebrates, (Eds) RM Dawley, JP Bogart. Bulletin 466, New York State Museum, Albany, New York, USA, pp. 289-301.

Graff JM, RS Thies, JJ Song, AJ Celeste, DA Melton. 1994. Studies with a *Xenopus* BMP receptor suggest that the ventral mesoderm-inducing signals override dorsal signals in vivo. Cell 79: 169-179.

Graff JM. 1997. Embryonic patterning: To BMP or not to BMP, that is the question. Cell 89: 171-174.

Green DM, SK Sessions. 1991. Amphibian cytogenetics and evolution. Academic Press, pp. 393-430.

Green DM. 1884. Sympatric hybridization and allozyme variation in the toads *Bufo americanus* and *B. fowleri* in southern Ontario. Copeia 1984: 18-26.

Green DM. 1983. Allozyme variation through a clinal hybrid zone between the toads *Bufo americanus* × *B. hemiophrys* in southeastern Manitoba. Herpetologica. 39: 28-40.

Green DM. 1985. Natural hybrids between the frogs *Rana cascadae* and *Rana pretiosa* (Anura, Ranidae). Herpetologica 41: 262-267.

Guerrini F, M Ragghianti, S Bucci, G Mancino, L Berger, H Hotz, T Uzzell, GD Guex. 1994. Repetitive DANN in the hemiclonal hybrid *Rana esculenta* and its parental species: Chromosomal location of two satellites. Zool. Poloniae 39: 503-504.

Guex GD, H Hoth, T Uzzell, RD Semlitsch, P Beerli, R Pascolini. 2001. Developmental disturbances in *Rana esculenta* tadpoles and metamorphs. Mitt. Mus. Nat. Kd. Berlin Zool. Reiche: 79-86.

Guex GD, H Hotz, SD Semlitsch. 2002. Deleterious alleles and differential viability in progeny of natural hemiclonal frogs. Evolution 56: 1036-1044.

Guillet F, C Aimar. 1971. Interaction nucléo-cytoplasmatique dans la realization d'un phenotype enzymatique (LDH) au cours de l'ontogenése, ches des embryons isogéniques obtenus par graffe nucléaire dans le gendre *Pleurodeles* (Amphibien-Urodéele). C. R. Acad. Sci. Paris Sreie D, 273: 2630-2633.

Günther R, J Plötner. 1994. Morphometric, enzymological and bioacoustic studies in Italian water frogs (Amphibia, Ranidae). Zool. Poloniae 39: 387-415.

Günther R. 1975a: Unterschungen der Meiose bei Mannchen von *Rana ridibunda* Pall., *Rana lessonae* Cam. und der Bastardform, "*Rana esculenta*" L. (Anura). Biol. Zntrbl. 94: 277-294.

Hall BK, MH Wake. 2000. Introduction: Larval development, evolution, and ecology. In: The Origin and Evolution of Larval Forms, (Eds) BK Hall, MH Wake. Academic Press, pp. 1-19.

Hanken J. 1999. Larvae in amphibian development and evolution. In: The Origin and Evolution of Larval Forms, (Eds) BK Hall, MH Wake, Academic Press, pp. 61-108.

Hardisty MW. 1966. The number of vertebrate primordial germ cells. Biol. Rev. 42: 265-287.

Harland R. 1994. The transforming growth factor β family and induction of the vertebrate mesoderm: Bone morphogenetic proteins are ventral inducers. Proc. Nat. Acad. Sci. USA 91: 10234-10246.

Harrison RG (Ed.). 1993. Hybrid Zones and the Evolutionary Process. Oxford University Press.

Hartley RS, JC Sible, AL Lewellyn, JL Maller. 1997. A role of cyclin E/cdk2 in the timing of the midblastula transition in *Xenopus* embryos. Dev. Biol. 188: 312-321.

Heasman J, A Crawford, K Goldstone, P Garner-Hamrick, B Gumbiner, P McCrea, C Kintner, CY Noro, C Wylie. 1994. Overexpression of cadherins and underexpression of beta-catenin inhibit dorsal mesoderm induction in early *Xenopus* embryos. Cell 79: 791-803.

Heasman J. 1997. Patterning the *Xenopus* blastula. Development 124: 4179-4191.

Hemati-Brivanlou A, OG Kelly, DA Melton. 1994. Follistatin, an antagonist of activin, is expressed in Spemann organizer and displays direct neutralizing activity. Cell 77: 283-295.

Hemati-Brivanlou A, DA Melton. 1992. A truncated activin receptor inhibits mesoderm induction and formation of axial structures in *Xenopus* embryos. Nature (London) 359: 609-614.

Hennen S. 1972. Morphological and cytological features of gene activity in an amphibian hybrid system. Dev. Biol. 29: 241-249.

Hennen S. 1973. Competence tests of early amphibian gastrula tissue containing nuclei of one species (*Rana palustris*) and cytoplasm of another (*Rana pipiens*). J. Embryol. Exp. Morph. 29: 529-538.

Heppich S, HG Tunner, J Greilhuber. 1982. Premeiotic chromosome doubling after genome elimination during spermatogenesis of the species hybrid *Rana esculenta*. Theor. Appl. Genet. 61: 101-104.

Hewitt MH. 1988. Hybrid zones—Natural laboratories for evolutionary studies. TREE 3: 158-167.

Hillis DM. 1988. Systematics of the *Rana pipiens* complex: Puzzle and paradigm. Ann. Rev. Ecol. Syst. 19: 39-63.

Hotz H, G Mancino, S Bucci-Innocenti, M Ragghianti, L Berger, T Uzzell. 1985. *Rana ridibunda* varies geographically in inducing clonal gametogenesis in interspecies hybrids. J. Exp. Zool. 236: 199-210.

Hotz H, T Uzzell, L Berger. 1994. Hemiclonal hybrid water frogs associated with the sexual host species *Rana perezi*. Zool. Poloniae 39: 243-266.

Joly P. 2001. The future of the selfish hemiclone: A Neodarwinian approach to water frog evolution. Mitt. Mus. Nat. Kd Berlin Zool. Reihe 77: 31-38.

Jones JM. 1972. Effects of thirty years hybridization on the toads *Bufo americanus* and *Bufo woodhousii fowleri* at Bloomington, Indiana. Evolution 27: 435-448.

Karakousis Y, P Kyriakopoulou-Sklavounoy. 1995. Genetic and morphological differentiation among populations of the green toad *Bufo viridis* from Greece. Biochem. Syst. Ecol. 23: 39-45.

Kawamura T, M Nishioka. 1986. Hybridization experiments among *Rana lessonae*, *Rana ridibunda* and *Rana esculenta*, with special reference to hybridogenesis. Sci. Rep. Lab. Amphibian Biol. Hiroshima Univ. 8: 117-271.

Kawamura T. 1984. Polyploidy in Amphibians. Zool. Sci. 1: 1-15.

Keller RE. 1991. Early embryonic development of *Xenopus laevis*. In: Methods in Cell Biology, (Eds) B Kay, HB Peng, Vol. 36, San Diego, Academic Press, pp. 61-113.

Key B. 2003. Molecular development. In: Reproductive Biology and Phylogeny of Anura, (Ed.) BGM Jamieson, Vol. 2, Academic Press, pp. 409-436.

Kimelman D, KJP Griffin. 1998. Mesoderm induction: A postmodern view. Cell 94: 419-421.

Kleckner N. 1996. Meiosis: How could it work? Proc. Nat. Acad. Sci. USA 93: 8167-8174.

Klotz C, MC Dabauvalle, M Paintrand, T Weber, M Bornrns, E Karsenti. Parthenogenesis in *Xenopus* eggs requires centrosomal integrity. J. Cell Biol. 110: 4-5-415.

Kobel HR, L Du Pasquier, RC Tinsley. 1981. Natural hybridization and gene introgression between *Xenopus gilli* and *Xenopus laevis laevis* (Anura: Pipidae). J. Zool. London 194: 317-322.

Kotlik P, V Zavadil. 1999. Natural hybrids between the newts *Triturus montandoni* and *T. vulgaris*: Morphological and allozyme data evidence of recombination between parental genomes. Folia Zoologica 48: 211-218.

Kraus F, MM Miyamoto. 1990. Mitochondrial genotype of a unisexual salamander of hybrid origin is unrelated to either of its nuclear haplotypes. Proc. Nat. Acad. Sci. USA 87: 2235-2238.

404

Lane MC, RE Keller. 1997. Microtubule disruption reveals that Spemann's organizer is subdivided into two subdomains by the vegetal alignment zone. Development 124: 895-906.

Liepins A, S Hennen. 1977. Cytochrome oxidase deficiency during development of amphibian nucleocytoplasmic hybrids. Dev. Biol. 57: 284-292.

Litvinchuk SN, LJ Borkin, JM Rosanov. 2003. On distribution of and hybridization between the newts *Triturus vulgaris* and *T. montandoni* in western Ukraine. Alytes 20: 161-168.

Lowcock LA, LE Licht, JP Bogart. 1987. Nomenclature in hybrid complexes of *Ambystoma* (Urodela: Ambystomatidae): No case for the erection of hybrid "species". Syst. Zool. 36: 328-336.

Madej Z. 1964. Studies on the fire bellied toad (*Bombina bombina* (Linnaeus, 1761)) and yellow bellied toad (*Bombina variegata* (Linnaeus, 1758)) of Upper Silesia and Moravian Gate. Acta Zool. Cracoviensia IX: 291-334.

Maller J, D Poccia, D Nishioka, P Kido, J Gerhart, H Hartman. 1976. Spindle formation and cleavage in *Xenopus* eggs injected with centriole containing fractions from sperm. Exp. Cell Res. 99: 285-294.

Mancino G, M Ragghianti, S Bucci-Innocenti. 1978. Experimental hybridization within the genus *Triturus* (Urodela: Salamandridae). Chromosoma 69: 27-46.

Mantovani B, M Passamonti, V Scali. 1999. Genomic evolution in parental hybrid taxa of the genus *Bacillus* (Insecta Phasmatodea). Ital. J. Zool. 66: 265-272.

Mantovani B, V Scali. 1992. Hybridogenesis and androgenesis in the stick insect *Bacillus rossius-grandii benazzi* complex (Insecta, Phasmatodea). Evolution 46: 783-796.

Marks SB, A Colazzo. 1998. Direct development in *Desmognathus aeneus* (Caudata: Plethodontidae). A staging table. Copeia 1998: 637-648.

McDarmid RW, R Altig. 1999. Tadpole. The Biology of Anuran Larvae. The University of Chicago Press, pp. 444.

McDowell N, JB Gurdon. 1999. Activin as a morphogen in Xenopus mesoderm induction. Semin. Cell Develop. Biol. 10: 311-317.

Michałowski J. 1964. Isolationmechanismen und Bastardirungmöglichkeiten bei den Amphibien. Biol. Abl. 85: 561-585.

Mikuliček P, P Kotlik. 2001. Two water frog populations from western Slovakia consisting of diploid females and diploid and triploid males of the hybridogenetic hybrid *Rana esculenta* (Amphibia, Anura). Mitt. Mus. Nat. Kd Berlin Zool. Reiche: 59-64.

Moon RT, D Kimelman. 1998. From cortical rotation to organizer gene expression: Toward a molecular explanation of axis specification in *Xenopus*. BioEssays 20: 536-545.

Moore JA. 1941. Developmental rate of hybrid frogs. J. Exp. Zool. 86: 405-422.

Moore JA. 1946. Studies in the development of frog hybrids. I. Embryonic development in the cross *Rana pipiens* ♀ × *Rana sylvatica* ♂. J. Exp. Zool. 101: 173-213.

Moore JA. 1947. Studies in the development of frog hybrids. II. Competence of the gastrula ectoderm of *Rana pipiens* ♀ × *Rana sylvatica* ♂. J. Exp. Zool. 105: 349-370.

Moore JA. 1948. Studies in the development of frog hybrids. III. Inductive ability of the dorsal lip region of *Rana pipiens* ♀ × *Rana sylvatica* ♂. J. Exp. Zool. 108: 127-147.

Moore JA. 1966. Hybridisation experiments involving *Rana dunni, Rana megapoda,* and *Rana pipiens.* Copeia 4: 673-675.

Nieuwkoop PD, J Faber. 1956. Normal Tables of *Xenopus laevis* (Daudin). Amsterdam, North Holland.

Ninomiya H, Q Zhang, RP Elinson. 2001. Mesoderm formation in *Eleuthero-dactylus coqui*: Body patterning in a frog with a large egg. Dev. Biol. 236: 109-123.

Nishioka M, H Ohtani. 1984. Hybridogenetic reproduction of allotriploids between Japanese and European pond frogs. Zool. Sci. 1: 291-326.

Ogielska M, L Jurgowiak. 1994. Exogastrulation in the progeny of water frog, *Rana esculenta* L. (Amphibia, Anura). Zool. Poloniae 39: 475-484.

Ogielska M, P Kierzkowski, M Rybacki. 2005. DNA content, erythrocyte size and genome composition of diploid and triploid water frogs belonging to the *Rana esculenta* complex (Amphibia, Anura). Can. J. Zool. 82: 1894-1901.

Ogielska M, A Kotusz. 2004. Pattern and rate of ovary differentiation wit h reference to somatic development in anuran amphibians. J. Morphol. 259: 41-54.

Ogielska M, E Wagner. 1990. Oogenesis and development of the ovary in European green frog, *Rana ridibunda* (Pallas). I. Tadpole stages until metamorphosis. Zool Jb. Anat. 120: 211-221.

Ogielska M, E Wagner. 1993. Oogenesis and ovary development in natural hybridogenetic water frog, *Rana esculenta* L. 1. Tadpole stages until metamorphosis. Zool. Jb. Physiol. 97: 349-368.

Ogielska M. 1991. Abnormal eye development in the progeny of the natural hybridogenetic frog, *Rana esculenta* L. (Amphibia, Anura). Zool. Anz. 226: 174-184.

Ogielska M. 1994a. Nucleus-like bodies in gonial cells of *Rana esculenta* (Amphibia, Anura) tadpoles—A putative way of chromosome elimination. Zool. Poloniae 39: 461-474.

Ogielska M. 1994b. *Rana esculenta* developmental syndrome: Fates of abnormal embryos from the first cleavage until spontaneous death. Zool. Poloniae 39: 447-459.

Ogielska M. 1995. Oogeneza i różnicowanie się jajnika u hybrydogenetycznego mieszańca, *Rana esculenta* L. (Oogenesis and ovary differentiation in the hybridogenetic hybrid, *Rana esculenta* L.). Acta Universitatis Wratislaviensis, Prace Zoologiczne, Vol. 32, pp. 1-42.

Ogielska, M, J Bartmańska. 1999: Development of testes and differentiation of germ cells in water frogs of the *Rana esculenta* complex (Amphibia, Anura). Amphibia-Reptilia 20: 251-263.

Ogielska-Nowak M. 1985. Stages of normal and spontaneous abnormal development of the natural hybrid, *Rana esculenta* L. (Amphibia, Anura). Zool. Poloniae 32: 37-61.

Ohler A. 1988. Developmental rate of *Rana synkl. esculenta* (Ranidae, Anura) embryos from different crosses: Consequences on the evolution of the populations. Alytes 7: 115-123.

Ohno S. 1970. Evolution by Gene Duplication. Springer Verlag, Berlin, Heidelberg, New York.

Ohtani H. 1993. Mechanism of chromosome elimination in the hybridogenetic spermatogenesis of allotriploid males between Japanese and European water frogs. Chromosoma 102: 158-162.

Otto SP, J Whitton. 2000. Polyploid incidence and evolution. Annu. Rev. Genet. 34: 401-437.

Phillips CA, T Uzzell, CM Spolsky, JM Serb, RE Szafoni, TR Pollowy. 1997. Persistent high levels of tetraploidy in salamanders of the *Ambystoma jeffersonianum* complex. J. Herpetol. 31: 530-535.

Piccolo S, E Agius, L Leyns, S Bhattacharyya, H Grunz, T Bouwmeester, EM De Robertis. 1999. The head inducer Cerberus is a multifunctional antagonist of Nodal, BPM and WNT signals. Nature (London) 397: 707-710.

Piccolo S, E Agius, B Lu, S Goodman, L Dale, EM de Robertis. 1997. Cleavage of chordin by Xolloid metalloprotease suggests a role for proteolytic processing in the regulation of Spemann organizer activity. Cell 91: 407-414.

Plénet S, A Pagano, P Joly, P Fouillet. 2000. Variation of plastic responses to oxygen availability within the hybridogenetic *Rana esculenta* complex. J. Evol. Biol. 13: 20-29.

Plötner J, C Grunwald. 1991. A mathematical model of the structure and dynamics of the *Rana ridibunda/esculenta-♂♂* populations (Anura, Ranidae). Z. Zool. Syst. Evolut. Forsch. 29: 201-207.

Plötner J, T Ohst. 2001. New hypothesis on the systematics of the western Palearctic water frog complex (Anura, Ranidae). Mitt. Mus. Nat. Kd. Berlin Zool. Reiche: 5-21.

Polls Pelaz M. 1994. Modes of gametogenesis among kleptons of the hybridogenetic water frog complex: An evolutionary synthesis. Zool. Poloniae 39: 123-138.

Ragghianti M, S Bucci, F Guerrini, G Mancini. 1999. Characterization of two repetitive DNA families (RrSS1 and Rana/Pol III) in the genomes of Palearctic green water frogs. Ital. J. Zool. 66: 255-263.

Reynhout JK, DL Kimmel. 1969. Chromosome studies of the lethal hybrid *Rana pipiens* ♀ × *Rana catesbeiana* ♂. Dev. Biol. 20: 501-517.

Richards CH, GW Nace. 1977. The occurrence of diploid ova in *Rana pipiens*. J. Heredity 68: 307-312.

Rose CS. 1999. Hormonal control in larval development and evolution— Amphibians. In: The Origin and Evolution of Larval Forms, (Eds) BK Hall, MH Wake. Academic Press, pp. 167-216.

Rosenkilde P, AP Ussing. 1996. What mechanisms control neoteny and regulate induced metamorphosis in urodeles? Int. Rev. Dev. Biol. 40: 665-673.

Rozenblut B. 2006. Wiek osiągania dojrzałości płciowej i rozwój jajnika u żab zielonych (*Rana esculenta* complex). (Age of sexual maturity and ovary development in water frogs (*Rana esculenta*-complex). Ph.D. thesis, Wroclaw University, Zoological Institute.

Ruiz i Altaba A, DA Melton. 1990. Axial patterning and the establishment of the polarity in the frog embryo. Trends Genet. 6: 57-64.

Ryan K, N Garret, A Mitchell, JB Gurdon, 1996. *Eomesodermin*, a key early gene in the frog embryo. Trends Genet. 6: 57-64.

Rybacki M, L Berger. 1994. Distribution and ecology of water frogs in Poland. Zool. Poloniae 39: 293-303.

Rybacki M, L Berger. 2001. Types of water frog populations (*Rana esculenta* complex) in Poland. Mitt. Mus. Nat. Kd Berlin Zool. Reiche: 51-57.

Rybacki M. 1994a. Diploid males of *Rana esculenta* from natural populations in Poland producing diploid spermatozoa. Zool. Poloniae 39: 517-518.

Rybacki M. 1994b. Water frogs (*Rana esculenta* complex) of the Bornholm Island, Denmark. Zool. Poloniae 39: 331-344.

Rybacki M. 1994c. Structure of water frog populations (*Rana esculenta* complex) of the Wolin Island, Poland. Zool. Poloniae 39: 345-364.

Schempp W, M Schmid. 1981. Chromosome banding in Amphibia. VI. BrdU replication patterns in Anura and demonstration of XX/XY sex chromosomes in *Rana esculenta*. Chromosoma 83: 697-710.

Schlosser G, H Kintner, RG Northcutt. 1999. Loss of ectodermal competence for lateral line placode formation in the direct developing frog *Eleuterodactylus coqui*. Dev. Biol. 213: 354-369.

Schmeller D, A Seitz, A Crivelli, A Pagano, M Veith. 2001. Inheritance of the water frog *Rana ridibunda* Pallas 1771–Is it Mendelian or hemiclonal? Mitt. Mus. Nat. Kd. Berlin Zool. Reiche: 39-42.

Schmeller D. 2004. Tying ecology and genetics of hemiclonally reproducing waterfrogs (Rana, Anura). Ann. Zool. Fen. 41: 681-687.

Schmidt BR. 1993. Are hybridogenetic frogs cyclical parthenogens? TREE 8: 271-273.

Schultz RJ. 1969. Hybridization, unisexuality, and polyploidy in the teleost *Poeciliopsis*. Science 157: 1564-1567.

Schultz RJ. 1977. Evolution and ecology of unisexual fishes. Evol. Biol. 10: 277-331.

Semlitsch RD, S Schmiedehausen, H Hotz, P Beerli. 1996. Genetic compatibility between sexual and clonal genomes in local populations of the hybridogenetic *Rana esculenta* complex. Evol. Ecol. 10: 531-543.

Sessions SK. 1982. Cytogenetics of diploid and triploid salamanders of the *Ambystoma jeffersonianum* complex. Chromosoma 84: 599-621.

Shi YB. 2000. Amphibian Metamorphosis. From Morphology to Molecular Biology. Wiley-Liss, pp. 288.

Sible JC, JA Anderson, AL Lewellyn, JL Maller. 1997. Zygotic transcription is required to block a maternal program of apoptosis in *Xenopus* embryos. Dev. Biol. 189: 335-346.

Simon, JC, F Delmotte, C Rispe, T Crease. 2003. Phylogenetic relationship between parthenogens and their sexual relatives: The possible routes to parthenogenesis in animals. Biol. J. Linnean Soc. London 79: 151-163.

Smith J. 1997. Brachyury and T-box genes. Curr. Opinion in Genet. Develop. 7: 474-480.

Smith JC, BMJ Price, JBA Green, D Weigl, BG Herman. 1991. Expression of a *Xenopus* homolog of *Brachyury* (T) is an immediate-early response to mesoderm induction. Cell 67: 79-87.

Smith JC, R White. 2003. Patterning the Xenopus embryo. In: Frontiers in Molecular Biology: Patterning in Molecular Biology, (Ed.) C Tickle, Oxford University Press, pp. 24-37.

Socha M. 2005. Struktura I reprodukcja mieszanych populacji żab zielonych *Rana ridibunda—Rana escu lenta* (Amphibia, Anura). (Structure and reproduction

408

of mixed populations *Rana ridibunda-Rana esculenta*). Ph.D. thesis, Wroclaw University, Zoological Institute.

Spolsky K, T Uzzell. 1986. Evolutionary history of hybridogenetic hybrid frog, *Rana esculenta*, as deduced from mtDNA analyses. Mol. Biol. Evol. 3: 44-56.

Spolsky KM, CA Phillips, T Uzzell. 1992. Gynogenetic reproduction in hybrid mole salamanders (genus *Ambystoma*). Evolution 46: 1935-1944.

Stearns T, M Kirchner. 1994. *In vitro* reconstruction of centrosome assembly and function: the role of γ-tubulin. Cell 76: 623-637.

Stöck M, WR Grosse. 1997. Erythrocyte size and polyploidy determination in green toads (*Bufo viridis* complex) from Middle Asia. Alytes 15: 72-90.

Stöck M, DK Lamatsch, C Steilein, JT Epplen, WR Grosse, R Hock, T Klapperstück, KP Lampert, U Scheer, M Schmid, M Schartl. 2002. A bisexually reproducing all-triploid vertebrate. Nature Genet. 30: 325-328.

Stöck M. 1997. Untersuchungen zur Morphologie di- und tetraploider Grüunkröten (Bufo viridis-Komplex) in Mittelasien (Amphibia: Anura: Bufonidae). Zool. Abh. St. Mus. Tierk. Dresden 49: 193-222.

Subtelny S. 1974. Nucleocytoplasmic interactions in development of amphibians. Int. Rev. Cytol. 39: 35-88.

Suomalainen E, A Saura, j Lokki. 1987. Cytology and Evolution in Parthenogenesis. CRC Press Inc., Boca Raton, Florida, USA, pp. 216.

Szymura JM, NH Barton. 1986. Genetic analysis of a hybrid zone between the fire-bellied toads, *Bombina bombina* and *B. variegata* near Cracovin southern Poland. Evolution 40: 1141-1159.

Szymura JM, NH Barton. 1991. The genetic structure of the hybrid zone between the fire-bellied toads, *Bombina bombina* and *B. variegata*: comparison between transects and between loci. Evolution 45: 237-261.

Szymura JM. 1993. Analysis of hybrid zones with *Bombina*. In: 1993. Hybrid Zones and the Evolutionary Process, (Ed.) RG Harrison, Oxford University Press, pp. 216-289.

Tata JR. 1993. Gene expression during metamorphosis: An ideal model for post-embryonic development. BioEssays 15: 239-248.

Thibaudeau G, R Altig. 1999. Endotrophic anurans. Development and evolution. In: Tadpole. The Biology of Anuran Larvae. The University of Chicago Press, pp. 170-188.

Tickle C, M Devey. 2003. Laying down the vertebrate body plan. In: Frontiers in Molecular Biology: Patterning in Molecular Biology, (Ed.) C Tickle. Oxford University Press, pp. 10-23.

Tinti F, V Scali. 1992. Genome exclusion and gametic DAPI-DNA content in the hybridogenetic *Bacillus rossius-grandii benazzi* complex (Insecta, Phasmatodea). Mol. Rep. Dev. 33: 235-242.

Tournier F, Y Bobinnec, A Debec, P Santamaria, M Bornens. 1999. Drosophila centrosomes are unable to trigger parthenogenetic development of *Xenopus* eggs. Biol. Cell. 91: 99-108.

Tournier F, M Bornens. 1994. Cell cycle regulation of centrosome function. In: Microtubules, (Ed.) JS Hyams, CW Lloyd. John Wiley-Liss and Sons, New York, pp. 303-324.

Tunner HG, MT Dobrowsky. 1976. Zur morphologischen, serologischen und enzymologischen Differenzierung von *Rana lessonae* und der hybridogenetischen *Rana esculenta* aus dem Seewinkel und dem Neusiedlersee (Österreich, Burgenland). Zool. Anz. 197: 6-22.

Tunner HG, S Heppich. 1981. premeiotic genome exclusion during oogenesis in the common edible frog, *Rana esculenta*. Naturwissenschaften 68: 207-208.

Tunner HG, S Heppich-Tunner. 1991. Genome exclusion and two strategies of chromosome duplication in oogenesis of a hybrid frog. Naturwissenschaften 78: 32-34.

Tunner HG, H Nopp. 1979. Heterosis in the common European water frog. Naturwissenschaften 66: 268-269.

Tunner HG. 1974. Die klonale Strukture einer Wasserfroschenpopulation. Z. Zool. Syst. Evol. Forsch. 12: 309-314.

Uzzell T, R Gunther, L Berger. 1977. *Rana ridibunda* and *Rana esculenta*: A leaky hybridogenetic system (Amphibia, Salientia). Proc. Acad. Nat. Sci. Philadelphia 128: 147-171.

Uzzell T, H Hotz, L Berger. 1980. Genome exclusion in gametogenesis by an interspecific *Rana* hybrid: evidence from electrophoresis of individual oocytes. J. Exp. Zool. 259: 214-251.

Uzzell T, H Hotz. 1979. Electrophoretic and morphological evidence for two forms of green frogs (*Rana esculenta* complex) in Peninsular Italy (Amphibia, Salientia). Mitt. Zool. Mus. Berlin 55: 13-27.

Uzzell T, HG Tunner. 1983. An immunological analysis of Spanish and French water frogs. J. Herpet. 17: 320-326.

Uzzell TM, L Berger. 1975. Electrophoretic phenotypes of *Rana ridibunda*, *Rana lessonae* and their hybridogenetic associate, *Rana esculenta*. Proc. Acad. Nat. Sci. Philadelphia 127: 13-24.

Uzzell TM. 1979. Immunological distances between the serum albumins of *Rana ridibunda* and *Rana lessonae*. Proc. Acad. Nat. Sci. Philadelphia 127: 12-24.

Uzzell TM. 1982. Immunological relationships of western Palearctic frogs (Salienta, Ranidae). Amphibia-Reptilia 3: 135-143.

Valée L. 1959. Recherches sur triturus blasii de L'Isle, hybride naturel de *Triturus cristatus* Laur. × *Triturus marmoratus* Latr. Lons-le-Saunier, Imprimerie Maurice Declume.

Vincent JP, G Oster, JC Gerhart. 1986. Kinetics of the grey crescent formation in the *Xenopus* egg: Displacement of the subcortical cytoplasm relative to the egg surface. Dev. Biol. 114: 484-500.

Vinogradov AE, LJ Borkin, R Günther, MJ Rosanov. 1990b. Two germ cell lineages with genomes of different species in one and the same animal. Hereditas 114: 245-251.

Vinogradov AE, LJ Borkin, MJ Rosanov. 1990a. Genome elimination in triploid and diploid *Rana esculenta* males: cytological evidence from DNA flow cytometry. Genome 33: 619-626.

Vogel P, PS Chen. 1976. Genetic control of LDH isoenzymes in the *Rana esculenta* complex. Experientia 32: 304-307.

Volpe EP. 1952. Physiological evidence for normal hybridization of *Bufo americanus* and *Bufo fowleri*. Evolution VI: 393-406.

Vrijenhoek RC. 1984. The evolution of clonal diversity in *Poeciliopsis*. In: Evolutionary Genetics of Fishes, (Ed.) BJ Turner, Plenum Press, New York, pp. 399-429.

Vrijenhoek RC. 1990. Genetic diversity and the ecology of asexual populations. In: Population Biology, (Eds) KK Wöhrmann, S Jain, Springer Verlag Berlin, pp. 175-197.

Wagner E, M Ogielska. 1990. Oogenesis and development of the ovary in European green frog, Rana ridibunda (Pallas). 2. Juvenile stages until adults. Zool. Jb. Anat. 120: 223-231.

Wagner E, M Ogielska. 1993. Oogenesis and ovary development in natural hybridogenetic water frog, Rana esculenta L. 2. After metamorphosis until adults. Zool. Jb. Physiol. 97: 349-368.

Wake DB, J Hanken. 1996. Direct development in the lungless salamanders: What are the consequences for developmental biology, evolution and phylogenesis? Int. J. Dev. Biol. 40: 859-869.

Winklbauer R, M Schürfeld. 1999. Vegetal rotation, a new gastrulation movement involved in the internalization of the mesoderm and endoderm in Xenopus. Development 126: 3703-3713.

Woodland HR, JB Gurdon. 1969. RNA synthesis in an amphibian nuclear-transplant hybrid. Dev. Biol. 20: 89-104.

Yamamoto TS, C Takagi, AC Hyodo, N Ueno. 2001. Suppression of head formation by Xmsx-1 through the inhibition of intercellular nodal signaling. Development 128: 2769-2779.

Yasuda GK, G Schubinger. 1992. Temporal regulation in the early embryo: Is MBT too good to be true? Trends Genet. 8: 124-127.

Zhang J, DW Hoston, ML King, C Payne. 1998. The role of maternal Veg-T in establishing the primary germ layers in Xenopus embryos. Cell 94: 515-524.

Zhang J, ML King. 1996. Xenopus Veg-T RNA is localized to the vegetal cortex during oogenesis and encodes a novel T-box transcription factor involved in mesodermal patterning. Development 122: 4119-4129.

Zimmerman LB, JM Jesus-Escobar, RM Harland. 1996. The Spemann organizer signal noggin binds and inactivates Bone Morphogenetic Protein 4. Cell 86: 599-606.

Index

Printed and bound by CPI Group (UK) Ltd, Croydon, CR0 4YY

21/10/2024

01777107-0020